Bioinformatics: Design and Implementation

Bioinformatics: Design and Implementation

Editor: Alexis White

MURPHY & MOORE
www.murphy-moorepublishing.com

Murphy & Moore Publishing,
1 Rockefeller Plaza,
New York City, NY 10020, USA

Visit us on the World Wide Web at:
www.murphy-moorepublishing.com

ISBN: 978-1-63987-075-2 (Hardback)

Cataloging-in-Publication Data

Bioinformatics : design and implementation / edited by Alexis White.
p. cm.
Includes bibliographical references and index.
ISBN 978-1-63987-075-2
1. Bioinformatics. 2. Biology--Data processing. 3. Computational biology.
4. Systems biology. I. White, Alexis.
QH324.2 .B56 2022
570.285--dc23

Table of Contents

Preface

Bioinformatics is an interdisciplinary field that focuses on developing methods and software tools for understanding large and complex biological data. It consolidates the disciplines such as biology, computer science, information engineering, mathematics, and statistics to analyze and interpret biological data. Some of the subfields of bioinformatics are computational biology, genetics, genomics, proteomics, transcriptomics, phylogenetics, metabolomics, systems biology, structural analysis, molecular modeling and pathway analysis. Bioinformatics is applied in various sectors such as molecular medicine, personalized medicine, preventative medicine, gene therapy, drug development, waste cleanup, etc. This book is a compilation of chapters that discuss the most vital concepts and emerging trends in bioinformatics. It elucidates new techniques and their applications in a multidisciplinary manner. For all readers who are interested in this field, the case studies included in this book will serve as an excellent guide to develop a comprehensive understanding.

The researches compiled throughout the book are authentic and of high quality, combining several disciplines and from very diverse regions from around the world. Drawing on the contributions of many researchers from diverse countries, the book's objective is to provide the readers with the latest achievements in the area of research. This book will surely be a source of knowledge to all interested and researching the field.

In the end, I would like to express my deep sense of gratitude to all the authors for meeting the set deadlines in completing and submitting their research chapters. I would also like to thank the publisher for the support offered to us throughout the course of the book. Finally, I extend my sincere thanks to my family for being a constant source of inspiration and encouragement.

Editor

1

Bioinformatics for Membrane Lipid Simulations: Models, Computational Methods, and Web Server Tools

S. W. Leong, T. S. Lim and Y. S. Choong

Abstract

Biological membranes are complex environments consisting of different types of lipids and membrane proteins. The structure of a lipid bilayer is typically difficult to study because the membrane liquid crystalline state is made up of multiple disordered lipid molecules. This complicates the description of the lipid membrane properties by the conformation of any single lipid molecule. Molecular dynamics (MD) simulations have been used extensively to investigate properties of membrane lipids, lipid vesicles, and membrane protein systems. All-atom membrane models can elucidate detailed contacts between membrane proteins and its surrounding lipids, while united-atom and coarse-grained description have allowed larger models and longer timescales up to microsecond mark to be probed. Additionally, membrane models with mixed phospholipids and lipopolysaccharide content have made it possible to model improved views of biological membranes. Here, we present an overview of commonly used lipid force fields by the biosimulation community, useful tools for membrane MD simulations, and recent advances in membrane simulations.

Keywords: all-atom (AT), coarse-grain (CG), force field, lipid bilayer, united-atom (UA)

1. Introduction

The biological membrane is a selective barrier delineating the boundaries of cells and organizing cellular organelles into compartments. One of the major functions of the membrane is to regulate movements of water, ions, and certain constituents in and out of the cell or organelle. Singer and Nicholson described the membrane as a "fluid mosaic" model in which lipids and proteins freely diffuse in the membrane plane [1]. While the biological membrane is actually made up of 1.5- to 4-fold proteins by weight, it is usually described as phospholipids arranged in a bilayer.

The lipid composition in membrane is highly specific with variation identified from membranes of different organisms, different cells of the same organism, and even different membranes of the same cell [2]. Generally, a lipid consists of a polar head group and hydrophobic tail region where a cell may have over a thousand different lipid species. Formation of bilayers and other lipid structures such as micelles occurs as the hydrophobic tails orientates away from the aqueous environment while the polar head groups interact favorably with water. Changes in the bilayer structures can affect the function of the membrane, membrane-embedded proteins and complexes. The surrounding environment can greatly affect the structure and functions of proteins, much like the effects of water on water-soluble proteins. The membrane produces a complex environment for embedded proteins, and membrane thickness, fluidity, charge, curvature, and phase have been demonstrated to play a role in protein structure and function determination [3].

The structure and properties of membranes are a complex matter and cannot be easily described by a single-lipid molecule alone. Molecular dynamics (MD) simulations offer a viable alternative to study the properties of membrane and lipid-forming structures such as vesicles. This review serves as an introduction to currently available membrane lipid force fields and recent advances in membrane simulations.

2. Force fields for lipid simulation

In general, all-atom (AT), united-atom (UA), and coarse-grained (CG) are the three-membrane lipid force fields. **Figure 1** illustrates the AT, UA and CG force field of a lipid by spherical representation.

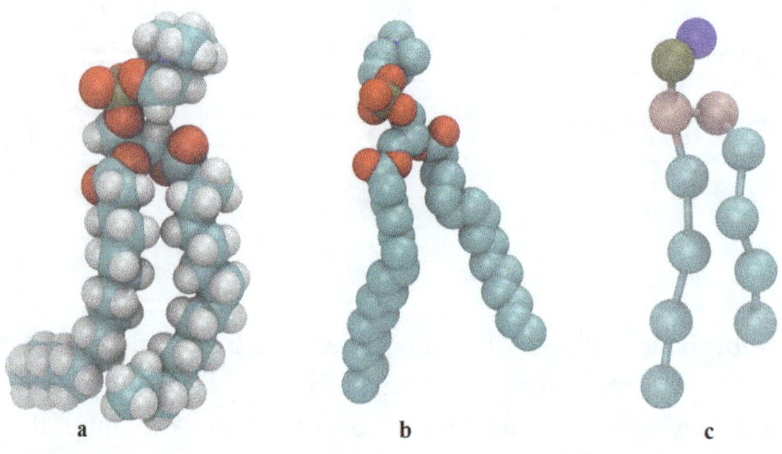

Figure 1. Representation of 1,2-dipalmitoyl-sn-glycero-3-phosphocholine (DPPC) with (a) atomistic (all-atom; AT), (b) united-atom (UA), and (c) coarse-grain (CG) force fields as van der Waals spheres.

2.1. All-atom (AT) MD simulation: details and details

AT MD simulation represents every atom in the system as a single interaction site. **Figure 2** shows the example of AT simulation system of *Salmonella enterica ser.* Typhi TolC protein in POPE. To date, Chemistry at HARvard Macromolecular Mechanics (CHARMM) and Assisted Model Building with Energy Refinement (AMBER) are the only fully AT force field parameterization available for lipids. Prior to the development of CHARMM36, CHARMM27r was widely used for membrane simulations [4, 5]. Simulations using CHARMM27r require a large positive surface tension (30–40 dyn/cm) to achieve the experimentally determined surface area per lipid (APL). However, theoretical considerations of self-assembled systems and macroscopic black lipid bilayers [6] indicate that the surface tension of bilayers are about zero to several dyn/cm even when undulations are taken into account [7]. Therefore, simulations of lipid bilayers using CHARMM27 and CHARMM27r shrinks to a near gel phase state without the use of surface tension. Additionally, CHARMM27 and CHARMM27r also failed to reproduce the experimental deuterium order parameters, S_{CD} in the glycerol and upper chain regions [4]. A wide range of glycerophospolipids exhibit splitting in the carbon 2 of the aliphatic chain and carbon 1 for glycerols, but this observation cannot be replicated when simulations were performed with CHARMM27 or CHARMM27r [8]. This may affect conclusions of interactions of lipids with surface-active agents drawn from MD simulations. Besides that, the area compressibility modulus, K_A, was underestimated, the head group region was underhydrated, the electron density in the bilayer midplane was underestimated while the frequency dependence of the ^{13}C NMR T_1 of the acyl chains near the head group was overestimated.

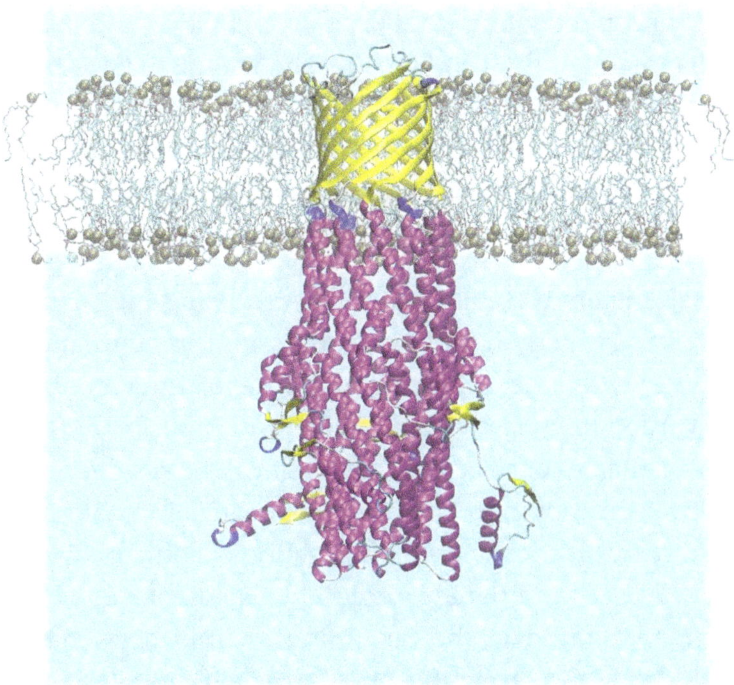

Figure 2. All-atom (AT) simulation of *Salmonella enterica ser.* Typhi TolC protein in POPE.

These major weaknesses of the CHARMM27r force field eventually motivated the development of CHARMM36. CHARMM36 corrected most of the existing problems with lipid force fields, most importantly allowing the simulation of lipid bilayer without the use of surface tension and improved the reproducibility of the experimental deuterium order parameters in the glycerol and upper chain regions of phospholipids [8]. A comparative study of lipid force fields by Piggot and colleagues found that CHARMM36 was the only force field that accurately reproduced the experimental order parameters of carbon 2 in both acyl chains of 1,2-dipalmitoyl-*sn*-glycero-3-phosphocholine (DPPC) and 1,2-dipalmitoyl-*sn*-glycero-3-phosphocholine (POPC) [9]. The other properties of the membrane were reasonably well reproduced. However, there are several limitations that have to be considered while pursuing CHARMM36. One is the approximate treatment of long-range Lennard-Jones (LJ) forces for bilayers. Simulations augmented with long-range LJ forces using 3D-isotropic periodic sum/discrete fast Fourier transform (3D-IPS/DFFT) may improve results for lipid monolayers but this increases surface tension, therefore making it less suited for bilayer simulations. The recommendation for bilayer simulations is to use particle-mesh Ewald (PME) with r_c = 10 or 12 Å and no long-range correction for the LJ term, but these setting will cause underestimation of the surface tension of lipid monolayers.

AMBER force fields were generally less used for membrane protein simulation due to the lack of a specific parameter set for lipids. However, the general AMBER force field (GAFF) which was originally parameterized for the simulation of arbitrary organic molecules with pre-existing AMBER force fields have been shown to reproduce lipid bilayer parameters satisfactorily [10]. GAFF has been tested on a range of lipids, including 1,2-dimyristoyl-d54-*sn*-glycero-3-phosphocholine (DMPC), 2-dioleoyl-*sn*-glycero-3-phosphocholine (DOPC), 1,2-dilauroyl-*sn*-glycero-3-phosphocholine (DLPC), DPPC, POPC, and 1-palmitoyl-2-oleoyl-*sn*-glycero-3-phosphoethanolamine (POPE) [11, 12]. The use of surface tension is necessary to achieve the correct APL for POPC [10], DOPC [13], and DMPC [12]. In order to better fit GAFF for lipid molecules, Dickson and colleagues set to re-parameterize the LJ terms for acyl chain carbons and hydrogens as the initial GAFF LJ terms were developed for proteins, nucleic acids, and small organic molecules [11].

In addition, torsion parameters were also re-optimized using high-level quantum chemical data with "Paramfit." This strategy has allowed a tensionless simulation of the lipids in the isothermal-isobaric (NPT) ensemble, achieving high level of agreement with experiment data for volume per lipid (V_L) and thickness value within 5% of experiment [11]. However, the APL for POPE lipids is lower than expected and thickness value was overestimated by 10% [11]. The force field has also been shown to be able to reproduce the data on large lipid bilayers containing 288 lipids, with little changes in the APL, V_L, and bilayer thickness [11]. The comparison with CHARMM27 and Berger force field also displayed GAFF's ability to reproduce the order asymmetry found at the beginning of lipid acyl chains [13]. Prior to the development of CHARMM36, GAFF was the only force field capable of capturing the carbon-2 deuterium order parameter splitting [13], although it has also produced lower surface APL and higher deuterium order parameters compared to experimental data [10, 12, 13]. Even so, compatibility of GAFF with previous AMBER force fields makes it a suitable choice for the simultaneous simulation of membrane, protein, and organic molecule simulation [13].

A recently developed Lipid14 force field with an updated Lipid11 head group and tail group charges and parameters enabled proper tensionless simulation of lipid bilayers in AMBER [14]. Lipid14 LJ and torsion parameters were modified to reproduce the experimental density, ρ and heat of vaporization, ΔH_{vap} of alkanes of different chain lengths by fitting the CH_2-CH_2-CH_2-CH_2 torsion to *ab initio* data using "Paramfit" and altering the LJ and torsion simultaneously. Lipid14 force field has a modular nature that allows new lipid species to be added into the force field by constructing them from head group and tail group "building blocks." Testing of Lipid14 on DLPC, DMPC, DPPC, DOPC, POPC, and POPE showed that APL for all simulations was within 3% of the experimental value, with the exception of POPE. The APL was closer to the older experimental value of 56.6 Å^2 [15] than the more current APL of 59–60 Å^2 [16]. The V_L was found to be within 5% of experimental value which was acceptable, although it may be considered a slight underestimation [14].

Furthermore, the isothermal area compressibility modulus, K_A falls close to the experiment, again with the exception of POPE, which has a higher K_A with a large standard deviation value. The authors suggested that implementation of other barostats into AMBER may improve K_A values as the Berendsen method for pressure control is not ideal for simulations in which volume fluctuations is an important parameter that is capable of influencing the outcome of the simulation [14]. The Luzzati thickness, D_B, which is calculated using the z-dimension of the simulation box and the integral probability distribution of the water density along the z-axis was slightly underestimated for saturated lipids, which implies more water penetration into the hydrophobic region of the bilayer of these lipids. Lipid14 was also able to reproduce the experimental order parameter trend, including the splitting of the sn-2 chain from the sn-1 chain and the drop at the carbon-9 and carbon-10 positions of POPC and POPE lipids, which is the *cis* double bond. Results from GPU repeats and CPU runs were also consistent.

Recent analysis applying full AT force field have reached timescales of up to tens of nanoseconds, with ambitious simulations pushing the microsecond mark, e.g. to characterize the interaction of the multiple sclerosis synthetic biomarker CSF114(Glc) with the membrane bilayer [17], probing of the huntingtin Htt17 membrane anchor on a POPC bilayer [18], mixing of lipids [19], and membrane-binding mechanism of the yeast Osh4 peripheral membrane protein [20]. Even though microsecond simulations have been reported, the accessibility of this approach is limited as many biologically relevant phenomena may require sampling time beyond the microsecond timescale and investigators may not have access to intensive computational resources. Given that lipid reorganization is quite slow, a properly equilibrated membrane may require 20–40 ns of simulation time. Convergence of membrane protein simulations still remains a question, as 100 ns of simulation may still be insufficient to fully describe e.g. rhodopsin loop fluctuations in a membrane [21]. Indeed, since all atomic-level interactions are retained and time-steps for integration of Newtonian motions are in the femtosecond range, AT simulations could be a time consuming and computationally expensive practice. The issue of convergence has been rectified in part by running several long simulations and using non-equilibrium sampling method or steered MD to calculate the observables of interest. A series of extended simulations (~100 ns) on the voltage sensor domain of potassium channels revealed the importance of lipid phosphates in accommodating the significantly charged S4 helix [22]. Using umbrella sampling, potential of mean force has been calculated for permeation and effect of a potassium ion on the second ion passing through gramicidin A channel [23].

Regardless, AT simulations still provided the highest level of detail and reliability when it comes to quantitative prediction of properties such as motional timescales or interaction strengths. Such details recently provided insightful interactions between protein and membrane, such as the rearrangement of amino acid side chains and local bilayer deformation due to hydrophobic mismatch and how the hydrophobic membrane layer accommodates charged arginine side chain of outer membrane phospholipase A [24, 25].

2.2. United-atom (UA) lipid models: the best of both worlds?

The UA representation of lipids simplifies the carbon tails of the lipid by associating the aliphatic carbon and its hydrogen atoms into a single particle. Because the non-polar hydrogen atoms are treated implicitly, the number of interaction sites per lipid can be reduced by two third. The computational costs for simulations of such membrane systems become relatively cheap as the 60% of the pairwise interactions in the membrane is reduced. The model lipid DPPC can be represented by 50 particles in UA force field, but needed 130 interaction sites in an AT force field. Since limited physical information may be collected from explicit acyl chain hydrogens, it became desirable to utilize UA force fields for membrane with an AT protein force field for membrane protein simulation [26, 27].

The UA lipid models parameterized by Berger et al. (1997) were one of the most popular lipid force field for lipids and were originally developed by Essex and colleagues [28] from the Optimized Potentials for Liquid Simulations (OPLSs) UA force field. Bonded parameters of the Berger lipids were obtained from the GROMOS87 force field (note: GROMOS is the GROningen Molecular Simulation package), the acyl chains used Ryckaert-Bellemans dihedral parameters whereas the van der Waals terms were from OPLS and atomic partial charges were from Chiu and colleagues' calculations [29]. Berger and colleagues further optimized the LJ parameters of the lipid tails based on thermodynamic data of pentadecane [30]. Berger lipid parameters were recommended if one desires maximal sampling due to its fast diffusion and good simulation efficiency [9]. For membrane protein simulations, Berger lipids are commonly used with OPLS and GROMOS [27]. It has also been demonstrated to be compatible with AMBER99SB and could give marginally better free energy calculations than the widely used OPLS/Berger combination [31]. While simulations of membrane proteins using such hybrid parameters have been validated by various groups [32–34], the combination of different force fields require care. Protein lipid interactions in Berger-OPLS and Berger-GROMOS have been found to be overestimated and result in drastic changes of lipid properties upon protein insertion [13, 27].

A CHARMM UA representation of the acyl chains is also available, and compatible for simulations with AT CHARMM protein force fields [26]. The CHARMM UA model was derived from C27 phospholipids, where explicit hydrogen atoms of the acyl chains were replaced with a UA representation. The force field, called C27-UA, retains the accuracy of the AT counterpart and provides a practical alternative when used in simulations of proteins and other compounds described by C27. C27-UA was parameterized by fitting to experimental data and AT simulations of liquid phase model systems (pentadecane for saturated hydrocarbon chains, *cis*-5-decene for monounsaturated chains and methyl hexanoate for the ester region). Simulation of POPC bilayer with C27-UA was comparable to C27 and reproduced experimental NMR and X-ray diffraction data, including electron density profile and carbon-deuterium order parameter. The free energy profiles of transfer of ethane, methanol, and water across a water-dodecane interface were identical in C27-UA and C27 simulation, suggesting that the force field is capable of simulating mixed system containing UA lipids and AT proteins and organic molecules described by the standard CHARMM force field. However, it also retained C27's feature of requiring a positive surface tension to be able to reproduce the experimental APL.

Several GROMOS parameters are used in membrane simulations, such as 43A1-S3 [35], 53A6 [36], and Kukol's modification of the 53A6 parameters [37]. 43A1-S3 is an extension and modification of the 43A1 force field designed to improve the properties of lipid membranes in simulations. 43A1-S3 employs charges from Chiu et al. [29] while van der Waals and dihedral parameters were modified from 43A1 to improve hydrocarbon and choline head group dynamics [38]. GROMOS 43A1-S3 accurately reproduces many properties of DPPC bilayers, including APL, lipid diffusion, and the order parameter [39]. The 53A6 parametrization also uses Chiu et al.'s charges [29] and LJ parameters of the choline methyls and phosphate ester oxygen atoms, thus providing good agreement for the APL, density profile, and order parameter for saturated and unsaturated acyl chain PC lipids [40, 41]. Even with improvements introduced, the isothermal area compressibility modulus was still overestimated. Meanwhile, Kukol reparametrize 53A6 following reports that the force field failed to reproduce DPPC satisfactorily [37]. He used Chiu's charges and increased the van der Waals radii of the carbonyl carbons of DPPC, DMPC, POPC, and POPG. He also employed non-standard GROMOS dihedral parameters for the double bond in the unsaturated lipids. This results in fairly good agreement with experimentally determined properties with the exception of order parameters for the *sn*-2 oleoyl chains. A comparative force field study by Piggot et al. recommended against using 43A1-S3 for POPC membranes and Kukol's 53A6 POPC parameters [9]. This is because 43A1-S3 could not reliably reproduce the drop in order parameter at the carbon 10 of oleoyl chains whereas Kukol's POPC parameters showed several disagreements with experimental value in terms of membrane thickness and order parameter [9].

With the advances in computational power, large-scale simulation projects using UA representation of the membrane region have been able to reach the microsecond mark. Access to such timescale allowed the observation of structural rearrangement and changes in hydrogen bonding pattern in an integral Kv1.2 channel embedded in a hyperpolarized membrane [42]. On a shorter timescale, CorA magnesium transport channel has been probed to undergo conformational changes to a putative open state in 110 ns [43].

2.3. Coarse-grained (CG): the need for speed

CG simulations are being widely used to investigate phenomenon occurring in timescales not accessible by AT simulation. In a CG simulation, 3–4 heavy atoms (non-H) are grouped together and represented by a single particle. For example, a DMPC lipid consisting of 130 atoms can be represented by 12 interaction sites [44]. The choline moiety is modeled with a single positively charged particle, the phosphate group with a negatively charged particle, the glycerol linkages with two nonpolar particles, while the lipid chains are modeled with 4 apolar particles each [45].

Early CG approaches were typically parameterized based on comparison to AT simulations by using inverted Monte Carlo schemes [46–48] or force matching [49], which aims to reproduce the structural details at a particular state for a particular system. Instead, the MARTINI (note: MARTINI is a CG force field developed by Marrink and coworkers [44]) CG force field calibrated the building blocks of biomolecules against thermodynamic data, particularly the oil/water partitioning coefficients so that there is no need to re-parameterize the model each time. A consistent, atomic level compatible CG force field will be beneficial for multi-scale applications. In MARTINI, an average of four heavy atoms were represented by a single interaction site, with the exception of ring structures which has 2 or 3 ring atoms mapped to a CG bead. The beads are then distinguished into four main types, i.e., polar, nonpolar, apolar, and charged. MARTINI was able to reproduce the properties of lipid bilayers on a semiquantitative level. These are including the APL, the distribution of groups across the membrane and the bending and area compression moduli.

MARTINI 2.0 further improved the stress profile across the lipid bilayer and its tendency to form pores. Additionally, the free energy of lipid desorption and lipid flip-flop across the bilayer agreed well with AT simulations and the condensing effect of cholesterol on the APL can be reproduced. MARTINI achieved 5- to 10-fold faster sampling of the configurational space of liquid hydrocarbons and the lipid tails inside a bilayer compared to AT force fields [44]. Among others, MARTINI has allowed applications in simulations of vesicle formation and fusion [50, 51], phase transition of lipid bilayers [52], and the structure and dynamics of membrane-protein assemblies [53, 54]. Currently, MARTINI has also been supplemented with an implicit solvent force field, named Dry MARTINI, which provided 1–2 order speed up and is expected to find application in simulations of large membranes containing millions of lipids [55].

The ELBA (an acronym for "electrostatic-based" by Orsi and Essex [56]) force field, so named because it is electrostatics based, offers another alternative for CG simulations of lipids and lipid bilayers. The ELBA model features two approaches in contrast to other CG methods. Notably, LJ interactions are treated using standard Lorentz-Berthelot mixing rules, similar to AT force fields. This is due to the explicit treatment of lipid electrostatics and water dipoles whereby a relative dielectric constant of unity ($\varepsilon_r = 1$) was used to model their interactions. In addition, a realistic dipolar water model based on a simpler soft sticky dipole potential, i.e. the Stockmayer potential is used in ELBA, which helped provide a correct diffusion coefficient of lipids in the liquid phase, a point often overestimated by other CG models. It also shows 15 and 200 times speed improvement over UA and AT models, respectively. Validation has been performed on DOPC, DOPE, and gel phase DSPC, whereby ELBA satisfactorily reproduced several fundamental experimental properties including APL, volume per lipid, curvature elastics constants, electrostatic potential distribution, electrostatic diffusion constant, and lipid diffusion coefficient. At the time of the development, ELBA was only available with the authors' in-house software, and it has since been implemented in LAMMPS [57]. However, there has yet to be a current parameterization for protein and other organic molecules, potentially limiting its use in mixed systems.

CG models for proteins are available and have been adopted into protein-bilayer and peptide-bilayer systems [58, 59]. The advantage in employing CG models is the improved speed and size. As CG force fields have a reduced degree of freedom and removed high-frequency motions such as hydrogen bond vibrations, integration time step could be pushed up to 20–40 fs and highly increases the accessible timescale by 100-fold. Therefore, it is possible to access phenomenon occurring at timescales not accessible to classical simulations, for example, membrane protein aggregation, demonstrated by the formation and domain-specific distribution of Ras proteins in plasma membranes [60]. CG simulation has also been used to self-assemble lipid bilayers around membrane proteins, providing information into protein positioning in the bilayer [61]. Comparison of membrane positioning for six proteins in the CG approach with available experimental data showed high qualitative similarity, even though there are discrepancies from different lipid composition used in the simulation and experiments [61].

Nevertheless, due to simplification of the representations, loss of details is inevitable. Two approaches have been employed to probe for detailed interactions while simulating in CG membrane models is by mixed AT-CG simulation and multiscaling. To elucidate small molecules, membrane peptides and proteins at high resolution, some approaches have used mixed CG-AT systems by simulating the ligand and protein atomistically in a CG membrane [62, 63]. Such mixed descriptions can be likened to quantum mechanical/molecular mechanical (QM/MM) simulations that have achieved considerable success in the past decade (see [64, 65] for reviews regarding usage of QM/MM with proteins). A major consideration in building a mixed AT-CG system is the parameterization and definition of the interactions at the AT-CG interface. As many CG force fields are derived from AT simulations, force-matching procedures can be used to derive effective pairwise CG force field from AT simulations. In AT-CG approach, all atoms simulation of the whole system is first carried out, the system will be subsequently divided into AT and CG parts whereby the most interesting part of the system will retain the atomistic details. The effective AT-CG force field is subsequently obtained by treating the AT and CG parts equally in the force matching procedure [63].

CHARMM-GUI PACE CG Builder [66] was developed for the modeling of large and complex biological system. It utilizes the mixed UA/CG where the PACE force field was used for protein in UA format and Martini CG force field for the water, ions and lipids. Thus, the number of atoms in a system can be reduced by the factor of 10. Analysis showed the PACE/MARTINI hybrid simulation has most of the proteins in the root means square deviation (RMSD) of less than 3 Å.

On the other hand, several methods have been developed for conversion of CG system to AT representation through fragment-based approach [67], simulated annealing [68], and force-matching [49]. These methods, termed the multiscale approach allows the use of CG simulations to explore membrane-lipid interactions, which, after a sufficient equilibration period is converted back to AT model simulation for detailed characterization [67].

3. Tools for membrane simulation setup and analysis

3.1. Automated setup of membrane simulation systems

The availability of many different force fields and parameters for a range of lipid molecules has made it easier to construct systems consisting of mixed lipids. Therefore, the lipid-converter tool is beneficial to easily adapt a system between force fields [69]. The lipid-converter tool can be used in command line by defining a PDB or Gromacs coordinate file and is also available as a web server [70]. The tool currently supports the Berger, GROMOS 43A1-S3, GROMOS 53A6, GROMOS 53A7, CHARMM36, OPLS-UA, and Lipid11. Thus, Stockholm lipids that are compatible with AMBER force field and used CHARMM nomenclature are also supported by this extension. Moreover, lipid converter may be useful to build non-conventional systems as it can generate asymmetric lipid distribution and even label leaflets in curved systems like vesicles.

Towards the end of automating the process of building heterogeneous membrane, the CHARMM-GUI Membrane Builder [71] is a useful tool to generate coordinates for membrane models and protein/membrane systems [72–74]. The membrane builder offers a selection of commonly used lipid models in addition to cholesterol which can be customized according to concentration, APL, hydration number, and thickness of the water layer [73].

Alternatively, a new web server MemGen is able to automatically set up lipid membrane simulation systems without restrictions in force fields, lipid types, or MD simulation software [75]. The user can upload one or more lipid structure files as well as amphiphilic molecules such as alcohol or detergent. A compact representation of each lipid aligned along the *z*-axis is generated by building GAFF topology of each lipid using ACPYPE, then applying simulated annealing with constant-force pulling on the head group and tail atoms as well as position-restraining potentials with Gromacs. The server subsequently hydrates the membrane with a number of water molecules, which can also be specified by the user. After the addition of counter ions or sodium chloride, a PDB format of the final structure is available for download. However, it must be noted that MemGen provides highly ordered, unphysical configurations which requires careful equilibration of at least 10 ns. In addition, it is also unable to produce asymmetric bilayers with different composition of lipids in the two monolayers.

iMembrane is another useful web-based tool which can predict the orientation of a membrane protein within the membrane [76]. Early approaches use a two-state membrane model or a simple hydrophobic slab to model the orientation of a membrane protein in the membrane. Scott and colleagues developed a CG MD to simulate membrane proteins in the presence of membrane lipids which self-assemble into a lipid bilayer [61]. Using the simulation results, iMembrane can predict the orientation of proteins of homologous structure or sequence. BLAST is first performed against the CGDB database for any input sequence or structure. Matches are subsequently realigned to the query using MUSCLE [77] for sequence realignment or "MAMMOTH" [78] for structure super-positioning. Residues in the query are then annotated as N (not in contact with the membrane), H (in contact with polar head group of the membrane lipids), or T (in contact with the lipid hydrophobic tails).

3.2. Tools for membrane MD simulation analysis

As MD simulations for membrane and membrane protein systems became widespread, many groups began developing tools to allow more efficient analysis of the MD trajectories. APL is an important indicator of the membrane phase and stability of the simulation. "GridMAT-MD" is a Perl program which can calculate the APL as well as thickness of a membrane [79]. For bilayer thickness calculation, the user can define a reference atom (such as the lipid phosphate, P atom) and the program first uses the upper leaflet as a reference and assign a paired lipid in the lower leaflet with the upper leaflet based on proximity in the x- and y-direction. The z-distance between the two points is calculated. The program then repeats the same step using the bottom leaflet as reference, and the two results are averaged, and written to a generic ASCII.dat file. Meanwhile, APL calculation of lipid-only systems can be as simple as taking the box size divided by the number of lipids in the upper or lower leaflet. Calculation of APL for membrane-protein systems are not as simple, and "GridMAT-MD" solves the problem by assigning protein atoms found within the lipid head groups to grid points then subtracts the total protein area from the size of the system. As of version 2.0, "GridMAT-MD" can now calculate the bilayer thickness and APL for multi ".pdb" or ".gro" files.

Mori and colleagues proposed a more sophisticated method for calculating the APL using Voronoi tessellation and Monte Carlo simulation [80]. Coordinates of center of mass for each lipid molecules and coordinates for protein atoms located between the maximum and minimum z-coordinates for the monolayer are projected onto the XY plane. Two-dimensional Voronoi analysis is subsequently performed for the lipids only. The APL for non-boundary lipids is the area of the Voronoi polygon where the lipid center of mass is located. The APL for boundary lipids can be determined by using a Monte Carlo integration method where the lipid region is probed by randomly shooting a pseudoparticle into the lipid Voronoi polygon. Thus, the APL for the boundary lipid is the product of the area of the Voronoi polygon, and the probability of the shot missing a protein atom. This method finds application in analysis of membrane-protein system and can differentiate between boundary and non-boundary lipids.

Not only that, analysis of other membrane properties became easier with the development of "Membrainy," an intelligent membrane analysis tool that can provide the calculation of various membrane-specific properties for planar bilayer trajectories [81]. This include APL, order parameter, head group orientation, lipid mixing/demixing entropy, time evolution of the transmembrane voltage, 2D surface map generation, gel percentage, membrane thickness, detection of lipid flip-flop and annular shell lipid analysis. While the program has been primarily designed for use with Gromacs MD package, it is also compatible with pdb trajectory from other MD packages. Currently, it is implemented with CHARMM36, Berger/GROMOS87 and Martini v2.0 force fields, but is also expandable to include other force fields and trajectory formats. Output graphs can be readable by the Grace plotting software.

Other than that, MEMBPLUGIN [82] is another tool to study the MD trajectories of membrane-protein and complex membrane structures. This is a plugin in Visual MD package to measure biophysical properties in the simulated membranes.

4. Towards realistic bilayer simulations

Beyond the improvements in computational power, force field developments, and CG methodologies, more accurate representations of the membrane continued to evolve. The biological membrane is a complex entity composed of numerous lipid species such as phosphatidylcholines, phosphatidylethanolamines, and phosphatidylserine. Other molecules such as cholesterols, sphingomyelins, and cardiolipins also play a role in regulating membrane structure and function. In the outer membrane of many species of Gram-negative bacteria, the presence of lipopolysaccharide in the upperleaflet modulates the insertion, folding, and dynamics of outer membrane proteins within the membrane. Available tools for generate mixed membrane bilayers are including CHARMM-GUI Membrane Builder [73] and MemBuilder [83]. CHARMM-GUI Membrane Builder and MemBuilder supports a total of 32 and 18 different lipids types, respectively.

Straatsma and Soares first reported the simulation of the outer membrane protein OprF in an asymmetric outer membrane with a lipopolysaccharide and phospholipids, describing the saccharides component using GLYCAM parameters [84]. Holdbrook et al. performed simulations of the *Haemphilus influenza* Hia autotransporter domain in LPS and a realistic outer membrane inner leaflet which comprises 1-myristoyl 2-palmitoleoyl phosphatidylethanolamine (DMPE) lipid [85]. Comparison with simulations of the autotransporter in simpler, single species DMPC lipid model showed that the DMPC membrane accurately replicated the membrane thickness of the outer membrane and reproduced similar dynamics of the protein in asymmetric LPS/MPoPE membrane [85]. The realistic bilayer, however, revealed a patch of positive lysine and arginine residues on the extracellular mouth of Hia that interact regularly with phosphate and sugar groups of the LPS and are suggested to anchor Hia within the outer membrane [85].

5. Conclusion

The continuous update and improvement of atomistic force fields expanded the types of lipid molecules which could be simulated and increased the accuracy to better match experimental data. Depending on the level of detail of the simulation, UA force fields are an excellent alternative to balance between accuracy and speed. By using CG force fields approaches, sampling and size limitations may be tackled efficiently. Ultimately, advances in computational power and hardware have improved the timescale and system size where MD can be employed. In membrane simulations, the microsecond mark has been reached and simulations are slowly becoming routine work to complement experimental results. In addition, various web server tools and useful analysis programs have been developed to aid membrane simulation analysis. Further advances in lipid force fields will make it possible to characterize membrane structures in greater time and physical scale.

Acknowledgements

This work was funded by the Malaysia Ministry of Higher Education Fundamental Research Grant Scheme (FRGS) (203/CIPPM/6711439) and the Higher Institutions Center of Excellence (HICoE) Grant. S. W. Leong would also like to thank the Malaysia Ministry of Higher Education for MyBrain Science scholarship.

Author details

S. W. Leong, T. S. Lim and Y. S. Choong*

*Address all correspondence to: yeesiew@usm.my

Institute for Research in Molecular Medicine, Universiti Sains Malaysia, Minden, Malaysia

References

[1] Singer SJ, Nicolson GL. The fluid mosaic model of the structure of cell membranes. Science. 1972;175:720–731. doi: 10.1126/science.175.4023.720

[2] van Meer G, Voelker DR, Feigenson GW. Membrane lipids: where they are and how they behave. Nat. Rev. Mol. Cell Biol. 2008;9:112–124. doi: 10.1038/nrm2330

[3] Lee AG. How lipids affect the activities of integral membrane proteins. Biochim. Biophys. Acta 2004;1666:62–87. doi: 10.1016/j.bbamem.2004.05.012

[4] Klauda JB, Brooks BR, MacKerell AD, Venable RM, Pastor RW. An *ab Initio* study on the torsional surface of alkanes and its effect on molecular simulations of alkanes and a DPPC bilayer. J. Phys. Chem. B. 2005;109:5300–5311. doi: 10.1021/jp0468096

[5] Klauda JB, Pastor RW, Brooks BR. Adjacent gauche stabilization in linear alkanes: implications for polymer models and conformational analysis. J. Phys. Chem. B. 2005;109:15684–15686. doi: 10.1021/jp0527608

[6] Tien HT, Diana AL. Bimolecular lipid membranes: a review and a summary of some recent studies. Chem. Phys. Lipids 1968;2:55–101. doi: 10.1016/0009-3084(68)90035-2

[7] Marsh D. Renormalization of the tension and area expansion modulus in fluid membranes. Biophys. J. 1997;73:865–869. doi: 10.1016/S0006-3495(97)78119-0

[8] Klauda JB, Venable RM, Freites JA, O'Connor JW, Tobias DJ, Mondragon-Ramirez C, Vorobyov I, MacKerell AD, Pastor RW, Connor JWO. Update of the CHARMM all-atom additive force field for lipids: validation on six lipid types. J. Phys. Chem. B 2010;114:7830–7843. doi: 10.1021/jp101759q

[9] Piggot TJ, Piñeiro Á, Khalid S. Molecular dynamics simulations of phosphatidylcholine membranes: a comparative force field study. J. Chem. Theory Comput. 2012;8:4593–4609. doi: 10.1021/ct3003157

[10] Jójárt B, Martinek TA, Jo ZS. Performance of the general AMBER force field in modeling aqueous POPC membrane bilayers. J. Comput. Chem. 2007;28:2051–2058. doi: 10.1002/jcc

[11] Dickson CJ, Rosso L, Betz RM, Walker RC, Gould IR. GAFFlipid: a General AMBER force field for the accurate molecular dynamics simulation of phospholipid. Soft Matter. 2012;8:9617–9627. doi: 10.1039/c2sm26007g

[12] Rosso L, Gould IR. Structure and dynamics of phospholipid bilayers using recently developed general all-atom force fields. J. Comput. Chem. 2007;29:24–37. doi: 10.1002/jcc.20675

[13] Siu SWI, Vácha R, Jungwirth P, Böckmann R. Biomolecular simulations of membranes: physical properties from different force fields. J. Chem. Phys. 2008;128:125103–125103. doi: 10.1063/1.2897760

[14] Dickson CJ, Madej BD, Skjevik ÅA, Betz RM, Teigen K, Gould IR, Walker RC. Lipid14: the AMBER lipid force field. J. Chem. Theory Comput. 2014;10:865–879. doi: 10.1021/ct4010307

[15] Rand RP, Parsegian VA. Hydration forces between phospholipid bilayers. BBA Rev. Biomembranes 1989;988:351–376. doi: 10.1016/0304-4157(89)90010-5

[16] Rappolt M, Hickel A, Bringezu F, Lohner K. Mechanism of the lamellar/inverse hexagonal phase transition examined by high resolution X-ray diffraction. Biophys. J. 2003;84:3111–3122. doi: 10.1016/S0006-3495(03)70036-8

[17] Bruno A, Scrima M, Novellino E, D'Errico G, D'Ursi AM, Limongelli V. The glycan role in the glycopeptide immunogenicity revealed by atomistic simulations and spectroscopic experiments on the multiple sclerosis biomarker CSF114(Glc). Sci. Rep. 2015;5:9200. doi: 10.1038/srep09200

[18] Côté S, Binette V, Salnikov ES, Bechinger B, Mousseau N. Probing the huntingtin 1-17 membrane anchor on a phospholipid bilayer by using all-atom simulations. Biophys. J. 2015;108:1187–1198. doi: 10.1016/j.bpj.2015.02.001

[19] Hong C, Tieleman DP, Wang Y. Microsecond molecular dynamics simulations of lipid mixing. Langmuir 2014;30:11993–12001. doi: 10.1021/la502363b

[20] Rogaski B, Klauda JB. Membrane-binding mechanism of a peripheral membrane protein through microsecond molecular dynamics simulations. J. Mol. Biol. 2012;423:847–861. doi: 10.1016/j.jmb.2012.08.015

[21] Grossfield A, Feller SE, Pitman MC. Convergence of molecular dynamics simulations of membrane proteins. Proteins 2007;40:31–40. doi: 10.1002/prot

[22] Andersson M, Freites JA, Tobias DJ, White SH. Structural dynamics of the S4 voltage-sensor helix in lipid bilayers lacking phosphate groups. J. Phys. Chem. B 2011;115:8732–8738. doi: 10.1021/jp2001964

[23] Li Y, Andersen OS, Roux B. Energetic of double ion occupancy in the gramicidin A channel. J. Phys. Chem. B 2010;114:13881–13888. doi: 10.1021/jp105820u

[24] Fleming PJ, Freites JA, Moon CP, Tobias DJ, Fleming KG. Outer membrane phospholipase A in phospholipid bilayers: a model system for concerted computational and experimental investigations of amino acid side chain partitioning into lipid bilayers. BBA Biomembranes 2012;1818:126–134. doi: 10.1016/j.bbamem.2011.07.016

[25] Sonntag Y, Musgaard M, Olesen C, Schiøtt B, Møller JV, Nissen P, Thøgersen L. Mutual adaptation of a membrane protein and its lipid bilayer during conformational changes. Nat. Commun. 2011;2:304. doi: 10.1038/ncomms1307

[26] Hénin J, Shinoda W, Klein ML. United-aom acyl chains for CHARMM phospholipids. J. Phys. Chem. B 2008;112:7008–7015. doi: 10.1021/jp800687p

[27] Tieleman DP, Maccallum JL, Ash WL, Kandt C, Xu Z, Monticelli L. Membrane protein simulations with a united-atom lipid and all-atom protein model: lipid-protein interactions, side chain transfer free energies and model proteins. J. Phys. Condens. Matter. 2006;18:S1221–S1234. doi: 10.1088/0953-8984/18/28/s07

[28] Essex JW, Hann MM, Richards WG. Molecular dynamics simulation of a hydrated phospholipid bilayer. Phil. Trans. R. Soc. B 1994;344:239–260. doi: 10.2307/55918

[29] Chiu SW, Clark M, Balaji V, Subramaniam S, Scott HL, Jakobsson E. Incorporation of surface tension into molecular dynamics simulation of an interface: a fluid phase lipid bilayer membrane. Biophys. J. 1995;69:1230–1245. doi: 10.1016/S0006-3495(95)80005-6

[30] Berger O, Edholm O, Jähnig F. Molecular dynamics simulations of a fluid bilayer of dipalmitoylphosphatidylcholine at full hydration, constant pressure, and constant temperature. Biophys. J. 1997;72:2002–2013. doi: 10.1016/S0006-3495(97)78845-3

[31] Cordomí A, Caltabiano G, Pardo L. Membrane protein simulations using AMBER force field and Berger lipid parameters. J. Chem. Theory Comput. 2012;8:948–958. doi: 10.1021/ct200491c

[32] Jerabek H, Pabst G, Rappolt M, Stockner T. Membrane-mediated effect on ion channels induced by the anesthetic drug ketamine. J. Am. Chem. Soc. 2010;132:7990–7997. doi: 10.1021/ja910843d

[33] Lemkul JA, Bevan DR. Characterization of interactions between PilA from *Pseudomonas aeruginosa* strain K and a model membrane. J. Phys. Chem. B 2011;115:8004–8008. doi: 10.1021/jp202217f

[34] Lensink MF, Govaerts C, Ruysschaert J-M. Identification of specific lipid-binding sites in integral membrane proteins. J. Biol. Chem. 2010;285:10519–10526. doi: 10.1074/jbc.M109.068890

[35] Braun AR, Sachs JN, Nagle JF. Comparing simulations of lipid bilayers to scattering data: the GROMOS 43A1-S3 force field. J. Phys. Chem. B 2013;117:5065–5072. doi: 10.1021/jp401718k

[36] Oostenbrink C, Villa A, Mark AE, Van Gunsteren WF. A biomolecular force field based on the free enthalpy of hydration and solvation: the GROMOS force-field parameter sets 53A5 and 53A6. J. Comput. Chem. 2004;25:1656–1676. doi: 10.1002/jcc.20090

[37] Kukol A. Lipid models for united-atom molecular dynamics simulations of proteins. J. Chem. Theory Comput. 2009;5:615–626. doi: 10.1021/ct8003468

[38] Scott WRP, Hünenberger PH, Tironi IG, Mark AE, Billeter SR, Fennen J, Torda AE, Huber T, Krüger P, van Gunsteren WF. The GROMOS biomolecular simulation program package. J. Phys. Chem. A 1999;103:3596–3607. doi: 10.1021/jp984217f

[39] Chiu SW, Pandit SA, Scott HL, Jakobsson E. An improved united atom force field for simulation of mixed lipid bilayers. J. Phys. Chem. B 2009;113:2748–2763. doi: 10.1021/jp807056c

[40] Poger D, Van Gunsteren WF, Mark AE. A new force field for simulating phosphatidylcholine bilayers. J. Comput. Chem. 2010;31:1117–1125. doi: 10.1002/jcc.21396

[41] Poger D, Mark AE. On the validation of molecular dynamics simulations of saturated and cis-monounsaturated phosphatidylcholine lipid bilayers: a comparison with experiment. J. Chem. Theory Comput. 2010;6:325–336. doi: 10.1021/ct900487a

[42] Bjelkmar P, Niemelä PS, Vattulainen I, Lindahl E. Conformational changes and slow dynamics through microsecond polarized atomistic molecular simulation of an integral Kv1.2 ion channel. PLoS Comput. Biol. 2009;5:e1000289. doi: 10.1371/journal.pcbi.1000289

[43] Chakrabarti N, Neale C, Payandeh J, Pai EF, Pomès R. An iris-like mechanism of pore dilation in the CorA magnesium transport system. Biophys. J. 2010;98:784–792. doi: 10.1016/j.bpj.2009.11.009

[44] Marrink SJ, Risselada HJ, Yefimov S, Tieleman DP, de Vries AH. The MARTINI force field: coarse grained model for biomolecular simulations. J. Phys. Chem. B 2007;111:7812–7824. doi: 10.1021/jp071097f

[45] Marrink SJ, de Vries AH, Mark AE. Coarse grained model for semiquantitative lipid simulations. J. Phys. Chem. B 2004;108:750–760. doi: 10.1021/jp036508g

[46] Shelley JC, Shelley MY, Reeder RC, Bandyopadhyay S, Klein ML. A coarse grain model for phospholipid simulations. J. Phys. Chem. B 2001;105:4464–4470. doi: 10.1021/jp010238p

[47] Lyubartsev AP. Multiscale modeling of lipids and lipid bilayers. Eur. Biophys. J. 2005;35:53–61. doi: 10.1007/s00249-005-0005-y

[48] Elezgaray J, Laguerre M. A systematic method to derive force fields for coarse-grained simulations of phospholipids. Comput. Phys. Commun. 2006;175:264–268. doi: 10.1016/j.cpc.2006.01.009

[49] Izvekov S, Voth GA. A multiscale coarse-graining method for biomolecular systems. J. Phys. Chem. B 2005;109:2469–2473. doi: 10.1021/jp044629q

[50] Marrink SJ, Mark AE. The mechanism of vesicle fusion as revealed by molecular dynamics simulations. J. Am. Chem. Soc. 2003;125:11144–11145. doi: 10.1021/ja036138+

[51] Marrink SJ, Mark AE. Molecular dynamics simulation of the formation, structure, and dynamics of small phospholipid vesicles. J. Am. Chem. Soc. 2003;125:15233–15242. doi: 10.1021/ja0352092

[52] Prates Ramalho JP, Gkeka P, Sarkisov L. Structure and phase transformations of DPPC lipid bilayers in the presence of nanoparticles: insights from coarse-grained molecular dynamics simulations. Langmuir 2011;27:3723–3730. doi: 10.1021/la200236d

[53] Sansom MSP, Scott KA, Bond PJ. Coarse-grained simulation: a high-throughput computational approach to membrane proteins. Biochem. Soc. Trans. 2008;36:27–32. doi: 10.1042/bst0360027

[54] Bucher D, Hsu Y-H, Mouchlis VD, Dennis EA, McCammon JA. Insertion of the Ca^{2+}-independent phospholipase A_2 into a phospholipid bilayer via coarse-grained and atomistic molecular dynamics simulations. PLoS Comput. Biol. 2013;9:e1003156. doi: 10.1371/journal.pcbi.1003156

[55] Arnarez C, Uusitalo JJ, Masman MF, Ingólfsson HI, de Jong DH, Melo MN, Periole X, de Vries AH, Marrink SJ. Dry Martini, a coarse-grained force field for lipid membrane simulations with implicit solvent. J. Chem. Theory Comput. 2015;11:260–275. doi: 10.1021/ct500477k

[56] Orsi M, Essex JW. The ELBA force field for coarse-grain modeling of lipid n
PLoS One 2011;6:e28637. doi: 10.1371/journal.pone.0028637

[57] Plimpton S. Fast parallel algorithms for short-range molecular dynamics. J. Comput. Phys. 1995;117:1–19. doi: 10.1006/jcph.1995.1039

[58] Monticelli L, Kandasamy SK, Periole X, Larson RG, Tieleman DP, Marrink SJ. The

MARTINI coarse-grained force field: extension to proteins. J. Chem. Theory Comput. 2008;4:819–834. doi: 10.1021/ct700324x

[59] Spijker P, van Hoof B, Debertrand M, Markvoort AJ, Vaidehi N, Hilbers PJ. Coarse grained molecular dynamics simulations of transmembrane protein-lipid systems. Int. J. Mol. Sci. 2010;11:2393–2420. doi: 10.3390/ijms11062393

[60] Janosi L, Li Z, Hancock JF, Gorfe AA. Organization, dynamics, and segregation of Ras nanoclusters in membrane domains. Proc. Natl. Acad. Sci. USA 2012;109:8097–8102. doi: 10.1073/pnas.1200773109

[61] Scott KA, Bond PJ, Ivetac A, Chetwynd AP, Khalid S, Sansom MSP. Coarse-grained MD simulations of membrane protein-bilayer self-assembly. Structure 2008;16:621–30. doi: 10.1016/j.str.2008.01.014

[62] Orsi M, Essex JW. Permeability of drugs and hormones through a lipid bilayer: insights from dual-resolution molecular dynamics. Soft Matter. 2010;6:3797–3797. doi: 10.1039/c0sm00136h

[63] Shi Q, Izvekov S, Voth G. Mixed atomistic and coarse-grained molecular dynamics: simulation of a membrane-bound ion channel. J. Phys. Chem. B 2006;110:15045–15048. doi: 10.1021/jp062700h

[64] Murphy RB, Philipp DM, Friesner RA. A mixed quantum mechanics/molecular mechanics (QM/MM) method for large-scale modeling of chemistry in protein environments. J. Comput. Chem. 2000;21:1442–1457. doi: 10.1002/1096-987X(200012)21:16<1442::AID-JCC3>3.0.CO;2-O

[65] Senn HM, Thiel W. QM/MM studies of enzymes. Curr. Op. Chem. Biol. 2007;11:182–187. doi: 10.1016/j.cbpa.2007.01.684

[66] Qi Y, Cheng X, Han W, Jo S, Schulten K, Im W. CHARMM-GUI PACE CG Builder for solution, micelle, and bilayer coarse-grained simulations. J. Chem. Inf. Model. 2014;54:1003–1009. doi: 10.1021/ci500007n

[67] Stansfeld PJ, Sansom MSPP. From coarse grained to atomistic: a serial multiscale approach to membrane protein simulations. J. Chem. Theory Comput. 2011;7:1157–1166. doi: 10.1021/ct100569y

[68] Rzepiela AJ, Schäfer LV, Goga N, Risselada HJ, De Vries AH, Marrink SJ. Reconstruction of atomistic details from coarse-grained structures. J. Comput. Chem. 2010;31:1333–1343. doi: 10.1002/jcc.21415

[69] Larsson P, Kasson PM. Lipid converter, a framework for lipid manipulations in molecular dynamics simulations. J. Membr. Biol. 2014;247:1137–1140. doi: 10.1007/s00232-014-9705-5

[70] Larsson P, Kasson P. Lipid-converter [Internet]. 2014. Available from: http://lipid-converter.appspot.com/ [accessed: 2016-02-16].

[71] Jo S, Lim JB, Klauda JB, Im W. CHARMM-GUI Membrane Builder [Internet]. 2009. Available from: http://www.charmm-gui.org/input/membrane [Accessed: 2016-02-16]

[72] Jo S, Kim T, Im W. Automated builder and database of protein/membrane complexes for molecular dynamics simulations. PLoS One 2007;2:e880. doi: 10.1371/journal.pone.0000880

[73] Jo S, Lim JB, Klauda JB, Im W. CHARMM-GUI membrane builder for mixed bilayers and its application to yeast membranes. Biophys. J. 2009;97:50–58. doi: 10.1016/j.bpj.2009.04.013

[74] Wu EL, Cheng X, Jo S, Rui H, Song KC, Davila-Contreras EM, Qi Y, Lee J, Monje-Galvan V, Venable RM, Klauda JB, Im W. CHARMM-GUI membrane builder toward realistic biological membrane simulations. J. Comput. Chem. 2014;35:1997–2004. doi: 10.1002/jcc.23702

[75] Knight CJ, Hub JS. MemGen: a general web server for the setup of lipid membrane simulation systems. Bioinformatics 2015;31:2897–2899. doi: 10.1093/bioinformatics/btv292

[76] Kelm S, Shi J, Deane CM. iMembrane: homology-based membrane-insertion of proteins. Bioinformatics 2009;25:1086–1088. doi: 10.1093/bioinformatics/btp102

[77] Edgar RC. MUSCLE: multiple sequence alignment with high accuracy and high throughput. Nucleic Acids Res. 2004;32:1792–1797. doi: 10.1093/nar/gkh340

[78] Ortiz AR, Strauss CEM, Olmea O. MAMMOTH (matching molecular models obtained from theory): an automated method for model comparison. Protein Sci. 2002;11:2606–2621. doi: 10.1110/ps.0215902

[79] Allen WJ, Lemkul JA, Bevan DR. GridMAT-MD: a grid-based membrane analysis tool for use with molecular dynamics. J. Comput. Chem. 2009;30:1952–1958. doi: 10.1002/jcc.21172

[80] Mori T, Ogushi F, Sugita Y. Analysis of lipid surface area in protein–membrane systems combining voronoi tessellation and monte carlo integration methods. J. Comput. Chem. 2012;33:286–293. doi: 10.1002/jcc.21973

[81] Carr M, MacPhee CE. Membrainy: a 'smart', unified membrane analysis tool. Source Code Biol. Med. 2015;10:1–10. doi: 10.1186/s13029-015-0033-7

[82] Guixa-Gonzalez R, Rodriguez-Espigares I, Ramirez-Anguita JM, Carrio-Gaspar P, Martinez-Seara H, Giorgino T, Selent J. MEMBPLUGIN: studying membrane complexity in VMD. Bioinformatics 2014;30:1478–1480. doi: 10.1093/bioinformatics/btu037

[83] Ghahremanpour MM, Arab SS, Aghazadeh SB, Zhang J, van der Spoel D. MemBuilder: a web-based graphical interface to build heterogeneously mixed membrane bilayers for the GROMACS biomolecular simulation program. Bioinformatics 2013;30:439–441. doi: 10.1093/bioinformatics/btt680

[84] Straatsma TP, Soares TA. Characterization of the outer membrane protein OprF of *Pseudomonas aeruginosa* in a lipopolysaccharide membrane by computer simulation. Proteins 2009;74:475–488. doi: 10.1002/prot.22165

[85] Holdbrook DA, Piggot TJ, Sansom MSP, Khalid S. Stability and membrane interactions of an autotransport protein: MD simulations of the Hia translocator domain in a complex membrane environment. Biochim. Biophys. Acta 2013;1828:715–723. doi: 10.1016/j.bbamem.2012.09.002

2

Application of Bioinformatics Methodologies in the Fields of Skin Biology and Dermatology

Sidra Younis, Valeriia Shnayder and
Miroslav Blumenberg

Abstract

Bioinformatics is a research field that uses computer-based tools to investigate life sciences questions employing "big data" results from large-scale DNA sequencing, whole genomes, transcriptomes, metabolomes, populations, and biological systems, which can only be comprehensively viewed *in silico*. The epidermis was among the earliest targets of bioinformatics studies because it represents one of the most accessible targets for research. An additional advantage of working with the epidermis is that the sample can even be recovered using tape stripping, an easy, noninvasive protocol. Consequently, bioinformatics methods in the fields of skin biology and dermatology generated a fairly large volume of bioinformatics data, which led us to originate the term "skinomics." Skinomics data are directed toward epidermal differentiation, malignancies, inflammation, allergens, and irritants, the effects of ultraviolet (UV) light, wound healing, the microbiome, stem cells, etc. Cultures of cutaneous cell types, keratinocytes, fibroblasts, melanocytes, etc., as well as skin from human volunteers and from animal models, have been extensively experimented on. Here, we review the development of the skinomics, its methodology, current achievements, and future potentials.

Keywords: epidermal differentiation, inflammation, microbiome, noninvasive, psoriasis

1. Introduction: both wide angle and focused view of skin biology

Bioinformatics is an umbrella term for a wide range of methodologies, studies that generate large datasets [1]. The term refers to the methodology, rather than a subject matter (akin to "microscopy"). Omics techniques, rather than focusing on a single protein, gene, metabolite, microorganism, etc., comprehensively deal with the entire collection of all proteins, the whole genome, the complete metabolic array, or the full microbiome of a given biological system. "Omics" methodology has reached nowadays full maturity and recognition of the research community. The methodology comprises very large datasets that require sophisticated *in silico* analyses. Accordingly, it uses, on the one hand, large central databanks, repositories of raw and "pre-processed" data, and complex suites of analysis programs developed by multidisciplinary teams that include statisticians, graphic designers, etc., and on the other hand, many individual laboratories and groups providing discrete pieces of the large "omics" puzzles and applying the algorithms to their specific objectives [2].

Bioinformatics approaches received a major impetus with the development of "omics" techniques. Arguably, DNA microarrays are the most widely used omics technology [3]. In microarrays, the DNA probes are immobilized on solid supports, and the samples, such as total bulk DNA or RNA from the specimen, are labeled and then hybridized to the arrays in order to measure individual genes. Requiring only minute amounts of input DNA or RNA, such microarrays probe simultaneously, in hugely parallel experiments, many genes, for example, all the genes in the human genome. These experiments create very large volumes of data. Strangely enough, microarrays allow not only a very broad but also a very detailed insight into the biological function, mechanisms, and diseases of interest to dermatology. Microarrays empower us to see both the "forest and the trees."

Bioinformatics is a very rapidly developing science that constantly improves its methodology, microarrays, sequencing aparati, data repositories, hardware, and software. To keep up with the field, we have found very useful the special database issue of the *Nucleic Acids Research* [4], which is published every January. In these issues, we find description of the functions and roles of various data repositories. Another invaluable resource is the *Bioconductor* [5] an ever-expanding collection of bioinformatics algorithms developed by computer scientists and programmers from all over the world. The *Bioconductor* analysis packages are freely available to all. They are well described and annotated, and usually the program developers are helpful in troubleshooting and even extensive hand-holding.

Directly accessible, skin was among the first organs analyzed using omics approaches. As a result, dermatology was one of the first medical disciplines to welcome and support omics results. Name "skinomics" has been proposed to designate specifically the bioinformatics studies in dermatology and skin biology [6]. The objectives of skinomics are to provide, enlarge, and buildup our knowledge of skin biology, to improve function of the healthy skin, and to assist in treating pathological skin conditions.

Skinomics studies focused to a significant extent, understandably, on skin cancers [6, 7]. For example, melanoma has been arguably the most studied skin disease. Microarray analyses identified markers of melanoma progression and of its metastatic potential. Similar studies targeted basal and squamous cell carcinomas. Specific for dermatology, noninvasive method using simple tape stripping can provide adequate material for transcriptional profiling of melanoma, psoriasis, and other skin diseases. The molecular changes in psoriatic plaques, that is, differences between uninvolved and involved skin and interestingly, the healed lesions of psoriatics, have also been defined using very large cohorts of patients. The psoriatic patients analyzed using this technology by now number in several hundreds [8]. Cutting-edge state-of-the-art international skinomics studies, involving almost 20 different countries, have characterized many of the psoriasis susceptibility loci in the human genome, and identified the genes with putative roles on the pathology of this disease. These genes represent potential targets for intervention [8, 9]. DNA microarrays have been used to follow the course of psoriasis treatment and to predict responses or resistance to specific treatment modalities. The characteristic changes in the microbiomes of psoriasis and of atopic dermatitis patients have been correlated with the progression of each disease.

Arguably the most frequent and continual methodology in skinomics is the use of DNA microarrays, such as those from Affymetrix and Illumina. The DNA microarrays are a perfect medium because they simultaneously measure the expression of the entire genome [10]. Printed cDNA arrays, originated by Brown at Stanford [11], are often homemade, inexpensive, and two color, that is, compare two samples on the same chip. They are easy to customize in-house for specific applications. The commercial synthetic oligonucleotide microarrays are pricier, but tend to be more reliable. The microarray community has established a set of guidelines known as "MIAMI" rules (minimal information about microarrays) to allow comparisons of data obtained using different microarrays, sample handling techniques, quality of data, etc.

In the field of skin biology, DNA microarrays have been used to identify the genes specific for epidermal stem cells. Moreover, the transcriptional changes occurring during the process of epidermal differentiation have been characterized. The consequences of epidermal and barrier disruption have been defined. Importantly, cultures of epidermal keratinocytes *in vitro* have been used by our group as well as many others in many studies because these cells respond to ultraviolet (UV) light, hormones and vitamins, inflammatory and immunomodulating cytokines, chemokines and growth factors, environmental toxins, microbes, physical injury, etc.

As the medical field presses forward in the direction of personalized medicine, we can anticipate that the skinomics approaches will be shortly applied at the bedside, directly to the personalized dermatology practice of the future.

2. Historical perspective

The first microarrays were developed at Stanford University by Dr. Pat Brown and his group [11]. Soon thereafter, they applied this methodology to skin biology [12]. Specifically, using the well-known model to achieve cell cycle synchronization by serum deprivation and then re-introduction of serum to the culture medium, Iyer et al. [12] characterized the timing and choreography of cell cycle gene regulation in synchronized cultures. Unexpectedly, Iyer et al. [12] also found that dermal fibroblasts respond to signals from serum by inducing specifically the wound-healing responses. In retrospect, this result makes perfect sense because dermal fibroblasts are not exposed to serum, except right after wounding, when they are required to mount an appropriate response.

Another Stanford team, the group of Dr. Paul Khavari, was the first to use microarrays in dermatology [13, 14]. They used microarrays to follow the outcomes of gene therapy for junctional epidermolysis bullosa, a lethal genetic skin disorder. The microarray analysis showed that the normal gene expression has not been completely reestablished, although the replacement of the affected gene restored cell growth and adhesion.

Subsequently, melanomas and skin carcinomas have been investigated using microarrays, as have certain inflammatory diseases, such as psoriasis and eczema, as well as responses to allergens and irritants, effects of UV light, skin aging, wound healing, keloid formation, etc. The large body of skinomics data was used in meta-analyses. For example, Dr. Noh and his team were the first to use meta-analysis of microarray data in dermatology [15]. Such meta-analyses include our own work as well [16, 17].

3. Noninvasive sample acquisition

A very important advantage of skin-oriented research is that the samples can be acquired from skin completely noninvasively and almost painlessly. Based on the work of Drs. Morhenn, Benson, and others, it was demonstrated that simple tape stripping provides sufficient quantity and quality of RNA for use in microarrays [18]. The methodology has been useful in studies of psoriasis, melanoma, etc. [19, 20]. Because of the noninvasive access to tissue, dermatology can be expected to lead further advance toward "omics" techniques. These will be directly applicable to the personalized medicine in the future.

4. Epidermal differentiation

Epidermal keratinocytes are "multifunctional" cells: on the one hand, they must differentiate through a tightly choreographed, multistage process in order to create cornified envelopes, unique three-dimensional structures in the stratum corneum (**Figure 1**); on the other hand, keratinocytes must respond to very many extracellular environmental stimuli, ranging from UV light and chemical irritants to bacteria and viruses. Keratinocytes also must communicate with nearby cells including other keratinocytes, melanocytes, dendritic cells, and others, both sending signals to these cells and receiving signals from them. As a result, keratinocytes have a large transcriptome, and they express relatively many genes. To determine which of the expressed genes are inherent to all keratinocytes, specific for the layers of epidermal differentiation and induced extracellularly, we compared the transcriptomes of harvested skin from human subjects in artificially and three-dimensionally cultured, differentiated, and reconstructed epidermis *in vitro* as well as in keratinocytes cultured as monolayers, and in nonkeratinocyte cell types.

Under all conditions, the keratinocytes express many proteases and protease inhibitors. Skin and the three-dimensional constructs, but not keratinocyte monolayers, express epidermal differentiation markers, including filaggrin, involucrin, loricrin, and other cornified envelope components. Skin specifically expresses a large number of transcription factors, cell surface receptors, and secreted proteins. Surprisingly, the mitochondrial genes were significantly suppressed in skin, which suggests a low metabolic rate. In culture, keratinocytes amply express the cell cycle and DNA replication machinery, and also integrins and extracellular matrix proteins. These results define the expressed and regulated genes in epidermal keratinocytes [21].

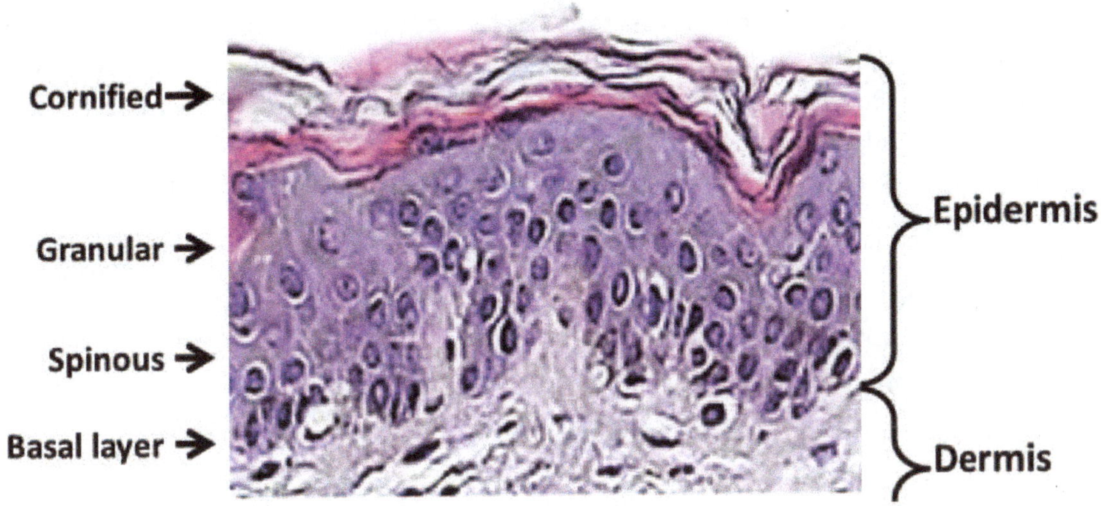

Figure 1. Cross-section through skin. Only the top portion of the dermis is shown, the rest of the dermis extends far below. The layers of the epidermis are marked on the left.

The regulatory circuits that control epidermal differentiation have been a focus of significant research efforts. Inhibition of Jun N-terminal kinase, JNK, in keratinocytes *in vitro* stimulates virtually all aspects of *in vivo* epidermal differentiation, including withdrawal from the cell cycle, cessation of motility, stratification, and production of cornified envelopes [22]. Inhibiting JNK also induces the expression of genes responsible for lipid and steroid metabolism, mitochondrial proteins, and histones. Simultaneously, the transcripts for basal cell markers are suppressed, including those for integrins, hemidesmosomes, and ECM components. We found that in the promoter sequences of JNK-regulated genes, the forkhead family binding sites and the c-Fos binding sites are overrepresented [22].

Vitamin D and calcium promote epidermal differentiation [23]. Specifically, kallikreins, serpins, and c-Fos were found to be vitamin D-responsive genes with roles in epidermal differentiation. Seo et al. [24] identified a subset of calcium-regulated genes in human keratinocytes [24]. Conversely, retinoic acid and its analogs inhibit differentiation of epidermal keratinocytes [25]. Microarray analysis identified Rho as another signaling molecule that suppresses differentiation-associated genes [26]. The papillomavirus type 16, HPV-16, E6 oncoprotein inhibits keratinocyte differentiation and suppresses transglutaminase, involucrin, elafin, and keratins [27]. Different classes of HPVs have different effects on cellular transcription patterns [28].

TGFβ promotes the basal, undifferentiated phenotype in keratinocytes via the SMAD4 transcription factor [29]. Interestingly, TGFβ-induced cell cycle arrest and migration genes via SMAD4, but not the epithelial–mesenchymal transition associated genes; these were not SMAD4-dependent, which suggest that a loss of SMAD4 in human carcinomas may interfere with the tumor-suppression, while maintaining the tumor-promoting responses to TGFβ.

To study epidermal differentiation transcriptome *in vivo,* we have separated the basal and suprabasal layers of skin and compared the transcriptomes of the two cell populations [28]. The human skin samples otherwise discarded after reduction mammoplasty are obtained usually within 2–6 h after surgery. The adipocytes and most of the dermis are physically removed leaving ~0.2 mm of mostly the epidermis. After enzymatic treatment, the epidermis is gently separated and single cell suspension derived using trypsin [16, 30, 31]. Magnetic beads attached to integrin β4 antibody are used to collect basal keratinocytes while the non-adherent, β4-cells represent the suprabasal cell populations. We disrupted the epidermal cells and isolated the RNA by using Trizol reagent. Next we use Qiashredders to homogenize cell extracts, remove DNA using on-column DNAse digestion RNeasy prepare the RNA using kits from Qiagen. As quality control, we visualize 28S and 18S ribosomal bands and check that the OD260/280 spectrophotometric ratio is at least 1.8. In the next step, ~5 μg of RNA is labeled according to the Affymetrix-suggested protocols.

5. The skin microbiome

Skin, our outermost layer, represents the first line of defense against pathogenic microbes. The intimate contact between the skin and the infectious microbial world has been known since the biblical days, was already discussed by Hippocrates, and has been studied for a very long time [32–34]. Microbes were perceived primarily as pathogens, which, fulfilling the Koch's postulates, can cause acne, impetigo, folliculitis, etc. As a result, skin microbes have been treated with disinfectants and antibiotics [35–37]. Knowledge of skin microorganisms was deficient because this knowledge was based on *in vitro* culturing of these microbes. While a few cutaneous bacteria and fungi could be grown in laboratories, the vast majority of micro-organisms known to reside on human skin was, and still is, uncultivatable [38]. Recent advances in large-scale DNA sequencing led to major breakthroughs in defining the cutaneous microbiome. Specifically, the 16S small subunit ribosomal RNA in prokaryotes and 18S in eukaryotes are encoded in genomes of all living organisms; genes encoding these RNAs are closely related so that a set of PCR primers can be used for multi-taxa amplification, providing unambiguous identification of individual species [39].

The major breakthrough in defining the cutaneous microbiome occurred in 2007 when Dr. M. Blaser's laboratory at the New York University, USA, published the molecular analysis of superficial skin bacterial biota of human forearm [40]. In this work, Gao et al. found that the cutaneous microbiome consists predominantly of six bacterial genera, *Propionibacterium, Corynebacterium, Staphylococcus, Streptococcus, Acinetobacter,* and *Finegoldia* were present in all subjects and represented 63% of all DNA clones analyzed. Approximately 8% of clones represented previously unknown organisms. Some 300 different bacteria inhabit our skin, although we humans demonstrate remarkable interpersonal variation in our cutaneous microbiota. Different body sites harbor specific microbial patterns that characteristically change in skin diseases [40]. Four phyla, *Actinobacteria, Firmicutes, Proteobacteria,* and *Bacteroidetes,* constitute vast majority of skin bacteria, while *Malassezia* dominates the skin fungi (**Figure 2**). The somewhat more complex fungal 18S RNA genes can be used to identify fungi, yeasts and other eukaryotes, and they have been used to confirm the abundance of *Malassezia* species on human skin [41, 42]. Current understanding of the cutaneous microbes has shown that they are, for the most part, commensal and beneficial, useful and protective, and only rarely dysbiosis and infection of pathogens occur.

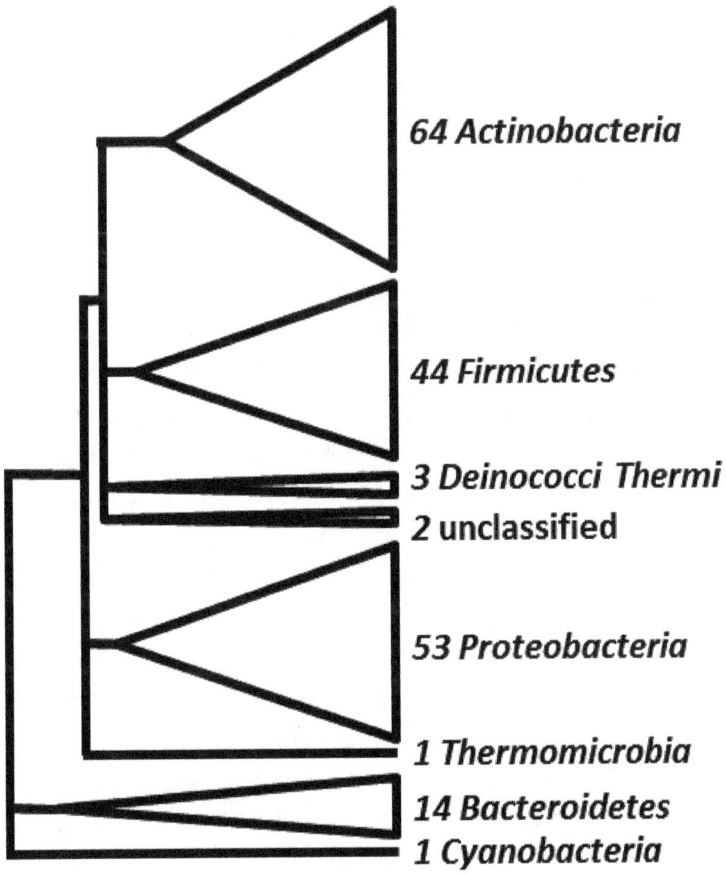

Figure 2. Bacterial populations in the cutaneous microbiota. Note the predominance of *Actinobacteria, Firmicutes,* and *Proteobacteria.* The numbers refer to different species, or "operational taxonomic units" detected.

Different body sites harbor different bacterial complexes. In a very significant microbiome analysis, Dr. Segre and her collaborators described microbiomes from 20 different body sites from 10 healthy individuals [43]. From such data, the authors were able to make several important conclusions regarding our cutaneous microbiome. Intrapersonal variation between symmetrical sites, that is, left vs. right forearms was much less than the interpersonal variation. Perhaps unexpectedly, the protected sites, such as inguinal and alar creases, were more related than were the freely exposed sites, such as forearms. Re-sampling resulted in higher intrapersonal similarity, suggesting a fairly high consistency of the microbiome during the 4–6 months duration of the study.

The types of the microbiota were classified according to the type of skin as sebaceous, moist, or dry [43]. These sites had different proportions of *Actinobacteria, Firmicytes, Proteobacteria,* and *Bacteroidetes*. In addition to sebaceous/moist/dry, it is possible to construct additional classifications of the skin microbiota: while overall *Corynebacteria* and *Propionibacteria* are the most common genera, *Actinobacteria* are especially common on the UV-exposed sites, glabella, nares, and occiput; these genera are not as common on sun-protected skin. Moreover, the *Proteobacteria* are particularly common on human arms, that is, on axillae, forearm, palms, and the interdigital space. Skin sites that are commonly subject to stretching and flexing, including fingers, toe webs, popliteal and antecubital fossa, inguinal crease, and occiput are particularly rich in *Staphylococci*. The gluteal crease and the toe webs are particularly rich in *Micrococci*. As suggested by Segre et al. [43], the human skin contains multiple and varied niches that are host to multiple and varied microbiota.

6. UV damage

UV light is a major environmental carcinogen. Photodamage of skin results in thinning, wrinkling, keratosis, and ultimately malignancy. Several groups, including ours, have analyzed the transcriptional responses to UV light in human epidermal keratinocytes [44–48]. Keratinocytes respond to UV by inducing a cell repair program to self-repair autonomously. However, the keratinocytes also must protect the underlying organism (**Figure 3**). The early transcriptional changes, that is, in the first two hours after UV illumination, contain expression of transcription factors, signal transducing, and cytoskeletal proteins resulting in a "paused" phenotype. This allows keratinocytes to assess the damage and commence repair functions. After 4–8 h post-irradiation, keratinocytes secrete signaling peptides, growth factors, cytokines, and chemokines; these serve to alert the underlying tissue to the UV damage. Subsequently, 16–24 h after treatment, the cornified envelope proteins are produced, as keratinocytes terminally differentiate. This has two beneficial effects: it boosts the stratum corneum, the protective inert layer of the epidermis, and also removes the cells containing potentially UV-damaged DNA, a carcinogenic threat [44]. The results from several laboratories were quite congruent, especially considering the differences in experimental approaches, countries of origin, and the time frames of the experiments [44–48].

These *in vitro* studies have been followed up by studies of UV-irradiated skin in human volunteers *in vivo* [49, 50]. *In vivo* in skin, markers of keratinocyte activation, such as keratin K6 and S100A proteins, were prominently induced by UV, as were the DNA repair proteins. Interestingly, keratinocytes exposed to gamma or X-rays irradiation produce similar transcriptional changes to those in the UV-treated cells [51, 52], specifically, inducing the genes involved in cell energy metabolism. An interesting study compared samples of lentigines with adjacent sun-exposed skin and with matched samples of sun-protected buttocks skin [53]; genes specifically upregulated in solar lentigo included melanocyte-related genes, genes related to fatty acid metabolism and genes related to inflammation.

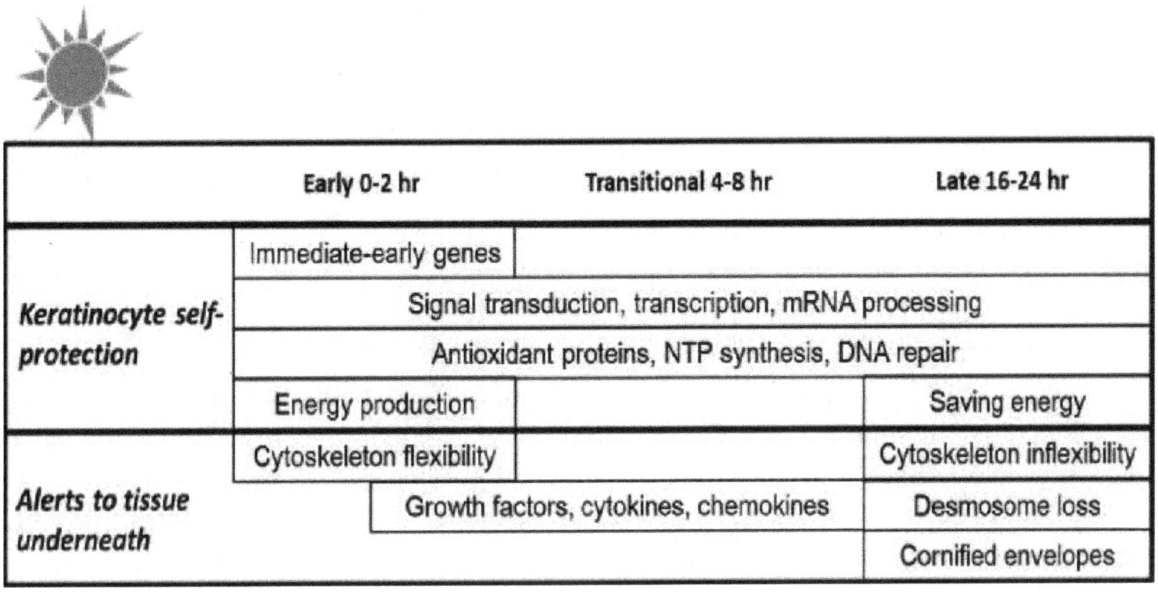

Figure 3. Transcriptional effects of UV light on epidermal keratinocytes. The time course was followed for the first 24 h post-illumination. The UV-treated keratinocytes repair the damage they suffered, as well as alert the underlying tissue that injury has occurred.

7. Skin aging

Skin is the most obvious yardstick of aging. Molecular comparisons of transcriptional profiles of young vs. aged and of sun-protected vs. sun-exposed skin indicate that photoaging and chronical aging, although partly overlapping, have different, characteristic features [54]. Genes associated with skin aging were identified by comparing foreskin keratinocytes from young, 3–4 years of age, old, 68–72 years, subjects [55]. A total 105 genes changed; for example, epidermal differentiation and keratinocyte activation markers were overexpressed in the aged skin, while the immune response, cell cycle, and extracellular matrix associated genes were overexpressed in keratinocytes from young skin. Proteomic profiling using two-dimensional gel electrophoresis to compare young and old foreskin samples identified additional markers of intrinsic aging, including aging-related posttranslational protein modification [56].

One of the hallmarks of aging skin is impaired wound healing. Using microarrays to compare gene expression in wounds of elderly vs. young humans, it was found that the differences seem to be related to regulation by estrogen [57]. These results suggest that estrogen has a profound influence on skin aging in general, and particularly in the context of wound healing. Another hallmark of aging is the graying of hair; analysis of differential gene expressions between pigmented, gray, and white scalp hair follicles identified close to 200 upregulated and as many downregulated genes in human gray hair. As expected, the melanogenesis and structural genes of the melanosome are overrepresented among the regulated genes [58].

8. Skinomics genome-wide association studies

Genome-wide association studies, GWAS, comprise examination of many common DNA polymorphisms in a large population cohort to detect association of polymorphisms with a given disease. Such polymorphisms can point to the genes where disease-causing mutations may map. GWAS are particularly useful in the analysis of diseases, such as psoriasis, which are common and with a strong genetic component. Psoriasis is a hyperproliferative autoimmune skin disease and involves keratinocytes and T cells [9]. A successful GWAS analysis of the psoriasis susceptibility loci in the human genome has been accomplished by an extensive multinational effort (**Figure 4**) and reported in a set of manuscripts [59–66]. A total to 36 loci have been associated with psoriasis in European populations, with additional ones detected in the Chinese population [67, 68]. Other skin diseases, for example, eczema, have also been studied using GWAS; eczema was associated several genetic loci, including major histocompatibility complex (MHC) on chromosome 6 and the epidermal differentiation complex, EDC, on chromosome 1. The human filaggrin gene, known to be associated with eczema, is encoded within the EDC.

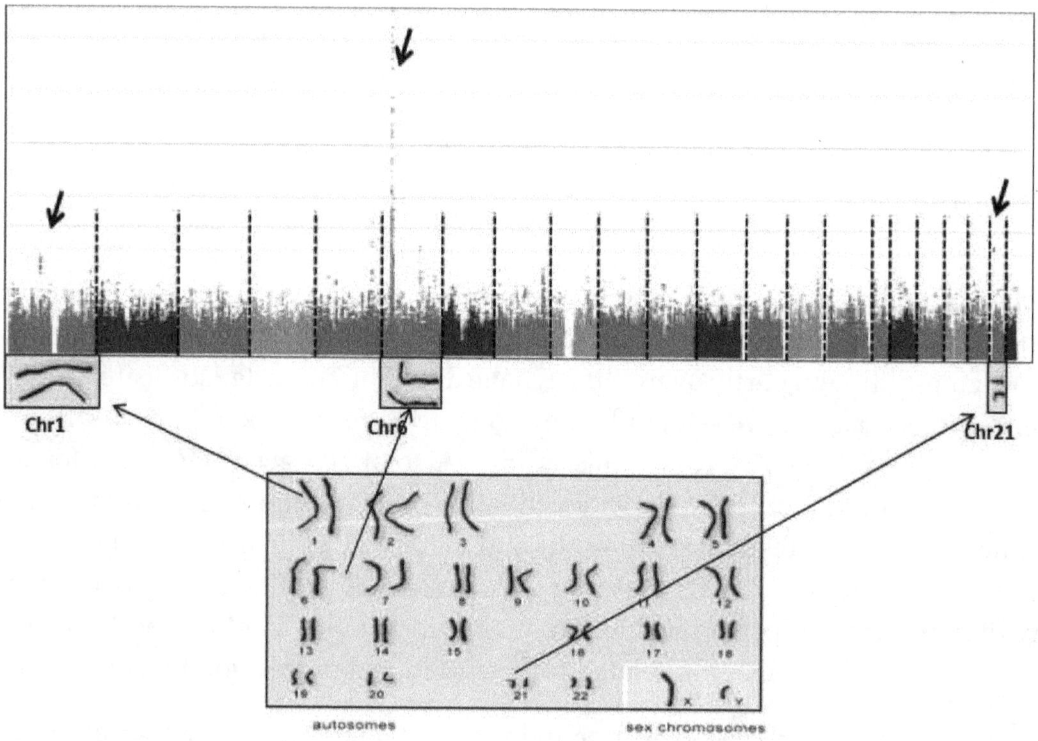

Figure 4. Genome-wide association studies of psoriasis. While many loci have been identified, and additional ones keep being reported, the most prominent ones on chromosomes 1, 6, and 21 are marked with arrows.

Carcinomas and melanoma were also analyzed using GWAS. Basal cell and squamous cell carcinomas have both joint and specific susceptibility loci [69]. Carcinoma GWAS loci are not associated with melanoma risks. GWAS also identified the loci important for human skin pigmentation [70].

9. Transcriptional studies of melanoma

Melanoma is one of the most aggressive human cancers, which is why it was among the earliest cancers studied using microarrays, already 1 year after the first report of cDNA microarrays [11, 71]. Great amount of attention has been devoted to melanoma diagnosis and the natural history of its progression. Multiple studies compared healthy melanocytes to melanoma tumor cells, other studies examined the differences between melanoma lines that differ in their metastatic potential [71–82]. Some of the differentially expressed genes are, as expected, encoded in the chromosomal regions identified by GWAS as commonly altered in melanomas. Osteopontin, for example, was identified as an overexpressed marker of invasive melanoma.

Microarrays have also been used to compare the transcriptomes of benign moles and melanomas. The metastatic samples exhibited two distinct patterns of gene expression, and these were similar to the microdissected nodular vs. flat components of large primary melanomas [74]. The epigenetic characteristics of melanomas, such as differences in the genome, its methylation, and the expression of miRNAs have also received significant attention. The abundance of primary, raw skinomics data enabled meta-analysis approaches, which consolidated the findings from many studies and established a specific "melanoma signature" gene sets; these could become potentially useful in detection, classification, and outcome prediction for melanomas [83].

10. Wound healing studies

A very active area of skinomics research relates to wound healing, a multi-step process involving coordinated and interacting regulatory pathways [84–86]. For example, we have demonstrated that microarrays can be utilized at the bedside to guide surgical debridement of non-healing wounds [87]. Specifically, the healing edges of ulcers express keratinocyte markers, whereas the non-healing ones express the dermal and inflammatory markers [88]. Diabetic foot ulcers and chronic venous ulcers have distinct transcriptional profiles [88, 89]. The microbiomes of chronic wounds and wound healing have also received attention [90].

11. Inflammation, cytokines, and chemokines

Psoriasis has been linked with multiple cytokines and chemokines, including IL-1, IL-12, IL-17, tumor necrosis factor (TNF)α, interferon (INF)-γ, and oncostatin M. Therefore, the transcriptional effects of these signaling proteins on epidermal keratinocytes in culture have been studied extensively [91–97]. The transcriptional profiling studies of the effects of corticosteroids, anti-inflammatory agents, demonstrated their inhibition of the TNFα, IFN-γ, and IL-1 pathways [98]. However, the anti-inflammatory effects of corticosteroids have characteristic choreography and phasing: the earliest are the anti-TNFα effects, clear already in the first hour. These are followed by the anti-IL-1 effects, peaking between 24 and 48 h. Finally, the anti-IFN-γ effects occur later, at 72 h.

In separate studies, comparisons of eczema and psoriasis showed that the expression of many antimicrobial proteins of the innate immune response genes are relatively decreased in eczema, which could explain the increased susceptibility to infection in eczema than in psoriasis [99–103].

12. Psoriasis transcriptome

The studies of psoriasis, a paradigmatic skin inflammatory disease, have provided several hundreds of patient samples from several laboratories [8, 9]. This provided data for meta-analysis of the psoriatic transcriptome [16]. Microarray data can be obtained from annotated and curated repositories. The two main data repositories that collect and annotate transcription profiling using microarrays are NIH-GEO [104] and ArrayExpress [105]. The two largely overlap, but have a few differences, that is, studies present in one but not the other collection. Additional datasets exist in proprietary databases, but we found that searching these is time consuming and usually unproductive.

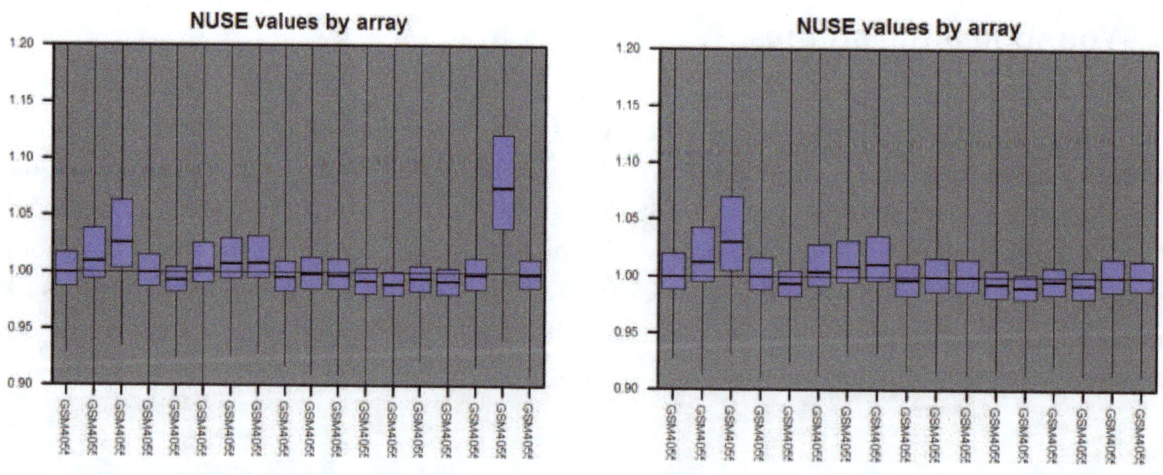

Figure 5. Quality control features of RMAExpress. Before removing a "bad" chip, left and after right.

We searched GEO Datasets for the key term "Psoriasis" and selecting "Homo sapiens" as the organism; from the results, we selected nine studies that met our criteria, namely comparing lesional and nonlesional psoriatic skin. These nine studies included seven Affymetrix-based and one each Sentrix and Illumina microarrays, with a combined total of 645 samples. The Affymetrix .CEL or .TXT files from these studies were downloaded and unzipped, then log2 transformed. Datasets obtained were combined and analyzed using RMAExpress for quality control [106]. RMAExpress is a program that can compute gene expression values from

Affymetrix chips using the Robust Multichip Average protocol; it is available [107]. RMA can also perform chip quality assessments.

We cannot overemphasize the importance of data normalization and selection, when conducting meta-analyses. One of the most is important, but often overlooked task is to perform quality control on the microarrays deposited in the databanks. The quality control graphs can be viewed in the RMAExpress program, the best routine being the normalized unscaled standard error (NUSE). The gene chips with NUSE medians more than 5% different from other chips should be ignored (**Figure 5**) [31]. These microarrays chips usually stem from poor quality input RNA.

In the case of non-Affymetrix studies, the simplest approach is to download the TXT data files directly from PubMed [108]. Conveniently, downloading all compressed files into same directory allows batch uncompressing of TXT files, which can be opened using Excel for further analysis. We often use AddIns DataLoader [109] in order to assign gene annotations to the expression data. A considerable stumbling block for meta-analysis is the merging of data from different microarray platforms. The difficulties include different number of genes on different platforms, different levels of redundancies in probing the genes, and inappropriate differences in identifying genes.

To harmonize gene IDs for various platforms we find useful BioMart [110], the downloaded files can be opened using Excel. We also find very useful the AddIns DataLoader [109] for combining similar data from different spreadsheets. Where the smaller arrays do not value for a given gene, we simply added 1. This does not affect the subsequent steps of analysis.

To select differentially expressed genes, we use RankProd, a nonparametric method [30]. The RankProd method combines different datasets thus increasing the number of differentially expressed genes identified. Here we present a simple simulated RankProd set of commands an imaginary analysis of three datasets with 5 + 5, 6 + 24, and 2 + 2 microarrays (psoriatic lesional + nonlesional [14]):

```
memory.size(max = FALSE) 1

memory.limit(size = 24000)

library (RankProd)

data (Your_txt_file)

n1 <- 5

n2 <- 5

n3 <- 6

n4 <- 24

n5 <- 2

n6 <- 2

cl <- rep(c(0,1,0,1,0,1), c(n1,n2,n3,n4,n5,n6)) 2
```

cl

rownames(Your_txt_file)= Your_txt_file [,1] 3

Your_txt_file = Your_txt_file [,-1]

origin <- c(rep(1,10), rep(2,30), rep(3,4)) 4

origin

RP.adv.out <- RPadvance(Your_txt_file, cl, origin, rand = 100) 5

plotRP(RP.adv.out, cutoff = 0.01) 6

topGene(RP.adv.out, cutoff = 0.01)

write.table(topGene(RP.adv.out, num.gene = 1000), row.names = TRUE, col.names = NA, file = "Your_txt_file.txt") 7.

To annotate the differentially expressed genes, we find extremely useful and convenient the program Database for Annotation Visualization and Integrated Discovery (DAVID) [31, 32, 111]. Starting with the uploaded list, DAVID make available "tables" containing details known about the genes, "charts" which contain over-represented pathways, ontological categories, etc., as well as "clusters" of such categories, eliminating some of the redundancies and overlaps. The transcription factor binding sites in the listed genes can also be evaluated using DAVID, although the oPOSSUM programs are much more comprehensive, sophisticated, and convenient [39, 40, 112]. Many microarray data clustering programs are available; our favorite is MEV [113]. Generally, we did not find them informative for the analysis of data of psoriasis patients.

13. Conclusions

One clear advantage in dermatology over other medical specialties is that in the clinic, the noninvasive, painless sampling using the tape stripping methods can provide high-quality samples for skinomics analysis. Already proven useful in diagnosis of certain diseases, the tape stripping will be also used in the microbiome analyses to provide samples of the cutaneous viruses, bacteria, and fungi, alerting to the presence of pathogens. Similar approaches can detect also cutaneous microbial imbalances. Moreover, disease treatments using microbes may be in our future! We cannot even imagine today the future developments in skinomics. In summary, great strides have been already achieved in skinomics, the omics technology applied in dermatology, and skin biology. Skinomics techniques will eventually provide individualized personalized treatments to dermatology patients. Exciting and wonderful times are ahead.

Author details

Sidra Younis[1,2], Valeriia Shnayder[1] and Miroslav Blumenberg[1*]

*Address all correspondence to: miroslav.blumenberg@nyumc.org

1 The R.O.Perelman Department of Dermatology and Department of Biochemistry and Molecular Pharmacology, NYU Langone Medical Center, New York, USA

2 Department of Biochemistry, Quaid-i-Azam University, Islamabad, Pakistan

References

[1] Pevsner J. Bioinformatics and Functional Genomics. Wiley-Blackwell; Hoboken, NJ, USA, 2015.

[2] Gligorijevic V, Przulj N. Methods for biological data integration: perspectives and challenges. J R Soc Interface. 2015;112:1–19.. pii:20150571. DOI: 10.1098/rsif.2015.0571.

[3] Bumgarner R. Overview of DNA microarrays: types, applications, and their future. Curr Protoc Mol Biol. 2013;Chapter(22):Unit 22.1. DOI: 10.1002/0471142727.mb2201s101.

[4] http://nar.oxfordjournals.org/content/vol36/suppl_1/index.dtl [Accessed: 2015].

[5] http://www.bioconductor.org/ [Accessed: 2015].

[6] Blumenberg M. Skinomics. J Invest Dermatol. 2005;124(4):viii–x.

[7] Blumenberg M. Skinomics: past, present and future for diagnostic microarray studies in dermatology. Expert Rev Mol Diagn. 2013;(8):885–94. DOI: 10.1586/14737159.2013.846827.

[8] Lowes MA, Suarez-Farinas M, Krueger JG. Immunology of psoriasis. Annu Rev Immunol. 2014;32:227–55. DOI: 10.1146/annurev-immunol-032713-120225.

[9] Bowcock AM, Krueger JG. Getting under the skin: the immunogenetics of psoriasis. Nat Rev Immunol. 2005;9:699–711.

[10] Quackenbush J, Hegde P, Qi R, Abernathy K, Gay C, Dharap S, et al. Genomics. Microarrays – guilt by association A concise guide to cDNA microarray analysis. Science. 2003;302:240–1.

[11] Schena M, Shalon D, Davis RW, Brown PO. Quantitative monitoring of gene expression patterns with a complementary DNA microarray. Science. 1995;270:467–70.

[12] Iyer VR, Eisen MB, Ross DT, Schuler G, Moore T, Lee JC, et al. The transcriptional program in the response of human fibroblasts to serum. Science. 1999;283:83–7.

[13] Robbins PB, Sheu SM, Goodnough JB, Khavari PA. Impact of laminin 5 beta3 gene versus protein replacement on gene expression patterns in junctional epidermolysis bullosa. Hum Gene Ther. 2001;12:1443–8.

[14] Hinata K, Gervin AM, Jennifer Zhang Y, Khavari PA. Divergent gene regulation and growth effects by NF-kappa B in epithelial and mesenchymal cells of human skin. Oncogene. 2003;22:1955–64.

[15] Noh M, Yeo H, Ko J, Kim HK, Lee CH. MAP17 is associated with the T-helper cell cytokine-induced down-regulation of filaggrin transcription in human keratinocytes. Exp Dermatol. 2010;19:355–62. Epub 2009 Jul 8.

[16] Mimoso C, Lee DD, Zavadil J, Tomic-Canic M, Blumenberg M. Analysis and meta-analysis of transcriptional profiling in human epidermis. Methods Mol Biol. 2014;1195:61–97. DOI: 10.1007/7651_2013_60.

[17] Blumenberg M. Profiling and metaanalysis of epidermal keratinocytes responses to epidermal growth factor. BMC Genomics. 2013;14:85. DOI: 10.1186/471-2164-14-85.

[18] Morhenn VB, Chang EY, Rheins LA. A noninvasive method for quantifying and distinguishing inflammatory skin reactions. J Am Acad Dermatol. 1999;41:687–92.

[19] Wong R, Tran V, Morhenn V, Hung SP, Andersen B, Ito E, et al. Use of RT-PCR and DNA microarrays to characterize RNA recovered by non-invasive tape harvesting of normal and inflamed skin. J Invest Dermatol. 2004;123:159–67.

[20] Wachsman W, Morhenn V, Palmer T, Walls L, Hata T, Zalla J, et al. Noninvasive genomic detection of melanoma. Br J Dermatol. 2011;164:797–806. DOI: 10.1111/j.365-2133.011.10239.x. Epub 2011 Mar 25.

[21] Gazel A, Ramphal P, Rosdy M, De Wever B, Tornier C, Hosein N, et al. Transcriptional profiling of epidermal keratinocytes: comparison of genes expressed in skin, cultured keratinocytes, and reconstituted epidermis, using large DNA microarrays. J Invest Dermatol. 2003;121:1459–68.

[22] Gazel A, Banno T, Walsh R, Blumenberg M. Inhibition of JNK promotes differentiation of epidermal keratinocytes. J Biol Chem. 2006;281:20530–41.

[23] Lu J, Goldstein KM, Chen P, Huang S, Gelbert LM, Nagpal S. Transcriptional profiling of keratinocytes reveals a vitamin D-regulated epidermal differentiation network. J Invest Dermatol. 2005;124:778–85.

[24] Seo EY, Namkung JH, Lee KM, Lee WH, Im M, Kee SH, et al. Analysis of calcium-inducible genes in keratinocytes using suppression subtractive hybridization and cDNA microarray. Genomics. 2005;86:528–38. Epub 2005 Aug 9.

[25] Baron JM, Heise R, Blaner WS, Neis M, Joussen S, Dreuw A, et al. Retinoic acid and its 4-oxo metabolites are functionally active in human skin cells in vitro. J Invest Dermatol. 2005;125:143–53.

[26] Grossi M, Hiou-Feige A, Tommasi Di Vignano A, Calautti E, Ostano P, Lee S, et al. Negative control of keratinocyte differentiation by Rho/CRIK signaling coupled with up-regulation of KyoT1/2 (FHL1) expression. Proc Natl Acad Sci U S A. 2005;102:11313–8. Epub 2005 Aug 1.

[27] Duffy CL, Phillips SL, Klingelhutz AJ. Microarray analysis identifies differentiation-associated genes regulated by human papillomavirus type 16 E6. Virology. 2003;314:196–205.

[28] Thomas JT, Oh ST, Terhune SS, Laimins LA. Cellular changes induced by low-risk human papillomavirus type 11 in keratinocytes that stably maintain viral episomes. J Virol. 2001;75:7564–71.

[29] Levy L, Hill CS. Smad4 dependency defines two classes of transforming growth factor {beta} (TGF-{beta}) target genes and distinguishes TGF-{beta}-induced epithelial–mesenchymal transition from its antiproliferative and migratory responses. Mol Cell Biol. 2005;25:8108–25.

[30] Radoja N, Gazel A, Banno T, Yano S, Blumenberg M. Transcriptional profiling of epidermal differentiation. Physiol Genomics. 2006;27:65–78.

[31] Lee DD, Zavadil J, Tomic-Canic M, Blumenberg M. Comprehensive transcriptional profiling of human epidermis, reconstituted epidermal equivalents, and cultured keratinocytes using DNA microarray chips. Methods Mol Biol. 2010;585:193–223.

[32] Matheson CD, Vernon KK, Lahti A, Fratpietro R, Spigelman M, Gibson S, et al. Molecular exploration of the first-century Tomb of the Shroud in Akeldama, Jerusalem. PLoS One. 2009;4:e8319. DOI: 10.1371/journal.pone.0008319.

[33] Marples MJ. Life on the human skin. Sci Am. 1969;220:108–15.

[34] Kugelman TP. Skin bacteria and their role in infection. J Am Med Assoc. 1966;195:146–7.

[35] Roth RR, James WD. Microbial ecology of the skin. Annu Rev Microbiol. 1988;42:441–64.

[36] Till AE, Goulden V, Cunliffe WJ, Holland KT. The cutaneous microflora of adolescent, persistent and late-onset acne patients does not differ. Br J Dermatol. 2000;142:885–92.

[37] Fredricks DN. Microbial ecology of human skin in health and disease. J Investig Dermatol Symp Proc. 2001;6:167–9.

[38] Grice EA, Segre JA. The skin microbiome. Nat Rev Microbiol. 2011;9:244–53.

[39] Woo PC, Lau SK, Teng JL, Tse H, Yuen KY. Then and now: use of 16S rDNA gene sequencing for bacterial identification and discovery of novel bacteria in clinical

microbiology laboratories. Clin Microbiol Infect. 2008;14:908–34. DOI: 10.1111/j.469-0691.2008.02070.x.

[40] Gao Z, Tseng CH, Pei Z, Blaser MJ. Molecular analysis of human forearm superficial skin bacterial biota. Proc Natl Acad Sci U S A. 2007;104:2927–32. Epub 2007 Feb 9.

[41] Paulino LC, Tseng CH, Blaser MJ. Analysis of Malassezia microbiota in healthy superficial human skin and in psoriatic lesions by multiplex real-time PCR. FEMS Yeast Res. 2008;8:460–71. Epub 2008 Feb 20.

[42] Paulino LC, Tseng CH, Strober BE, Blaser MJ. Molecular analysis of fungal microbiota in samples from healthy human skin and psoriatic lesions. J Clin Microbiol. 2006;44:2933–41.

[43] Grice EA, Kong HH, Conlan S, Deming CB, Davis J, Young AC, et al. Topographical and temporal diversity of the human skin microbiome. Science. 2009;324:1190–2.

[44] Li D, Turi TG, Schuck A, Freedberg IM, Khitrov G, Blumenberg M. Rays and arrays: the transcriptional program in the response of human epidermal keratinocytes to UVB illumination. Faseb J. 2001;15:2533–5.

[45] Sesto A, Navarro M, Burslem F, Jorcano JL. Analysis of the ultraviolet B response in primary human keratinocytes using oligonucleotide microarrays. Proc Natl Acad Sci U S A. 2002;99:2965–70.

[46] Murakami T, Fujimoto M, Ohtsuki M, Nakagawa H. Expression profiling of cancer-related genes in human keratinocytes following non-lethal ultraviolet B irradiation. J Dermatol Sci. 2001;27:121–9.

[47] Takao J, Ariizumi K, Dougherty, II, Cruz PD, Jr. Genomic scale analysis of the human keratinocyte response to broad-band ultraviolet-B irradiation. Photodermatol Photoimmunol Photomed. 2002;18:5–13.

[48] Howell BG, Wang B, Freed I, Mamelak AJ, Watanabe H, Sauder DN. Microarray analysis of UVB-regulated genes in keratinocytes: downregulation of angiogenesis inhibitor thrombospondin-1. J Dermatol Sci. 2004;34:185–94.

[49] Enk CD, Shahar I, Amariglio N, Rechavi G, Kaminski N, Hochberg M. Gene expression profiling of in vivo UVB-irradiated human epidermis. Photodermatol Photoimmunol Photomed. 2004;20:129–37.

[50] Enk CD, Jacob-Hirsch J, Gal H, Verbovetski I, Amariglio N, Mevorach D, et al. The UVB-induced gene expression profile of human epidermis in vivo is different from that of cultured keratinocytes. Oncogene. 2006;25:2601–14.

[51] Koike M, Shiomi T, Koike A. Identification of skin injury-related genes induced by ionizing radiation in human keratinocytes using cDNA microarray. J Radiat Res (Tokyo). 2005;46:173–84.

[52] Lamartine J, Franco N, Le Minter P, Soularue P, Alibert O, Leplat JJ, et al. Activation of an energy providing response in human keratinocytes after gamma irradiation. J Cell Biochem. 2005;95:620–31.

[53] Aoki H, Moro O, Tagami H, Kishimoto J. Gene expression profiling analysis of solar lentigo in relation to immunohistochemical characteristics. Br J Dermatol. 2007;156:1214–23. Epub 2007 Apr 5.

[54] Robinson MK, Binder RL, Griffiths CE. Genomic-driven insights into changes in aging skin. J Drugs Dermatol. 2009;8:s8–11.

[55] Lener T, Moll PR, Rinnerthaler M, Bauer J, Aberger F, Richter K. Expression profiling of aging in the human skin. Exp Gerontol. 2006;41:387–97. Epub 2006 Mar 10.

[56] Laimer M, Kocher T, Chiocchetti A, Trost A, Lottspeich F, Richter K, et al. Proteomic profiling reveals a catalogue of new candidate proteins for human skin aging. Exp Dermatol. 2010;19:912–8. DOI: 10.1111/j.600-0625.2010.01144.x.

[57] Hardman MJ, Ashcroft GS. Estrogen, not intrinsic aging, is the major regulator of delayed human wound healing in the elderly. Genome Biol. 2008;9:R80. DOI: 10.1186/gb-2008-9-5-r80. Epub May 13.

[58] Peters EM, Liezmann C, Spatz K, Ungethum U, Kuban RJ, Daniltchenko M, et al. Profiling mRNA of the graying human hair follicle constitutes a promising state-of-the-art tool to assess its aging: an exemplary report. J Invest Dermatol. 2013;133:1150–60. DOI: 10.038/jid.2012.462. Epub Dec 13.

[59] Nair RP, Duffin KC, Helms C, Ding J, Stuart PE, Goldgar D, et al. Genome-wide scan reveals association of psoriasis with IL-23 and NF-kappaB pathways. Nat Genet. 2009;41:199–204. Epub 2009 Jan 25.

[60] Zhang XJ, Huang W, Yang S, Sun LD, Zhang FY, Zhu QX, et al. Psoriasis genome-wide association study identifies susceptibility variants within LCE gene cluster at 1q21. Nat Genet. 2009;41:205–10. DOI: 10.1038/ng.310. Epub 2009 Jan 25.

[61] de Cid R, Riveira-Munoz E, Zeeuwen PL, Robarge J, Liao W, Dannhauser EN, et al. Deletion of the late cornified envelope LCE3B and LCE3C genes as a susceptibility factor for psoriasis. Nat Genet. 2009;41:211–5. Epub 2009 Jan 25.

[62] Strange A, Capon F, Spencer CC, Knight J, Weale ME, Allen MH, et al. A genome-wide association study identifies new psoriasis susceptibility loci and an interaction between HLA-C and ERAP1. Nat Genet. 2010;42:985–90. Epub 2010 Oct 17.

[63] Stuart PE, Nair RP, Ellinghaus E, Ding J, Tejasvi T, Gudjonsson JE, et al. Genome-wide association analysis identifies three psoriasis susceptibility loci. Nat Genet. 2010;42:1000–4. Epub 2010 Oct 17.

[64] Ellinghaus E, Ellinghaus D, Stuart PE, Nair RP, Debrus S, Raelson JV, et al. Genome-wide association study identifies a psoriasis susceptibility locus at TRAF3IP2. Nat Genet. 2010;42:991–5. Epub 2010 Oct 17.

[65] Sun LD, Cheng H, Wang ZX, Zhang AP, Wang PG, Xu JH, et al. Association analyses identify six new psoriasis susceptibility loci in the Chinese population. Nat Genet. 2010;42:1005–9. Epub 2010 Oct 17.

[66] Huffmeier U, Uebe S, Ekici AB, Bowes J, Giardina E, Korendowych E, et al. Common variants at TRAF3IP2 are associated with susceptibility to psoriatic arthritis and psoriasis. Nat Genet. 2010;42:996–9. Epub 2010 Oct 17.

[67] Tsoi LC, Spain SL, Knight J, Ellinghaus E, Stuart PE, Capon F, et al. Identification of 15 new psoriasis susceptibility loci highlights the role of innate immunity. Nat Genet. 2012;44:1341–8. DOI: 10.038/ng.2467. Epub 012 Nov 11.

[68] Julia A, Tortosa R, Hernanz JM, Canete JD, Fonseca E, Ferrandiz C, et al. Risk variants for psoriasis vulgaris in a large case–control collection and association with clinical subphenotypes. Hum Mol Genet. 2012;21:4549–57. Epub 2012 Jul 19.

[69] Nan H, Xu M, Kraft P, Qureshi AA, Chen C, Guo Q, et al. Genome-wide association study identifies novel alleles associated with risk of cutaneous basal cell carcinoma and squamous cell carcinoma. Hum Mol Genet. 2011;20:3718–24. DOI: 10.1093/hmg/ddr287. Epub 2011 Jun 23.

[70] Nan H, Kraft P, Qureshi AA, Guo Q, Chen C, Hankinson SE, et al. Genome-wide association study of tanning phenotype in a population of European ancestry. J Invest Dermatol. 2009;129:2250–7. DOI: 10.1038/jid.2009.62. Epub Apr 2.

[71] DeRisi J, Penland L, Brown PO, Bittner ML, Meltzer PS, Ray M, et al. Use of a cDNA microarray to analyse gene expression patterns in human cancer. Nat Genet. 1996;14:457–60.

[72] Dooley TP, Curto EV, Davis RL, Grammatico P, Robinson ES, Wilborn TW. DNA microarrays and likelihood ratio bioinformatic methods: discovery of human melanocyte biomarkers. Pigment Cell Res. 2003;16:245–53.

[73] Hoek K, Rimm DL, Williams KR, Zhao H, Ariyan S, Lin A, et al. Expression profiling reveals novel pathways in the transformation of melanocytes to melanomas. Cancer Res. 2004;64:5270–82.

[74] Haqq C, Nosrati M, Sudilovsky D, Crothers J, Khodabakhsh D, Pulliam BL, et al. The gene expression signatures of melanoma progression. Proc Natl Acad Sci U S A. 2005;102:6092–7. Epub 2005 Apr 15.

[75] Becker B, Roesch A, Hafner C, Stolz W, Dugas M, Landthaler M, et al. Discrimination of melanocytic tumors by cDNA array hybridization of tissues prepared by laser pressure catapulting. J Invest Dermatol. 2004;122:361–8.

[76] Gallagher WM, Bergin OE, Rafferty M, Kelly ZD, Nolan IM, Fox EJ, et al. Multiple markers for melanoma progression regulated by DNA methylation: insights from transcriptomic studies. Carcinogenesis. 2005;26:1856–67. Epub 2005 Jun 15.

[77] Nambiar S, Mirmohammadsadegh A, Doroudi R, Gustrau A, Marini A, Roeder G, et al. Signaling networks in cutaneous melanoma metastasis identified by complementary DNA microarrays. Arch Dermatol. 2005;141:165–73.

[78] Busam KJ, Zhao H, Coit DG, Kucukgol D, Jungbluth AA, Nobrega J, et al. Distinction of desmoplastic melanoma from non-desmoplastic melanoma by gene expression profiling. J Invest Dermatol. 2005;124:412–8.

[79] Talantov D, Mazumder A, Yu JX, Briggs T, Jiang Y, Backus J, et al. Novel genes associated with malignant melanoma but not benign melanocytic lesions. Clin Cancer Res. 2005;11:7234–42.

[80] Zhou Y, Dai DL, Martinka M, Su M, Zhang Y, Campos EI, et al. Osteopontin expression correlates with melanoma invasion. J Invest Dermatol. 2005;124:1044–52.

[81] Winnepenninckx V, Van den Oord JJ. Gene expression profiling of primary cutaneous melanoma. Verh K Acad Geneeskd Belg. 2007;69:23–45.

[82] Winnepenninckx V, Lazar V, Michiels S, Dessen P, Stas M, Alonso SR, et al. Gene expression profiling of primary cutaneous melanoma and clinical outcome. J Natl Cancer Inst. 2006;98:472–82.

[83] Liu W, Peng Y, Tobin DJ. A new 12-gene diagnostic biomarker signature of melanoma revealed by integrated microarray analysis. PeerJ. 2013;1:e49. DOI:10.7717/peerj.49. Print 2013.

[84] Tomic-Canic M, Brem H. Gene array technology and pathogenesis of chronic wounds. Am J Surg. 2004;188:67–72.

[85] Nuutila K, Siltanen A, Peura M, Bizik J, Kaartinen I, Kuokkanen H, et al. Human skin transcriptome during superficial cutaneous wound healing. Wound Repair Regen. 2012;20:830–9. DOI: 10.1111/j.524-475X.2012.00831.x. Epub 2012 Oct 19.

[86] Deonarine K, Panelli MC, Stashower ME, Jin P, Smith K, Slade HB, et al. Gene expression profiling of cutaneous wound healing. J Transl Med. 2007;5:11.

[87] Brem H, Stojadinovic O, Diegelmann RF, Entero H, Lee B, Pastar I, et al. Molecular markers in patients with chronic wounds to guide surgical debridement. Mol Med. 2007;13:30–9.

[88] Charles CA, Tomic-Canic M, Vincek V, Nassiri M, Stojadinovic O, Eaglstein WH, et al. A gene signature of nonhealing venous ulcers: potential diagnostic markers. J Am Acad Dermatol. 2008;59:758–71. DOI: 10.1016/j.jaad.2008.07.018. Epub Aug 20.

[89] Brem H, Tomic-Canic M. Cellular and molecular basis of wound healing in diabetes. J Clin Invest. 2007;117:1219–22.

[90] Grice EA, Segre JA. Interaction of the microbiome with the innate immune response in chronic wounds. Adv Exp Med Biol. 2012;946:55–68.

[91] Finelt N, Gazel A, Gorelick S, Blumenberg M. Transcriptional responses of human epidermal keratinocytes to Oncostatin-M. Cytokine. 2005;31:305–13.

[92] Molenda M, Mukkamala L, Blumenberg M. Interleukin IL-12 blocks a specific subset of the transcriptional profile responsive to UVB in epidermal keratinocytes. Mol Immunol. 2006;43:1933–40. Epub 2006 Feb 8.

[93] Gazel A, Rosdy M, Bertino B, Tornier C, Sahuc F, Blumenberg M. A characteristic subset of psoriasis-associated genes is induced by oncostatin-M in reconstituted epidermis. J Invest Dermatol. 2006;126:2647–57.

[94] Banno T, Gazel A, Blumenberg M, Adachi M, Mukkamala L. Effects of tumor necrosis factor-{alpha} (TNF{alpha}) in epidermal keratinocytes revealed using global transcriptional profiling. J Biol Chem. 2004;279:32633–42. Epub 2004 May 15.

[95] Banno T, Gazel A, Blumenberg M. The use of DNA microarrays in dermatology research. Retinoids. 2004;20:1–4.

[96] Banno T, Adachi M, Mukkamala L, Blumenberg M. Unique keratinocyte-specific effects of interferon-gamma that protect skin from viruses, identified using transcriptional profiling. Antivir Ther. 2003;8:541–54.

[97] Haider AS, Peters SB, Kaporis H, Cardinale I, Fei J, Ott J, et al. Genomic analysis defines a cancer-specific gene expression signature for human squamous cell carcinoma and distinguishes malignant hyperproliferation from benign hyperplasia. J Invest Dermatol. 2006;126:869–81.

[98] Stojadinovic O, Lee B, Vouthounis C, Vukelic S, Pastar I, Blumenberg M, et al. Novel genomic effects of glucocorticoids in epidermal keratinocytes: inhibition of apoptosis, interferon-gamma pathway, and wound healing along with promotion of terminal differentiation. J Biol Chem. 2007;282:4021–34. Epub 2006 Nov 9.

[99] Nomura I, Goleva E, Howell MD, Hamid QA, Ong PY, Hall CF, et al. Cytokine milieu of atopic dermatitis, as compared to psoriasis, skin prevents induction of innate immune response genes. J Immunol. 2003;171:326–9.

[100] Nomura I, Gao B, Boguniewicz M, Darst MA, Travers JB, Leung DY. Distinct patterns of gene expression in the skin lesions of atopic dermatitis and psoriasis: a gene microarray analysis. J Allergy Clin Immunol. 2003;112:1195–202.

[101] de Jongh GJ, Zeeuwen PL, Kucharekova M, Pfundt R, van der Valk PG, Blokx W, et al. High expression levels of keratinocyte antimicrobial proteins in psoriasis compared with atopic dermatitis. J Invest Dermatol. 2005;125:1163–73.

[102] Ogawa K, Ito M, Takeuchi K, Nakada A, Heishi M, Suto H, et al. Tenascin-C is upregulated in the skin lesions of patients with atopic dermatitis. J Dermatol Sci. 2005;40:35–41.

[103] Guttman-Yassky E, Lowes MA, Fuentes-Duculan J, Whynot J, Novitskaya I, Cardinale I, et al. Major differences in inflammatory dendritic cells and their products distinguish atopic dermatitis from psoriasis. J Allergy Clin Immunol. 2007;119:1210–7.

[104] http://www.ncbi.nlm.nih.gov/pubmed/ [Accessed: 2015].

[105] http://www.ebi.ac.uk/arrayexpress/ [Accessed: 2015].

[106] Eisen MB, Spellman PT, Brown PO, Botstein D. Cluster analysis and display of genome-wide expression patterns. Proc Natl Acad Sci U S A. 1998;95:14863–8.

[107] http://rmaexpress.bmbolstad.com/ [Accessed: 2015].

[108] http://www.ncbi.nlm.nih.gov/pubmed/ [Accessed: 2015].

[109] http://www.add-ins.com [Accessed: 2015].

[110] http://www.biomart.org/biomart/martview/39658a1f84f2d1822db6e184a5cc356e/ [Accessed: 2015].

[111] http://david.abcc.ncifcrf.gov [Accessed: 2015].

[112] http://www.cisreg.ca/cgi-bin/oPOSSUM/opossum/ [Accessed: 2015].

[113] http://www.tm4.org/ [Accessed: 2015].

3

Bioinformatics Tools and Genomic Resources Available in Understanding the Structure and Function of *Gossypium*

Venkateswara R. Sripathi, Ramesh Buyyarapu, Siva P. Kumpatla, Abreeotta J. Williams, Seloame T. Nyaku, Yonathan Tilahun, Venu Kalavacharla and Govind C. Sharma

Abstract

Cotton is economically and evolutionarily important crop for its fiber. In order to improve fiber quality and yield, and to exploit the natural genetic potential inherent in genotypes, understanding genome structure and function of cultivated cotton is important. In order to achieve this, a functional understanding of bioinformatics resources such as databases, software solutions, and analysis tools is required. But currently, there are very few unified reports on bioinformatics tools and even fewer repositories to access cotton genomic information. Also, resourceful developers and bioinformatics scientists actively addressing complex genomic challenges in cotton genomes are much in need. The primary goal of this chapter is to provide a review of such tools and resources for analyzing the structure and function of the cotton genome with preferential emphasis on this complex and economically important plant species. This discourse begins with a description of concurrent advances in high-throughput genome sequencing and bioinformatics analyses and focuses on four major sections covering bioinformatics tools and resources for analysis of: (1) genomes; (2) transcriptomes; (3) small RNAs; and (4) epigenomes. In each section, recent advances in cotton have been discussed. Cotton genome sequencing and annotation efforts are outlined within these sections. This review discusses the availability of genome information of both diploid and tetraploid species that have impelled cotton genome research into the post-genomics era, opening new avenues for exploring regulatory mechanisms associated with fine-tuning of gene expression of fiber-related genes. Finally, the potential impacts of these rapid advances, especially the challenges in handling and analyzing the large datasets are discussed.

Keywords: genome, transcriptome, epigenome, sequencing, cotton fiber

1. Introduction

Cotton is an economically and evolutionarily important crop species. Along with cotton improvement, which has progressed impressively with conventional and molecular breeding approaches, the genomic approaches utilizing next generation sequencing (NGS) technologies have enhanced our ability to understand and utilize the genetic potential of crop species [1]. The early sequencing efforts in cotton (*Gossypium* spp.) are mostly limited to diploid species such as *G. raimondii*, as its genome is structurally less complex. Now, DNA sequencing has become a routine tool in cotton genetic research. Sequencing genomic, transcriptomic, and regulatory regions of a plant species and comparing their patterns will provide a better understanding of genome architecture, genetic variation, gene identification, regulation, and its expression [2]. Understanding cotton genome, transcriptome, and regulatory molecules (small RNAs and epigenetic modulators) will therefore provide deeper insights into the structure and function of the genome [3]. Dissecting the complexity associated with tetraploid cotton genome and narrow genetic variability is more challenging and requires efficient methodologies. Comparative genome and transcriptome analysis of cultivated cotton and its progenitors will aid in the identification of novel single-nucleotide polymorphisms (SNPs), copy number variations (CNVs), transcripts, transcript quantification, and alternative splice junctions at the transcriptome level and the role of these elements in regulating cotton fiber quality and yield [4]. Similarly, transcriptome profiling studies are powerful in unravelling the underlying mechanisms involved in gene expression associated with the sub-genomes of a plant species. In addition, the role of small RNAs and epigenetic modulators is increasingly evident in determining the genome landscapes of plant species including cotton.

DNA sequencing technologies have evolved tremendously over two decades from Bacterial artificial chromosome (BAC)-based cloning to single-molecule sequencing [5, 6]. Concomitantly, the computational tools for addressing the problems in understanding sequence data have also evolved from sequence alignment algorithms such as Needleman–Wunsch and Smith–Waterman to assembly tools such as Burrows-Wheeler Alignment (BWA) and Bowtie [7]. As described here, a wide range of publicly available bioinformatics tools and some commercially available sequence analysis tools such as CLCBio Genomics Workbench, Strand NGS, Geneious, Laser gene suite, and NextGENe are being utilized [8].

2. Genome analysis in cotton

Approximately 50 naturally occurring cotton species are available, including 45 diploid (2n = 2x = 26) and five allotetraploid species (2n = 4x = 52) each with a haploid chromosome number of 13 [9]. Based on meiotic pairing and chromosome size, diploid species (2n = 26) have been placed into genomic groups A, B, C, D, E, F, G, or K. Of these, A-genome sources, *G. herba-*

ceum ($2n = 2x = A_1A_1 = 26$) and *G. arboreum* ($2n = 2x = A_2A_2 = 26$), and D-genome sources, *G. raimondii* ($2n = 2x = D_5D_5 = 26$), and *G. gossypioides* ($2n = 2x = D_6D_6 = 26$) are considered as closest progenitor species to cultivated allotetraploid cotton [10]. Diploid genome sizes of A-G and K vary significantly between species ranging from ~900 to ~2800 Mb. Two natural allotetraploid ($2n = 4x = A_tA_tD_tD_t = 52$) species, *G. hirsutum* and *G. barbadense,* are derived from a complex inter-specific hybridization process between two diploid species carrying A-genome and D-genomes [11]. However, the sequencing of polyploid plant genomes has been a tedious task due to repeat regions, whole-genome duplication events, and chromosomal rearrangements that have occurred in the process of evolution [12]. Effective strategies must be developed and employed for sequencing such complex genomes. In order to reduce such complexity, closest diploid progenitor species carrying D_5 (*G. raimondii*; 880 Mb), A_2 (*G. arboreum*; 1700 Mb), and A_1 (*G. herbaceum*; 1700 Mb) genomes can be sequenced and compared against cultivated tetraploid species carrying A_tD_t genome (*G. hirsutum*; ~2400 Mb and *G. barbadense*; ~2500 Mb), the estimated genome sizes [13] of the respective *Gossypium* species are given in parenthesis. The whole-genome sequencing efforts in cotton resulted in assembling ~775 Mb (~88%) of *G. raimondii* and ~1694 Mb (~90%) of *G. arboreum* genomes [14, 15]. Also, ~2.3 Gb (~90%) and ~2.5 Gb (~88%) of the genomes have been assembled into 13 pseudochromosomes each from A_t and D_t sub-genomes of *G. hirsutum* [16, 17] and *G. barbadense,* [18, 19] respectively. Presently, *G. herbaceum* (A_1) genome sequencing efforts are in progress with a partial genome assembled up to 1.2 Gb (~70%) [20].

Transposable elements (TEs) are abundant in plant genomes and are highly variable and often deleterious, as they undergo massive amplification within the genome. The role of TEs has been implicated in gene mutation, whole-genome duplication, chromosomal rearrangements and novel gene formation through genetic and epigenetic changes. They can alter gene expression and phenotypes by establishing and modifying gene regulatory networks. This is accomplished by inducing changes in genetic and epigenetic mechanisms [21]. The variation in genome structure and organization, even in closely related species, is primarily due to TEs and whole-genome duplication (WGD). This may be as a result of a combination of non-random events such as small RNA silencing and epigenetic mechanisms [22] and random events such as natural selection and adaptation [21].With the availability of allotetraploid and diploid cotton genomes, comparative genomics has been used to analyze and understand the structural variation and role of TEs in evolution [23]. Their study identified ~57, ~68, and ~67% of TEs in D_5, A_2, and A_tD_t genomes of cotton, respectively. Long terminal repeats (LTRs) have contributed significantly in whole-genome duplication and evolution of domesticated cotton. Though the overall TE content in A_2- and A_tD_t-genomes has been found to be similar, the frequency of LTR-gypsy and LTR-copia-type elements varied significantly. Terminal repeat retrotransposons in miniature (TRIMs) are a small, ubiquitous, conserved, poorly characterized, and scarcely reported group of repeats. TRIMs are often derived from partial deletion of long terminal repeat retrotransposons found in genic regions and in gene-body methylation within the genome. TRIMS are often targeted by small RNAs (sRNAs) of 21–24 nt in length. Screening for TRIMs in land plants is critical to understanding differences in selection pressure and evolutionary relationships among the clades. Using high-throughput sequencing followed by bioinformatics analysis, 145 unique families of TRIMs have been identified after screening 48 plant genomes [24].

Genetic variation is an important element for crop improvement. An understanding of the genetic and genomic relationships of cotton species and cultivars is critical for further utilization of diversity in the development of improved cultivars with favorable alleles [25]. Allelic variations within a genome of the species can be classified into three major groups at DNA level: microsatellites, insertions/deletions, and single-nucleotide polymorphisms (SNPs) [26]. Molecular markers serve as efficient tools for genome characterization, understanding the genetic complex traits, marker-assisted selection (MAS) and for map-based cloning in breeding programs. Several molecular marker technologies have been used to study the genetic diversity and relationships of *Gossypium* species [27]. However, cotton crop improvement is limited by its narrow genetic base and limited variation among the cultivated cotton cultivars. Genetic variation at molecular level in cotton was previously characterized using isozyme/allozyme markers [28]; using non-coding genomic markers such as restriction fragment length polymorphisms (RFLPs) [29]; amplified fragment length polymorphisms (AFLPs) [30]; microsatellites [31, 32]; single-nucleotide polymorphisms [33] in *G. hirsutum* and its related species.

Simple sequence repeat (SSR) markers are widely used in many plant and animal genomes due to their abundance, hypervariability, and suitability for high-throughput analysis. Development of SSR markers using molecular methods is time consuming, laborious, and expensive. Use of computational approaches to mine ever-increasing sequences such as expressed sequence tags (ESTs) in public databases permits rapid and economical discovery of SSRs [34]. SSR mining programs such as Repeat Pattern tool kit; SSR Finder; Advanced Content Matching Engine for Sequences (ACMES); Spectral-repeat finder; Adplot; REPEATS and other programs are routinely used to mine EST databases, genome survey sequences, and other nucleotide databases [35]. In cotton, SSR markers mined from diploid species such as *G. arboreum* were also successfully employed to understand the structural variation in tetraploid cultivars [32]. SSR markers had been extensively used for many genetic mapping, quantitative trait loci (QTL), and trait mapping experiments for favorable characteristics such as fiber quality, higher yield [36], pathogen resistance [37, 38], and other important traits in cotton. With the advent of next generation sequencing technologies, identification of allelic variation at single-nucleotide level and their application in crop genetics is becoming a common practice.

SNP markers are becoming the "markers of choice" due to their abundance in the genomes, amenability to automation, and high-throughput genotyping capability. In cotton, using available EST or transcriptome sequences, gene-specific SNPs had been characterized [39]; however, these initial efforts and methods had limited application for genome wide SNP marker development in cotton species. Tetraploid nature, highly complex, and repetitive genome of cultivated cotton species poses significant challenges for genome wide SNP marker development. These complications usually result in high number of false positive SNPs, especially when they are developed from sequences that were not thoroughly characterized [25]. Though there is considerable genetic variation across the cultivated cotton species, the narrow germplasm base within each of the cultivated tetraploid species: *G. hirsutum* and *G. barbadense* had made the discovery of useful SNP markers more difficult. Using highly conservative parameters such as minimum coverage of 8× at each SNP and 20% minor allelic frequency, a total of 11,834 and 1679 non-genic SNPs were previously identified between accessions of *G. hirsutum* and *G. barbadense* in genome reduction assemblies, respectively, by Byers et al. [40] As a part of the same study, an additional 4327 genic SNPs were also identified between accessions of *G. hirsutum* in the EST assembly. The transcriptome sequencing has been extended to SNP marker discovery [41]. Using oligonucleotide microarrays, SNP markers in seven differentially expressed EXPANSIN transcripts in early cotton fiber development have been identified [39]. The variant analysis in cotton revealed 27,956 indels and 149,616 SNPs from 268,786 EST assembled contigs [42]. Over 1000 SNPs from 92 single-copy polymorphic loci have been characterized in tetraploid cotton [43].

Despite significant success in cotton breeding and genetic improvement for fiber-related traits, the genome-wide physical or high-density genetic maps are scarcely available in cotton due to its large genome size. Approximately 5000 markers are needed to fully saturate the cotton genome [44]. A diverse list of markers associated with cotton is available in the cotton marker database (CMD) [45]. CottonGen is a comprehensive database that contains genetic, breeding, and genomic data in cotton including 49 genetic maps, ~24,000 markers, ~1000 quantitative trait loci (QTL) linked to more than 30 agronomic traits, ~18,000 genes/transcripts and ~460,000 expressed sequence tags (ESTs) [46]. Cotton QTLdb contains a total of 2274 QTLs that have been identified in intraspecific populations [47]. Recently, using genotype-by-sequencing method ultra-dense inter-specific genetic map that comprised of ~5 million SNPs distributed unevenly across 26 linkage groups has been constructed in allotetraploid cotton [4].

Currently, there is a growing repertoire of available NGS platforms that support whole-genome sequencing, re-sequencing, and exome sequencing such as NextSeq, HiSeq and HiSeq-X series (Illumina, Inc.), and IonProton (Life technologies). While ABI3500 series (Applied Biosystems), MiniSeq and MiSeq (Illumina, Inc.), and Ion S5 and Ion PGM (Life technologies) are suitable for amplicon sequencing. Extensive genomic information related to gene regulation, genetic, and epigenetic codes are available at GenBank (NCBI), EBI (EMBL) and DDBJ (NIG). In addition, raw sequencing data obtained from NGS platforms as a result of sequencing/re-sequencing efforts has been reported at NCBI Sequence Read Archive (SRA). NCBI-Genome hosts information related to genomes including sequences, physical and genetic maps, chromosomes, assemblies, and annotations. Single-nucleotide polymorphisms (SNPs) and multiple small-scale variations that include insertions/deletions, microsatellites and non-polymorphic variants are also available at dbSNP. The databases available for repeats are Repbase, RepeatsDB, Dfam, P-MITE and the PGSB Repeat Database, while the information associated with plant *cis*-acting regulatory DNA elements in promoter regions are accessible by PLACE and PlantCARE.

Assembler category	Assembly program	References
Overlap layout consensus (OLC)-based	Celera assembler, Arachne, CAP/PCAP/CAP3, CABOG, MIRA, Newbler/GS de novo assembler, Edena	[49–53]
Eulerian-based/*De Bruijn* graph (DBG)-based	Euler, Velvet, AllPaths, ABySS, Trans-ABySS, SOAP and SOAPdenovo, miraEST, Oases and Rnnotator	[54–57]
Greedy assemblers	SSAKE, VCAKE, and SHARCGS	[58, 59]

Table 1. Genome and transcriptome assemblers.

Three major categories of assemblers the widely used in whole-genome and transcriptome assembly are as follows: Overlap Layout Consensus (OLC)-based, Eulerian-based/*De Bruijn* Graph (DBG)-based and Greedy assemblers. A table of various types of available sequence assembly. Tools are summarized in **Table 1** [48].

Bioinformatics utility	Programs	References
Mapping to reference genome	SAMtools, Picard, BEDtools, BWA, BWT, Bowtie2, CLC Genomics Workbench	[77–79]
Whole-genome alignment	LASTZ, AVID, VISTA	[80, 81]
Multiple sequence alignment	MUSCLE; MEGA5	[82]
Phylogenetic analysis	EMBOSS; PAML	[83]
Homology-based search	NCBI BLASTN, BLASTP	
Repeat regions	RepeatMasker, Tandem Repeat Finder, MGEScanLTR, TransposonPSI, MITE Digger,	[84–86]
Gene structure	Augustus, GeneMark, FGENESH, GLAD, GeneWise, GenScan, GlimmerHMM, HMMER, GeneID, SNAP, and GLEAN	[87–90]
Gene annotation	BLAST2GO, BLAT, PASA	[91, 92]
Orthologous sequence search	OrthoMCL	[93]
Paralogous sequence search	Mcscan, McscanX	[94]
Transcriptome tools	Trinity, Tuxedo, TopHat, Cufflinks	[95, 96]
Pseudogenes	Pseudopipe	[97]
Single-nucleotide polymorphisms detection	GATK pipeline, SnpEff	[98, 99]
Copy number variation detection	CNVKit	[100]
Linkage mapping	MapMaker, JoinMap, MapManager QTX	[101, 102]
QTL mapping	MapQTL, Windows QTL Cartographer	[103, 104]

Table 2. Commonly used bioinformatics programs in structural, functional and regulatory genome analyses of cotton.

The cotton genomes, D_5 and A_2, were assembled using SOAPdenovo, while A_tD_t utilized overlap-layout-consensus (OLC) assembly followed by SOAPdenovo to achieve a coverage of over 100×. Other tools commonly used in analyzing the structural and functional analysis in cotton genomes are summarized in **Table 2**.

3. Transcriptome analysis in cotton

The RNA-sequencing (RNA-Seq) and analysis research has exploded in parallel to the genome sequencing methodologies. RNA-Seq has been widely adopted due to its high accuracy in characterization and quantification of transcriptomes [60]. The major objectives of transcriptomics are to catalog the transcripts of all species, to predict the genes with biological functions and to quantify the divergence in gene expression during varied spatial, temporal, and ecological conditions. In the beginning, RNA-Seq was focused on deciphering the transcriptomes of model species, but it was quickly extended to non-model species due to the portability of the method because RNA-Seq analysis can be achieved with or without a reference genome. There was a huge shift in characterizing the high-resolution transcriptomes in plant species from EST-sequencing to microarray analysis and currently to RNA-Seq. In plants, RNA-Seq was first adopted for sequencing the transcriptome of *Arabidopsis* using massive parallel 454 sequencing platform [61]. Since then, transcriptomes of several plant species under differing genetic and physiological states have been sequenced including cotton [62]. The currently available NGS platforms that support whole-transcriptome or total RNA/mRNA analysis are GS FLX+ System (Roche 454), SOLiD and IonProton (Applied Biosystems), MiSeq, HiSeq series and NextSeq (Illumina, Inc.). The platforms that support targeted RNA analysis are GS Junior, IonTorrent PGM, and MiniSeq.

Several independent studies used RNA-Seq analysis to determine spatial-, temporal-, tissue-, genotype- and genome-specific, and stress-induced (abiotic and biotic) expression in both diploid and tetraploid cottons. Some recent studies are discussed here. Several groups have sequenced the cotton fiber cDNA or ESTs [63, 64] and transcriptome or mRNA [65, 66] using RNA-Seq, while others utilized microarrays for analysis [67, 68]. Recently, 40,976, 41,330, 66,434, and 80,876 high-confidence protein-coding genes have been predicted in *G. raimondii*, *G. arboreum*, *G. hirsutum* and *G. barbadense*, respectively. Cotton fibers are unique have the most elongated cells in plants, and they account to 40–50% of the whole transcriptome in Upland cotton [69].

Differential gene expression analyses in tetraploid and diploid cottons have been carried out [65]. Two separate studies in root transcriptome analyses of *G. hirsutum* revealed 519 and 1530 differentially expressed transcripts between well-watered and water-deficit conditions, respectively [66, 70]. The comparative transcriptome analysis of developing cotyledon and embryo axis in cotton revealed 17,384 differentially expressed unigenes between two tissues. Of these, ~8000 unigenes were down-regulated and ~10,000 unigenes were up-regulated in cotyledons [71]. Transcriptome analysis of interspecific hybrid F1, synthetic and natural allopolyploid cotton revealed homeolog expression bias (relative contribution of homeologs towards gene expression) and expression level dominance bias (over-all expression from both homeologs) towards A-genome in diploid and natural allopolyploid cotton but not in synthetic cotton [72]. Using RNA-Seq, the global transcriptome profiles of developing cotton fibers from four wild and five domesticated cottons were compared, and this study identified over 5000 differentially expressed genes during the primary and secondary cell wall synthesis between wild and domesticated cottons [73]. Comparative RNA-Seq analysis of *G. hirsutum* from young, mature, and different senescence stages of leaves identified 3624 differentially expressed genes during leaf senescence [74]. Comparative transcriptome profiling of *G. hirsutum* and *G. davidsonii* (diploid wild species) further characterized 4744 and 5337 differentially expressed salt-stress responsive genes from roots and leaves of *G. davidsonii* [75]. The potential role of signalling pathways, ethylene pathway in fiber elongation, and receptor-like kinases (RLKs) in cell wall integrity have been proposed in determining fiber quality using comparative RNA-Seq analyses of near isogenic lines (NILs) in *G. hirsutum* [76].

The major mRNA repositories are Crops-ESTdb, NCBI dbEST, UniGene, RefSeq, and GEO; plant-specific microarray databases are plant expression database (PLEXdb) and plant co-expression database (PLANEX). Tools for RNA-Seq expression analysis are R/bioconductor packages such as RSEM, DEGseq, DESeq, edger, and baySeq. The open access expVIP is a web-based tool for visualizing, analyzing and comparing RNA-Seq data for conducting gene expression analysis in diploid and polyploid plant species. Single-nucleotide variant analysis in RNA-Seq dataset can be performed by SAMtools followed by Genome Analysis Toolkit (GATK). RobiNA is a web-based tool for accessing and comparing EST, microarray, and RNA-Seq databases.

Though several OLC-based, DBG-based, and greedy assemblers discussed above have been utilized in RNA-Seq analyses, Trinity is the most promising tool in generating transcriptome assembly. The Tuxedo pipeline contains a series of tools including Bowtie and TopHat for aligning the RNA-Seq reads, Cufflinks for assembling the mapped reads, Cuffdiff for identifying differentially expressed genes and CummeRbund for visualizing differentially expressed genes and transcripts [96]. The parameters such as false discovery rate (FDR) ≤ 0.001, fold change (log2FC) ≥ 2, and P-value ≤ 0.05 are considered in comparative transcriptome profiling using RNA-Seq [105]. Using Trinity and Tuxedo, novel genes, and splice variants can be determined. In RNA-Seq analysis, FPKM (Fragments per kilobase of transcript per million mapped reads) and RPKM (Reads per kilobase of transcript per million mapped reads) values are often used in quantifying gene expression. Express tool has been used in detecting transcript abundance in *G. hirsutum*.

Additionally, to annotate gene functions, homology-based methods such as BLASTX, BLAST2GO, and BLAT have been used. PASA and EVM have been used in annotating 3' and 5' UTR regions and alternative splice events in the transcript assemblies. Functional annotation of transcript assembled fragments (TAFs) can be done using BLASTX (E-value 1×10^{-6}) or BLASTP (E-value 1×10^{-5}) against protein databases non-redundant (NR), SwissProt, and TrEMBL. Gene ontology annotation is conducted by using BLAST2GO and AgriGO analysis. InterPro is used to annotate motifs and domains by comparing TAFs with publicly available databases such as Pfam, PRINTS, PROSITE, ProDom, and SMART. The databases used for predicting pathways are KEGG, PANTHER, Pathguide, PlantCyc, BioCyc, and MetaCyc. Using BLASTP (E-value =< 1e-5) and OrthoMCL gene clusters between sub-genomes in tetraploid cotton have been classified. FASTQC and FASTX toolkit are widely used in filtering low quality and determining the quality of RNA-Seq reads. Reads contaminated with adapters are removed using Trimmomatic software. HMMER has been used in identifying transcription factor (TF) gene families in tetraploid cotton using the PlnTFDB. Using D_5 genome as a reference, homeologous genes/syntenic blocks between A_t and D_t sub-genomes have been identified using MCscanX. These tools were summarized in **Table 2.**

The comparative transcriptome analysis of *G. hirsutum* with its progenitors ameliorates the complexities associated with the co-existence of A_t and D_t genome transcripts in allotetraploid species. Moreover, comparative transcriptome mapping of sub-genomes leads to identification of high-confidence transcriptional modules that are evolutionarily conserved and are specific to the genus *Gossypium*. Identifying sub-genome specific transcripts and analyzing their role in cotton gene regulation would help in developing novel biomarker tools associated with various complex polygenic traits in cotton. Alternative splicing in eukaryotes results in transcriptome and proteome diversity. The presence of abundant splice variants and novel transcriptionally active regions (nTARs) has been identified in *Arabidopsis* and rice using RNA-Seq, but majority of these novel transcripts did not overlap with known protein-coding genes and open reading frames (ORFs), suggesting their potential role in post-transcriptional gene regulation and novel transcript/gene formation [105]. Identification of alternative splice events and nTARs in tetraploid cotton is computationally challenging due to ploidy, duplication, chromosomal rearrangements, and presence of homeologous bias.

4. Small RNA analysis in cotton

Napoli et al. (1990) first identified the phenomenon of RNA interference (RNAi) and referred to it as co-suppression in plants [106]. Hamilton & Baulcombe (1999) reported a group of antisense RNAs that mediate in post-transcriptional gene silencing (PTGS) in plants that target both cellular and viral mRNAs [107]. The world of sRNAs exploded with the discovery of miRNAs (microRNA) in *C. elgans* [108]. In plants, miRNAs were first identified in *A. thaliana* and studied extensively in other plant species using comparative genomic, cloning, or sequencing approaches [109]. Unlike in animals, plant miRNAs are mainly found in intergenic and intronic regions [110]. MiRNA genes are mostly localized and form clusters and they transcribe together as a single transcriptional unit [111]. MiRNA discovery has gained momentum in diverse plant species. The majority of plant miRNAs identified to date negatively regulate the target gene expression at the post-transcriptional level. MiRNAs regulate environmental stress response, metabolic processes, organogenesis, growth and development [112]. The miRNAs associated with seed-specific transcription factors have been identified and were found to influence differentiation and developmental timing [113].

The reduction in cotton fiber quality and yields is mainly attributed to several biotic and abiotic factors, and these complex traits are also regulated by small RNAs (sRNAs) which are mostly by miRNAs. Recent advances in cotton genomics have provided an impetus to pursue the discovery of novel miRNAs in the cotton genome [3, 114, 115]. MiRNAs and small interfering RNAs (siRNAs) are the two major classes of endogenous small non-coding RNAs that regulate the gene expression in plants. Earlier reports suggest that miRNAs mediate a variety of functional roles such as developmental timing, cell proliferation, differentiation, morphogenesis, defense response, and signal transduction [116–118]. Cotton miRNA research is limited due to the lack of genomic information of cultivated cotton until recently. Cotton miRNAs have been identified in *G. hirsutum* (A_tD_t), *G. barbadense* (A_tD_t), *G. herbaceum* (A_1), *G. arboreum* (A_2), and *G. raimondii* (D_5). Besides their role in fiber growth, development, initiation, and elongation, some miRNAs have been implicated in biotic and abiotic stress responses in cotton. Moreover, many cotton-specific miRNAs have been identified, but their functions need experimental validation. Recent studies that use qRT-PCR and degradome sequencing of cotton fiber RNA suggest that miRNAs play an important role in cotton fiber development. Similarly, Wang et al. (2012) identified 73 miRNAs that belong to 49 families in *G. arboreum* using homology-based approach [115]. Zhang et al. (2013) identified 65 novel miRNAs and their candidate gene targets in *G. hirsutum* using transcript sequences of *G. raimondii* [116]. MiRNA was also analyzed in salt tolerance response in *G. hirsutum*. The miRNVL5 precursor was first discovered in *Arabidopsis* and later found in cotton. *G. arboreum* miRNA was studied in response to Cotton Leaf Curl Disease [118].

NGS technologies that are mentioned in transcriptome analysis are also applicable in small RNA sequencing. MiRNAs are highly conserved across the species and can be predicted by using homology-based methods. Various resources commonly used in analyses of small RNAs are described in **Table 3**.

Purpose	Tool	References
miRNA precursors	RNAfold, mfold	[123]
miRNA databases	miRBase, MicroRNAdb	[124]
miRNA prediction	miRCat, miREAP, miRPlant, miRDeep-P, miRNAKey	[125–128]
Gene targets for miRNAs	miRanda and psRNATarget	[129, 130]
lncRNA homologs	RefSeq, NONCODE, lncRNADB, PLncDB, PNRD, PlncRNADB, and PLNlncRbase	[131–134]
Epigenomic databases	pENCODE, GEO, NGSmethDB, MethBase, plant methylome-db	[135–137]
ChIP-Enriched region identification	HOMER	[138]
Methylation detection	CyMATE, MeQA, MEDIPS, FASTmC	[139, 140]
Bisulphite data analysis	Methylpipe	[141]

Table 3. Bioinformatics tools used in regulatory analysis of cotton.

The other classes of small RNAs identified in *Gossypium* include trans-acting small interfering RNA (TasiRNA), long intergenic noncoding RNA (lincRNA), and long noncoding natural antisense transcript loci (lncNAT). Long noncoding RNAs (lncRNAs) are 200 bp in length, rich in repetitive sequences, preferentially expressed in a tissue-, genome- and lineage-specific manner, are poorly conserved and participate in various regulatory processes. Higher overall methylation levels are exhibited by lncRNAs when compared with protein-coding genes, while their expression is less affected by gene-body methylation. The lncRNAs play a pivotal role in regulating lignin metabolism, cotton fiber initiation, and elongation. There are two main classes of lncRNAs, long intergenic noncoding RNAs (lincRNAs), and long noncoding natural antisense transcript (lncNATs), which are structurally similar but vary in number and length of exons and transcripts. In cotton (*Gossypium spp.*), 30550 lincRNA and 4718 lncNAT loci have been reported. However, homeolog expression bias of lncRNAs has been identified in subgenomes of polyploid species when compared to wild parents [119]. In a similar study, 3510 lincRNAs and 2486 lncNATs that play an integral role in fiber initiation and elongation have been identified in *G. arboreum* using strand-specific RNA sequencing (ssRNA-Seq) of cotton fibers and leaves [120]. A preferential expression of lncRNAs was observed (~50%) than for protein coding genes (~20%) during rapid fiber elongation, thus establishing their function in fiber development. The direct role of 21–22 nt siRNA derived from GhMML3 gene (MYB-MIXTA-like transcription factor 3) that encodes a lncNAT, implicated in fiber development has been investigated in *G. hirsutum* [119].

5. Epigenome analysis in cotton

Post-transcriptional gene silencing (PTGS) is mainly associated with the regulation of gene expression directly by targeting the mRNA, while transcriptional gene silencing (TGS) is mostly coordinated by epigenetic modifications such as DNA methylation, histone modifications (methylation and acetylation), chromatin remodelling factors, polycomb group (PcG), and trithorax group (trxG) of proteins [121]. These epigenetic modulators can be identified by using homology-based searching in Upland cotton and its closest progenitor species, which will aid in understanding the epigenetic landscape of Upland cotton. The role of DNA methylation has been implicated in plant growth and development, by maintaining active and inactive chromatin states and in gene silencing by employing canonical and non-canonical RNA-directed DNA methylation (RdDM) pathways [121]. For eliciting both the pathways, core RNAi machinery including DICER-LIKE (DCL), RNA-DEPENDENT RNA POLYMERASE (RDR), ARGONAUTE (AGO) proteins, RNA polymerases (Pol IV and V), and plant specific-DNA methylases are required [122].

In plants, DNA methylation occurs in both CG and non-CG (CHG and CHH) contexts, where "H" denotes A, T, or C. Although DNA is primarily methylated in CG context in both plants and mammals, the non-CG contexts are abundant in plants when compared to mammals [142]. The CG, CHG methylation contexts are symmetrical and are maintained by DNA replication, while CHH context is established by *de novo* DNA methylation. The DNA methylation within the genome can occur in promoter, gene-body, transposable elements, and repeat regions [143]. Previous reports suggest that 35–43%, 30–41%, and 30–32% of loci were methylated in *A. thaliana* ecotypes [144], *Brassica oleracea* accessions [145], and *G. hirsutum* accessions [146], respectively. The role of DNA methylation in fiber growth and gene silencing has been proposed in cotton. It has also been suggested that CHH methylation may have a role in developmental timing in cotton. Annual fluctuations in DNA demethylases and methyltransferases have shown opposite trends in their abundance in cotton ovules. Also, substantial changes in CHH methylation was noticed in the promoter regions of three key genes ETHYLENE RESPONSIVE FACTOR 6 (ERF6), SUPPRESSION OF RVS 161 DELTA 4 (SUR4) and 3-KETOACYL-COA SYNTHASE 13 (KCS13) that regulate cotton fiber growth. Development of homozygous RNAi lines, specifically targeting demethylases and methyltransferases, will aid in determining DNA methylation patterns in cotton fiber growth and development [147].

The variation in epigenetic modifications has been observed in spontaneous reciprocal hybrids of *Rosa sect. Caninae* and *Rubigineae* when compared to their parental species using cDNA–AFLPs and methylation-sensitive amplified fragment length polymorphisms (MSAPs), suggesting the biased contribution of DNA methylation from parents to polyploid hybrids [148]. Similarly, genotype-specific and tissue-specific variations in DNA methylation have been identified in cotton by using MSAP with methylation insensitive enzyme (BsiSI). Their results suggest that CHG methylation is more diverse than the other two contexts (CG and CHH) and this work established a relationship between DNA methylation and fiber development but failed to establish the correlation between epigenetic regulation and fiber quality. To achieve this, a robust study involving diverse genotypes/accessions and tissues is required. Cytosine methylation is among one of the key epigenetic regulatory processes that likely silence duplicated genes in polyploid crop species such as cotton. The levels, patterns, and diversity of methylation-polymorphism in CG context have been investigated in 20 accessions of Upland cotton (*G. hirsutum*) to identify 32 methylation-polymorphic cytosine sites using MSAP [63]. The introgression of exotic DNA fragments from wild parents (*G. bickii*) into cultivated Upland cotton (*G. hirsutum*) has been investigated using AFLP and MSAP analysis to identify ~2000 genomic and ~800 epigenetic sites [66]. This study identified ~0.5% of alien DNA segments and ~0.7% of genetic variation in the genomes of introgression lines. Though the overall methylation content is close to the wild parent, the context-specific methylation varied significantly in introgression lines [149].

Using methylcytosine-sequencing (MethylC-Seq) and RNA-sequencing (RNA-Seq) analysis, variation in CHH methylation during ovule, and fiber development was reported in cotton [150]. Further, this study suggested that CHH hypermethylation triggered RdDM-dependent methylation in promoters, while RdDM-independent methylation occurred in TEs and nearby genes, thus facilitating ovule and fiber development in cotton. Moreover, the contribution of CHG and CHH methylation in genic regions towards homeologous expression patterns in A_t, or D_t subgenomes have been proposed [150]. Evolution is a very slow process when it is purely dependent on mutation and recombination. In plants, it is believed that there are additional forces that accelerated the process of evolution, which include interspecific hybridization. However, the mechanism of acceleration is less understood. The possible underlying mechanism has been proposed in wheat [151]. It was suggested that the interaction of alien nuclear DNA with cytoplasmic macromolecules as a result of interspecific hybridization can potentially trigger a network of epigenetic changes in nuclear DNA, thus altering the expression of genes and genetic pathways associated with physiological processes, which may serve as a possible source of variation that facilitate evolutionary process. This idea sheds light into development of epigenomic-segregating lines in crop plants, including cotton [151].

Screening the core RNAi machinery of important DNA methylases such as DOMAINS REARRANGED METHYLASE 1/2 (DRM1/2), CHROMOMETHYLASE 2/3 (CMT 2/3), and demethylases such as ROS1 and DEMETER (DME) in Upland cotton, and its closest progenitor species will aid in elucidating the key regulatory mechanisms in whole-genome duplication, chromosomal rearrangements, dosage compensation, and evolutionary advantage of being polyploids. Chromosome painting techniques such as fluorescence *in situ* hybridization (FISH), variations of FISH and genomic *in situ* hybridization (GISH) are not only useful in determining chromosome structure and distribution of hetero/euchromatin but also in understanding epigenomic patterns and evolution of polyploid species [152].

Although several databases and repositories are available for storing, cataloging and normalizing animal, and mammalian epigenomic data, they are limited for plants. The popular resource is NCBI epigenomics sample browser to access the diverse collection of epigenomic data sets including genome-wide DNA methylation maps and histone modifications. However, only *A. thaliana* dataset is currently available in the sample browser. Other than NCBI sample browser, the popular visualization tools for epigenomic data are as follows: UCSC epigenome browser, Ensembl encyclopedia of DNA elements (ENCODE), and WashU epigenome browser. A plant ENCODE (pENCODE) collects and curates epigenomic data from a wide range of plant species to compare, annotate, and understand mechanisms behind plant evolution. Epigenomics of Plants International Consortium (EPIC) provides a platform to share the protocols, methods, and results across the research labs to address basic questions of genetics and genome regulation beyond routine whole-genome sequencing. GEO is one of the seminal repositories for hosting epigenome data from various resources including animals (human) and plants (*Arabidopsis* and Maize). NGSmethDB and MethBase specifically store whole-genome single-base resolution data obtained from whole-genome bisulphite-sequencing (WGBS/BS-Seq) from diverse organisms. Currently, very few plant-specific epigenome databases are publicly available and include plant methylome-db, *Arabidopsis* Epigenome Browser, and Tomato Epigenome Database. Though the cotton-specific epigenome database is not publicly available, methylomes of evolutionarily close relatives such as *Ricinus communis* and *Theobroma cacao* are available at plant methylomes-db with user restrictions [84]. Epigenetic resources and tools were summarized in **Table 3**.

The wide applications of chromatin Immunoprecipitation-sequencing (ChIP-Seq) analysis include understanding protein-DNA interactions, histone modifications, chromatin states, and DNA methylation. Methylation and acetylation marked peaks can be identified using spatial clustering for identification of ChIP-enriched regions (SICER), and later peaks can be annotated by using HOMER (motif search algorithm). Other peak-finding algorithms, Model-based Analysis of ChIP-Seq (MACS), and PeakSeq have been extensively used in analysis. BS-Seq analysis helps in identification of genome-wide differentially methylated regions (DMRs) along with tools such as Bismark, BSMAP, and BS-Seeker can be used. Cytosine Methylation Analysis Tool for Everyone (CyMATE) is a plant-specific tool for analyzing BS-Seq data. A web-based tool, QUMA (quantification tool for methylation analysis) is suitable for BS-Seq analysis, primarily for estimating CG methylation. MeDIP analysis helps in detecting methylated cytosines in both CG and non-CG contexts and tools such as MeQA and MEDIPS can be used. FASTmC Webtool is available for comparative analysis of CG, CHG, and CHH DNA methylation levels in non-model species. Methpipe is useful in analyzing bisulphite-sequencing data including BS-Seq, WGBS, and reduced representation bisulfite sequencing (RRBS). Integrative epigenome analysis can be done using methylPipe and compEpiTools. Differentially methylated regions (DMRs) have been identified in *Arabidopsis* by using methylpipe. ChIP-Seq analysis in R (CSAR) accurately account the protein bound regions in the genome in plants. PolyCat determines mapping bias between genomes that may have resulted due to SNPs; hence, it can be used in comparing allopolyploid, *G. hirsutum* against diploid *G. raimondii* or *G. arboreum*. Moreover, it accepts data from diverse NGS platforms including RNA-Seq, BS-Seq, ChIP-Seq, and SNP analysis [153].

6. Future direction

The reduction in cost of sequencing per base, increase in throughput and read size, and availability of better algorithms have significantly facilitated the integration of one or more genomic methods to address biological questions. For instance, continued progress in detection of SNPs and CNVs from sequencing data and combining this with genotyping-by-sequencing methods will be helpful in crop improvement. Rapid advances in genomics, bioinformatics, and computational biology in the past two decades have already facilitated generation and screening of large datasets, and thus have ushered us into the "Big Data" era of genomics. The plant science community still lacks full-fledged computational infrastructure, data curation, and novel tools to extract information from the massive datasets. It should be realized that progress in analyzing polyploid genomes will continually need improvements because the nature of the data generated from high-throughput technologies is voluminous, heterogeneous, and often in unstructured forms. The concepts of parallel processing, cloud computing, and machine learning offer promising solutions for managing and analyzing large datasets. The other important limiting factors are trained professionals to access and analyze these resources. Nonetheless, progress is being made as we enter the burgeoning age of "Big Data" to tackle genomes of widely grown and utilized crops such as cotton using bioinformatics tools.

Acknowledgements

The authors acknowledge Dr. Avinash Sreedasyam of HudsonAlpha Institute of Biotechnology and Mr. Joshua Reid at Alabama A&M University for reviewing this book chapter. Also authors would like to thank Dr. Ibrokhim Abdurakhmonov and Ms. Ivona Lovric for their valuable suggestions and comments in improving this book chapter.

Author details

Venkateswara R. Sripathi[1*], Ramesh Buyyarapu[2], Siva P. Kumpatla[2], Abreeotta J. Williams[1], Seloame T. Nyaku[3], Yonathan Tilahun[4], Venu Kalavacharla[5] and Govind C. Sharma[1]

*Address all correspondence to: v.sripathi@aamu.edu

1 Center for Molecular Biology, Alabama A&M University, Normal, AL, USA

2 Dow AgroSciences, Indianapolis, IN, USA

3 Crop Science Department, College of Basic and Applied Sciences, University of Ghana, Legon-Accra, Ghana

4 Cooperative Extension, School of Agriculture & Applied Sciences, Langston University, Langston, OK, USA

5 Molecular Genetics & EpiGenomics Laboratory, College of Agriculture & Related Sciences, Delaware State University, Dover, DE , USA

References

[1] Varshney RK, Terauchi R, McCouch SR. Harvesting the promising fruits of genomics: applying genome sequencing technologies to crop breeding. PLoS Biology. 2014;12:e1001883. doi:10.1371/journal.pbio.1001883

[2] Joosen RV, Ligterink W, Hilhorst HW, Keurentjes JJ. Advances in genetical genomics of plants. Current Genomics. 2009;10:540-549. doi:10.2174/138920209789503914

[3] Abdurakhmonov IY, Ayubov MS, Ubaydullaeva KA, Buriev ZT, Shermatov SE, Ruziboev HS, Shapulatov UM, Saha S, Ulloa M, Yu JZ, Percy RG. RNA interference for functional genomics and improvement of cotton (*Gossypium* sp.). Frontiers in Plant Science. 2016;7:202. doi:10.3389/fpls.2016.00202

[4] Wang S, Chen J, Zhang W, Hu Y, Chang L, Fang L, Wang Q, Lv F, Wu H, Si Z, Chen S. Sequence-based ultra-dense genetic and physical maps reveal structural variations of allopolyploid cotton genomes. Genome Biology. 2015;16:1-8. doi:10.1186/s13059-015-0678-1

[5] Pareek CS, Smoczynski R, Tretyn A. Sequencing technologies and genome sequencing. Journal of Applied Genetics. 2011;52:413-435. doi:10.1007/s13353-011-0057-x

[6] Buyyarapu R, Kantety RV, John ZY, Xu Z, Kohel RJ, Percy RG, Macmil S, Wiley GB, Roe BA, Sharma GC. BAC-pool sequencing and analysis of large segments of A12 and D12 homoeologous chromosomes in Upland cotton. PloS One. 2013;8:e76757. doi:10.1371/journal.pone.0076757

[7] Shang J, Zhu F, Vongsangnak W, Tang Y, Zhang W, Shen B. Evaluation and comparison of multiple aligners for next-generation sequencing data analysis. BioMed Research International. 2014;2014:309650. doi:10.1155/2014/309650

[8] Shang J, Zhu F, Vongsangnak W, Tang Y, Zhang W, Shen B. Evaluation and comparison of multiple aligners for next-generation sequencing data analysis. BioMed Research International. 2014;2014:309650. doi:10.1155/2014/309650

[9] Fryxell P. A revised taxonomic interpretation of *Gossypium* L. (Malvaceae). Rheedea. 1992;2:108-165

[10] Wendel JF, Cronn RC. Polyploidy and the evolutionary history of cotton. Advances in Agronomy. 2003;78:13986. doi:10.1016/S0065-2113(02)78004-8

[11] Wendel JF, Schnabel A, Seelanan T. Bidirectional interlocus concerted evolution following allopolyploid speciation in cotton (*Gossypium*). Proceedings of the National Academy of Sciences of the United States of America. 1995;1995:92280-92284. doi: 10.1073/pnas.92.1.280

[12] Bancroft I, Morgan C, Fraser F, Higgins J, Wells R, Clissold L, Baker D, Long Y, Meng J, Wang X, Liu S. Dissecting the genome of the polyploid crop oilseed rape by transcriptome sequencing. Nature Biotechnology. 2011;29:762-766. doi:10.1038/nbt.1926

[13] Hendrix B, Stewart JM. Estimation of the nuclear DNA content of *Gossypium* species. Annals of Botany. 2005;95:789-797. doi:10.1093/aob/mci078

[14] Wang K, Wang Z, Li F, Ye W, Wang J, Song G, Yue Z, Cong L, Shang H, Zhu S, Zou C. The draft genome of a diploid cotton *Gossypium raimondii*. Nature Genetics. 2012;44:1098-103. doi:10.1038/ng.2371

[15] Li F, Fan G, Wang K, Sun F, Yuan Y, Song G, Li Q, Ma Z, Lu C, Zou C, Chen W. Genome sequence of the cultivated cotton *Gossypium arboreum*. Nature Genetics. 2014;46:567-572. doi:10.1038/ng.2987

[16] Li F, Fan G, Lu C, Xiao G, Zou C, Kohel RJ, Ma Z, Shang H, Ma X, Wu J, Liang X. Genome sequence of cultivated upland cotton (*Gossypium hirsutum* TM-1) provides insights into genome evolution. Nature Biotechnology. 2015;33:524-30. doi:10.1038/nbt.3208

[17] Zhang T, Hu Y, Jiang W, Fang L, Guan X, Chen J, Zhang J, Saski CA, Scheffler BE, Stelly DM, Hulse-Kemp AM. Sequencing of allotetraploid cotton (*Gossypium hirsutum* L. acc. TM-1) provides a resource for fiber improvement. Nature Biotechnology. 2015;33:531-537. doi:10.1038/nbt.3207

[18] Liu X, Zhao B, Zheng HJ, Hu Y, Lu G, Yang CQ, Chen JD, Chen JJ, Chen DY, Zhang L, Zhou Y. *Gossypium barbadense* genome sequence provides insight into the evolution of extra-long staple fiber and specialized metabolites. Scientific Reports. 2015;5:14139. doi: 10.1038/Srep14139

[19] Yuan D, Tang Z, Wang M, Gao W, Tu L, Jin X, Chen L, He Y, Zhang L, Zhu L, Li Y. The genome sequence of Sea-Island cotton (*Gossypium barbadense*) provides insights into the allopolyploidization and development of superior spinnable fibres. Scientific Reports. 2015;5:17662. doi:10.1038/Srep17662

[20] Sripathi VR. Towards understanding the genome of diploid cotton, *Gossypium herbaceum* using deep sequencing. In: Plant and Animal Genome XXIII Conference; 10-1437 January 2015; San Diego, CA, USA.

[21] Wei L, Cao X. The effect of transposable elements on phenotypic variation: insights from plants to humans. Science China Life Sciences. 2016;59:24-37. doi:10.1007/s11427-015-4993-2

[22] Springer NM, Lisch D, Li Q. Creating order from chaos: epigenome dynamics in plants with complex genomes. The Plant Cell. 2016;28:314-325. doi:10.1105/tpc.15.00911

[23] Wang K, Huang G, Zhu Y. Transposable elements play an important role during cotton genome evolution and fiber cell development. Science China Life Sciences. 2015;59:112-121. doi:10.1007/s11427-015-4928-y

[24] Gao D, Li Y, Do Kim K, Abernathy B, Jackson SA. Landscape and evolutionary dynamics of terminal repeat retrotransposons in miniature in plant genomes. Genome Biology. 2016;17:1-7. doi:10.1186/s13059-015-0867-y

[25] Buyyarapu R, Kumaptla S, Elango N, Chen W, Greene T. KASPar: An Efficient and Cost-effective Technology for SNP Genotyping in Cotton. In: Proceedings of the Beltwide Cotton Conferences; 2010; New Orleans.

[26] Mammadov J, Aggarwal R, Buyyarapu R, Kumpatla S. SNP markers and their impact on plant breeding. International Journal of Plant Genomics. 2012;2012:728398. doi: 10.1155/2012/728398

[27] Zhang HB, Li Y, Wang B, Chee PW. Recent advances in cotton genomics. International Journal of Plant Genomics. 2008;2008:742304. doi:10.1155/2008/742304

[28] Wendel JF, Brubaker CL, Percival AE. Genetic diversity in *Gossypium hirsutum* and the origin of Upland cotton. American Journal of Botany. 1992;1992:1291-1310.

[29] Reinisch AJ, Dong JM, Brubaker CL, Stelly DM, Wendel JF, Paterson AH. A detailed RFLP map of cotton, *Gossypium hirsutum* × *Gossypium barbadense*: chromosome organization and evolution in a disomic polyploid genome. Genetics. 1994;138:829-847.

[30] Abdalla AM, Reddy OU, El-Zik KM, Pepper AE. Genetic diversity and relationships of diploid and tetraploid cottons revealed using AFLP. Theoretical and Applied Genetics. 2001;102:222-229. doi:10.1007/s001220051639

[31] Liu S, Cantrell RG, McCarty JC, Stewart JM. Simple sequence repeat–based assessment of genetic diversity in cotton race stock accessions. Crop Science. 2000;40:1459-1469. doi:10.2135/cropsci2000.4051459x

[32] Buyyarapu R, Kantety RV, Yu JZ, Saha S, Sharma GC. Development of new candidate gene and EST-based molecular markers for *Gossypium* species. International Journal of Plant Genomics. 2011;2011:894598. doi:10.1155/2011/894598

[33] Hulse-Kemp AM, Lemm J, Plieske J, Ashrafi H, Buyyarapu R, Fang DD, Frelichowski J, Giband M, Hague S, Hinze LL, Kochan KJ. Development of a 63K SNP array for cotton and high-density mapping of intra-and inter-specific populations of *Gossypium* spp.G3: Genes| Genomes| Genetics. 2015;2015:g3-115. doi:10.1534/g3.115.018416

[34] Kumpatla SP, Mukhopadhyay S. Mining and survey of simple sequence repeats in 2 expressed sequence tags of dicotyledonous species. Genome. 2005;48:985-998. doi:3 10.1139/g05-060

[35] Sharma PC, Grover A, Kahl G. Mining microsatellites in eukaryotic genomes. Trends in Biotechnology. 2007;25:490-498. doi:10.1016/j.tibtech.2007.07.013

[36] Shen X, Guo W, Lu Q, Zhu X, Yuan Y, Zhang T. Genetic mapping of quantitative trait loci for fiber quality and yield trait by RIL approach in Upland cotton. Euphytica. 2007;155:371-380. doi:10.1007/s10681-006-9338-6

[37] Yang C, Guo W, Li G, Gao F, Lin S, Zhang T. QTLs mapping for Verticillium wilt resistance at seedling and maturity stages in *Gossypium barbadense* L. Plant Science. 2008;174:290-298. doi:10.1016/j.plantsci.2007.11.016

[38] Shen X, Van Becelaere G, Kumar P, Davis RF, May OL, Chee P. QTL mapping for resistance to root-knot nematodes in the M-120 RNR Upland cotton line (*Gossypium hirsutum* L.) of the Auburn 623 RNR source. Theoretical and Applied Genetics. 2006;113:1539-1549. doi:10.1007/s00122-006-0401-4

[39] An C, Saha S, Jenkins JN, Scheffler BE, Wilkins TA, Stelly DM. Transcriptome profiling, sequence characterization, and SNP-based chromosomal assignment of the EXPANSIN genes in cotton. Molecular Genetics and Genomics. 2007;278:539-553. doi:10.1007/s00438-007-0270-9

[40] Byers RL, Harker DB, Yourstone SM, Maughan PJ, Udall JA. Development and mapping of SNP assays in allotetraploid cotton. Theoretical and Applied Genetics. 2012;124:1201-1214. doi:10.1007/s00122-011-1780-8

[41] Ashrafi H, Hulse-Kemp AM, Wang F, Yang SS, Guan X, Jones DC, Matvienko M, Mockaitis K, Chen ZJ, Stelly DM, Van Deynze A. A long-read transcriptome assembly of cotton (*Gossypium hirsutum*) and intraspecific SNP discovery. The Plant Genome. 2015;8:1-14. doi:10.3835/plantgenome2014.10.0068

[42] Xie F, Sun G, Stiller JW, Zhang B. Genome-wide functional analysis of the cotton transcriptome by creating an integrated EST database. PloS One. 2011;6:e26980. doi: 10.1371/journal.pone.0026980

[43] Van Deynze A, Stoffel K, Lee M, Wilkins TA, Kozik A, Cantrell RG, John ZY, Kohel RJ, Stelly DM. Sampling nucleotide diversity in cotton. BMC Plant Biology. 2009;9:125. doi: 10.1186/1471-2229-9-125

[44] Lacape JM, Nguyen TB, Thibivilliers S, Bojinov B, Courtois B, Cantrell RG, Burr B, Hau B. A combined RFLP SSR AFLP map of tetraploid cotton based on a *Gossypium hirsutum×Gossypium barbadense* backcross population. Genome. 2003;46:612-626. doi:2 10.1139/G03-050

[45] Blenda A, Scheffler J, Scheffler B, Palmer M, Lacape JM, John ZY, Jesudurai C, Jung S, Muthukumar S, Yellambalase P, Ficklin S. CMD: a cotton microsatellite database resource for *Gossypium* genomics. BMC Genomics. 2006;7:132. doi:10.1186/1471-2164-6 7-132

[46] Yu J, Jung S, Cheng CH, Ficklin SP, Lee T, Zheng P, Jones D, Percy RG, Main D. CottonGen: a genomics, genetics and breeding database for cotton research. Nucleic Acids Research. 2013;2013:gkt1064. doi:10.1093/nar/gkt1064

[47] Said JI, Knapka JA, Song M, Zhang J. Cotton QTLdb: a cotton QTL database for QTL analysis, visualization, and comparison between *Gossypium hirsutum* and *G. hirsutum×G. barbadense* populations. Molecular Genetics and Genomics. 2015;290:1615-1625. doi:10.1007/s00438-015-1021-y

[48] Schulz MH, Zerbino DR, Vingron M, Birney E. Oases: robust de novo RNA-seq assembly across the dynamic range of expression levels. Bioinformatics. 2012;28:1086-1092. doi:16 10.1093/bioinformatics/bts094

[49] Pevzner PA, Tang H, Waterman MS. An Eulerian path approach to DNA fragment assembly. Proceedings of the National Academy of Sciences. 2001;98:9748-9753. doi:19 10.1073/pnas.171285098

[50] Batzoglou S, Jaffe DB, Stanley K, Butler J, Gnerre S, Mauceli E, Berger B, Mesirov JP, Lander ES. ARACHNE: a whole-genome shotgun assembler. Genome Research. 2002;12:177-189. doi:10.1101/gr.208902

[51] Huang X, Madan A. CAP3: A DNA sequence assembly program. Genome Research. 1999;9:868-877. doi:10.1101/gr.9.9.868.

[52] Chevreux B. MIRA: an automated genome and EST assembler. 2007.

[53] Margulies M, Egholm M, Altman WE, Attiya S, Bader JS, Bemben LA, Berka J, Braverman MS, Chen YJ, Chen Z, Dewell SB. Genome sequencing in microfabricated high density picolitre reactors. Nature. 2005;437:376-380. doi:10.1038/nature03959

[54] Chaisson M, Pevzner P, Tang H. Fragment assembly with short reads. Bioinformatics. 2004;20:2067-2074. doi:10.1093/bioinformatics/bth205

[55] Zerbino DR, Birney E. Velvet: algorithms for de novo short read assembly using de Bruijn graphs. Genome Research. 2008;18:821-829. doi:10.1101/gr.074492.107

[56] Simpson JT, Wong K, Jackman SD, Schein JE, Jones SJ, Birol I. ABySS: a parallel assembler for short read sequence data. Genome Research. 2009;19:1117-1123. doi: 10.1101/gr.089532.108

[57] Li R, Yu C, Li Y, Lam TW, Yiu SM, Kristiansen K, Wang J. SOAP2: an improved ultrafast tool for short read alignment. Bioinformatics. 2009;25:1966-1967. doi:10.1093/bioinformatics/btp336

[58] Warren RL, Sutton GG, Jones SJ, Holt RA. Assembling millions of short DNA sequences using SSAKE. Bioinformatics. 2007;23:500-501. doi:10.1093/bioinformatics/btl629

[59] Dohm JC, Lottaz C, Borodina T, Himmelbauer H. SHARCGS, a fast and highly accurate short-read assembly algorithm for de novo genomic sequencing. Genome Research. 2007;17:1697-1706. doi:10.1101/gr.6435207

[60] Lee S, Seo CH, Alver BH, Lee S, Park PJ. EMSAR: estimation of transcript abundance from RNA-seq data by mappability-based segmentation and reclustering. BMC Bioinformatics. 2015;16:278. doi:10.1186/s12859-015-0704-z

[61] Weber AP, Weber KL, Carr K, Wilkerson C, Ohlrogge JB. Sampling the *Arabidopsis* transcriptome with massively parallel pyrosequencing. Plant Physiology. 2007;144:32-42. doi:10.1104/pp.107.096677

[62] Zhu YN, Shi DQ, Ruan MB, Zhang LL, Meng ZH, Liu J, Yang WC. Transcriptome analysis reveals crosstalk of responsive genes to multiple abiotic stresses in cotton (*Gossypium hirsutum* L.). PLoS One. 2013;8:e80218. doi:10.1371/journal.pone.0080218

[63] Lacape JM, Claverie M, Vidal RO, Carazzolle MF, Pereira GA, Ruiz M, Pré M, Llewellyn D, Al-Ghazi Y, Jacobs J, Dereeper A. Deep sequencing reveals differences in the transcriptional landscapes of fibers from two cultivated species of cotton. PloS One. 2012;7:e48855. doi:10.1371/journal.pone.0048855

[64] Li X, Yuan D, Zhang J, Lin Z, Zhang X. Genetic mapping and characteristics of genes specifically or preferentially expressed during fiber development in cotton. PloS One. 2013;8:e54444. doi:10.1371/journal.pone.0054444

[65] Paterson AH, Wendel JF, Gundlach H, Guo H, Jenkins J, Jin D, Llewellyn D, Showmaker KC, Shu S, Udall J, Yoo MJ. Repeated polyploidization of *Gossypium* genomes and the evolution of spinnable cotton fibres. Nature. 2012;492:423-427. doi:10.1038/nature11798

[66] Bowman MJ, Park W, Bauer PJ, Udall JA, Page JT, Raney J, Scheffler BE, Jones DC, Campbell BT. RNA-Seq transcriptome profiling of Upland cotton (*Gossypium hirsutum* L.) root tissue under water-deficit stress. PloS One. 2013;8:e82634. doi:10.1371/journal.pone.0082634

[67] Rapp RA, Haigler CH, Flagel L, Hovav RH, Udall JA, Wendel JF. Gene expression in developing fibres of Upland cotton (*Gossypium hirsutum* L.) was massively altered by domestication. BMC Biology. 2010;8:139. doi:10.1186/1741-7007-8-139

[68] Hinchliffe DJ, Turley RB, Naoumkina M, Kim HJ, Tang Y, Yeater KM, Li P, Fang DD. A combined functional and structural genomics approach identified an EST-SSR marker with complete linkage to the Ligon lintless-2 genetic locus in cotton (*Gossypium hirsutum* L.). BMC Genomics. 2011;12:445. doi:10.1186/1471-2164-12-445

[69] Haigler CH, Betancur L, Stiff MR, Tuttle JR. Cotton fiber: a powerful single-cell model for cell wall and cellulose research. Frontiers in Plant Science. 2012;3:104. doi:10.3389/fpls.2012.00104

[70] Park W, Scheffler BE, Bauer PJ, Campbell BT. Genome-wide identification of differentially expressed genes under water deficit stress in Upland cotton (*Gossypium hirsutum* L.). BMC Plant Biology. 2012;12:90. doi:10.1186/1471-2229-12-90

[71] Jiao X, Zhao X, Zhou XR, Green AG, Fan Y, Wang L, Singh SP, Liu Q. Comparative transcriptomic analysis of developing cotton cotyledons and embryo axis. PloS One. 2013;8:e71756. doi:10.1371/journal.pone.0071756

[72] Yoo MJ, Szadkowski E, Wendel JF. Homoeolog expression bias and expression level dominance in allopolyploid cotton. Heredity. 2013;110:171-180. doi:10.1038/hdy.2012.94

[73] Yoo MJ, Wendel JF. Comparative evolutionary and developmental dynamics of the cotton (*Gossypium hirsutum*) fiber transcriptome. PLoS Genetics. 2014;10:e1004073. doi:10.1371/journal.pgen.1004073

[74] Lin M, Pang C, Fan S, Song M, Wei H, Yu S. Global analysis of the *Gossypium hirsutum* L. Transcriptome during leaf senescence by RNA-Seq. BMC Plant Biology. 2015;15:43. doi:10.1186/s12870-015-0433-5

[75] Zhang X, Fan S, Song M, Pang C, Wei H, Wang C, Yu S. Functional characterization of GhSOC1 and GhMADS42 homologs from Upland cotton (*Gossypium hirsutum* L.). Plant Science. 2016;242:178-186. doi:10.1016/j.plantsci.2015.05.001

[76] Islam MS, Fang DD, Thyssen GN, Delhom CD, Liu Y, Kim HJ. Comparative fiber property and transcriptome analyses reveal key genes potentially related to high fiber strength in cotton (*Gossypium hirsutum* L.) line MD52ne. BMC Plant Biology. 2016;16:36. doi:10.1186/s12870-016-0727-2

[77] Langmead B, Salzberg SL. Fast gapped-read alignment with Bowtie 2. Nature Methods. 2012;9:357-359. doi:10.1038/nmeth.1923

[78] Li H, Durbin R. Fast and accurate short read alignment with Burrows–Wheeler transform. Bioinformatics. 2009;25:1754-1760. doi:10.1093/bioinformatics/btp324

[79] Quinlan AR, Hall IM. BEDTools: a flexible suite of utilities for comparing genomic features. Bioinformatics. 2010;26:841-2. doi:10.1093/bioinformatics/btq033

[80] Frazer KA, Pachter L, Poliakov A, Rubin EM, Dubchak I. VISTA: computational tools for comparative genomics. Nucleic Acids Research. 2004;32:W273-W279. doi:10.1093/nar/gkh458

[81] Bray N, Dubchak I, Pachter L. AVID: a global alignment program. Genome Research. 2003;13:97-102. doi:10.1101/gr.789803

[82] Edgar RC. MUSCLE: multiple sequence alignment with high accuracy and high throughput. Nucleic Acids Research. 2004;32:1792-1797. doi:10.1093/nar/gkh340

[83] Olson SA. Emboss opens up sequence analysis. Briefings in Bioinformatics. 2002;3:87-91. doi:10.1093/bib/3.1.87

[84] Bedell JA, Korf I, Gish W. MaskerAid: a performance enhancement to RepeatMasker. Bioinformatics. 2000;16:1040-1041. doi:10.1093/bioinformatics/16.11.1040

[85] Benson G. Tandem repeats finder: a program to analyze DNA sequences. Nucleic Acids Research. 1999;27:573-580.

[86] Yang G. MITE Digger, an efficient and accurate algorithm for genome wide discovery of miniature inverted repeat transposable elements. BMC Bioinformatics. 2013;14:186. doi:10.1186/1471-2105-14-186

[87] Stanke M, Morgenstern B. AUGUSTUS: a web server for gene prediction in eukaryotes that allows user-defined constraints. Nucleic Acids Research. 2005;33:W465-W467. doi: 10.1093/nar/gki458

[88] Salamov AA, Solovyev VV. Ab initio gene finding in *Drosophila* genomic DNA. Genome Research. 2000;10:516-522. doi:10.1101/gr.10.4.516

[89] Hu Y, Comjean A, Perkins LA, Perrimon N, Mohr SE. GLAD: an online database of gene list annotation for *Drosophila*. Journal of Genomics. 2015;3:75-81. doi:10.7150/jgen.12863

[90] Mathé C, Sagot MF, Schiex T, Rouzé P. Current methods of gene prediction, their strengths and weaknesses. Nucleic Acids Research. 2002;30:4103-4117. doi:10.1093/nar/gkf543

[91] Conesa A, Götz S, García-Gómez JM, Terol J, Talón M, Robles M. Blast2GO: a universal tool for annotation, visualization and analysis in functional genomics research. Bioinformatics. 2005;21:3674-3676. doi:10.1093/bioinformatics/bti610

[92] Haas BJ, Salzberg SL, Zhu W, Pertea M, Allen JE, Orvis J, White O, Buell CR, Wortman JR. Automated eukaryotic gene structure annotation using EVidenceModeler and the program to assemble spliced alignments. Genome Biology. 2008;9:R7. doi:10.1186/gb-2008-9-1-r7

[93] Li L, Stoeckert CJ, Roos DS. OrthoMCL: identification of ortholog groups for eukaryotic genomes. Genome Research. 2003;13:2178-2189. doi:10.1101/gr.1224503

[94] Wang Y, Li J, Paterson AH. MCScanX-transposed: detecting transposed gene duplications based on multiple colinearity scans. Bioinformatics. 2013;29:1458-1460. doi: 10.1093/bioinformatics/btt150

[95] Haas BJ, Papanicolaou A, Yassour M, Grabherr M, Blood PD, Bowden J, Couger MB, Eccles D, Li B, Lieber M, MacManes MD. De novo transcript sequence reconstruction from RNA-seq using the Trinity platform for reference generation and analysis. Nature Protocols. 2013;8:1494-1512. doi:10.1038/nprot.2013.084

[96] Trapnell C, Roberts A, Goff L, Pertea G, Kim D, Kelley DR, Pimentel H, Salzberg SL, Rinn JL, Pachter L. Differential gene and transcript expression analysis of RNA-seq experiments with TopHat and Cufflinks. Nature Protocols. 2012;7:562-578. doi:10.1038/nprot.2012.016

[97] Zhang Z, Carriero N, Zheng D, Karro J, Harrison PM, Gerstein M. PseudoPipe: an automated pseudogene identification pipeline. Bioinformatics. 2006;22:1437-1439. doi:10.1093/bioinformatics/btl116

[98] McKenna A, Hanna M, Banks E, Sivachenko A, Cibulskis K, Kernytsky A, Garimella K, Altshuler D, Gabriel S, Daly M, DePristo MA. The genome analysis toolkit: a MapReduce framework for analyzing next-generation DNA sequencing data. Genome Research. 2010;20:1297-1303. doi:10.1101/gr.107524.110

[99] Cingolani P, Platts A, Wang LL, Coon M, Nguyen T, Wang L, Land SJ, Lu X, Ruden DM. A program for annotating and predicting the effects of single nucleotide polymorphisms, SnpEff: SNPs in the genome of *Drosophila melanogaster* strain w1118; iso-2; iso-3. Fly. 2012;6:80-92. doi:10.4161/fly.19695

[100] Talevich E, Shain AH, Botton T, Bastian BC. CNVkit: copy number detection and visualization for targeted sequencing using off-target reads. bioRxiv. 2014;1:010876. doi:10.1101/010876

[101] Lander ES, Green P, Abrahamson J, Barlow A, Daly MJ, Lincoln SE, Newburg L. MAPMAKER: an interactive computer package for constructing primary genetic linkage maps of experimental and natural populations. Genomics. 1987;1:174-181. doi:10.1016/0888-7543(87)90010-3

[102] Manly KF, Cudmore Jr RH, Meer JM. Map Manager QTX, cross-platform software for genetic mapping. Mammalian Genome. 2001;12:930-932. doi:10.1007/s00335-001-1016-3

[103] Van Ooijen JW. MapQTL® 5. Software for the Mapping of Quantitative Trait Loci in Experimental Populations. Wageningen: Kyazma B; 2004.

[104] Basten CJ, Weir BS, Zeng ZB. QTL Cartographer, version 1.17. Raleigh, NC: Department of Statistics, North Carolina State University; 2004.

[105] Leng X, Jia H, Sun X, Shangguan L, Mu Q, Wang B, Fang J. Comparative transcriptome analysis of grapevine in response to copper stress. Scientific Reports. 2015;5:17749. doi:10.1038/srep17749

[106] Napoli C, Lemieux C, Jorgensen R. Introduction of a chimeric chalcone synthase gene into petunia results in reversible co-suppression of homologous genes in trans. The Plant Cell. 1990;2:279-289. doi:10.1105/Tpc.2.4.279

[107] Hamilton AJ, Baulcombe DC. A species of small antisense RNA in posttranscriptional gene silencing in plants. Science. 1999;286:950-952. doi:10.1126/science.286.5441.950

[108] Reinhart BJ, Slack FJ, Basson M, Pasquinelli AE, Bettinger JC, Rougvie AE, Horvitz HR, Ruvkun G. The 21-nucleotide let-7 RNA regulates developmental timing in *Caenorhabditis elegans*. Nature. 2000;403:901-906. doi:10.1038/35002607

[109] Voinnet O. Origin, biogenesis, and activity of plant microRNAs. Cell. 2009;136:669-687. doi:10.1016/j.cell.2009.01.046

[110] Jones-Rhoades MW, Bartel DP, Bartel B. MicroRNAs and their regulatory roles in plants. Annual Review of Plant Biology. 2006;57:19-53. doi:10.1146/annurev.arplant.57.032905.105218

[111] Combier JP, Frugier F, De Billy F, Boualem A, El-Yahyaoui F, Moreau S, Vernié T, Ott T, Gamas P, Crespi M, Niebel A. MtHAP2-1 is a key transcriptional regulator of symbiotic nodule development regulated by microRNA169 in Medicago truncatula. Genes & Development. 2006;20:3084-3088. doi:10.1101/gad.402806

[112] Jones-Rhoades MW, Bartel DP. Computational identification of plant microRNAs and their targets, including a stress-induced miRNA. Molecular Cell. 2004;14:787-799. doi:10.1016/j.molcel.2004.05.027

[113] Lee H, Yoo SJ, Lee JH, Kim W, Yoo SK, Fitzgerald H, Carrington JC, Ahn JH. Genetic framework for flowering-time regulation by ambient temperature-responsive miRNAs in *Arabidopsis*. Nucleic Acids Research. 2010;2010:gkp1240. doi:10.1093/nar/gkp1240

[114] Kim VN. MicroRNA biogenesis: coordinated cropping and dicing. Nature Reviews Molecular Cell Biology. 2005;6(5):376-385. doi:10.1038/nrm1644

[115] Wang M, Wang Q, Wang B. Identification and characterization of microRNAs in Asiatic cotton (*Gossypium arboreum* L.). PLoS One. 2012;7:e33696. doi:10.1371/journal.pone.0033696

[116] Zhang H, Yang JH, Zheng YS, Zhang P, Chen X, Wu J, Xu L, Luo XQ, Ke ZY, Zhou H, Qu LH. Genome-wide analysis of small RNA and novel microRNA discovery in human acute lymphoblastic leukemia based on extensive sequencing approach. PLoS One. 2009;4:e6849. doi:10.1371/journal.pone.0069743

[117] Gao S, Yang L, Zeng HQ, Zhou ZS, Yang ZM, Li H, Sun D, Xie F, Zhang B. A cotton miRNA is involved in regulation of plant response to salt stress. Scientific Reports. 2016;6:19736. doi:10.1038/Srep19736

[118] Khan MA, Shahid AA, Rao AQ, Bajwa KS, Muzaffar A, Rehman Samiullah T, Nasir IA, Husnain T. Molecular and biochemical characterization of cotton epicuticular wax in defense against cotton leaf curl disease (CLCuD). Iranian Journal of Biotechnology. 2016;13(4):3-9. doi:10.15171/ijb.1234

[119] Wang M, Yuan D, Tu L, Gao W, He Y, Hu H, Wang P, Liu N, Lindsey K, Zhang X. Long noncoding RNAs and their proposed functions in fibre development of cotton (*Gossypium* spp.). New Phytologist. 2015;207(4):1181-1197. doi:10.1111/nph.13429

[120] Zou C, Wang Q, Lu C, Yang W, Zhang Y, Cheng H, Feng X, Prosper MA, Song G. Transcriptome analysis reveals long noncoding RNAs involved in fiber development in cotton (*Gossypium arboreum*). Science China Life Sciences. 2016;59:164-171. doi: 10.1007/s11427-016-5000-2

[121] Matzke MA, Kanno T, Matzke AJ. RNA-directed DNA methylation: the evolution of a complex epigenetic pathway in flowering plants. Annual Review of Plant Biology. 2015;66:243-267. doi:10.1038/nrg3683

[122] Chen T, Li E. Structure and function of eukaryotic DNA methyltransferases. Current Topics in Developmental Biology. 2004;60:55-89. doi:10.1016/S0070-2153(04)60003-2

[123] Zuker M. Mfold web server for nucleic acid folding and hybridization prediction. Nucleic Acids Research. 2003;31:3406-3415. doi:10.1093/nar/gkg595

[124] Griffiths-Jones S, Grocock RJ, Van Dongen S, Bateman A, Enright AJ. miRBase: microRNA sequences, targets and gene nomenclature. Nucleic Acids Research. 2006;34:D140-D144. doi:10.1093/nar/gkj112

[125] Moxon S, Schwach F, Dalmay T, MacLean D, Studholme DJ, Moulton V. A toolkit for analysing large-scale plant small RNA datasets. Bioinformatics. 2008;24:2252-2253. doi: 10.1093/bioinformatics/btn428

[126] An J, Lai J, Sajjanhar A, Lehman ML, Nelson CC. miRPlant: an integrated tool for identification of plant miRNA from RNA sequencing data. BMC Bioinformatics. 2014;15:275. doi:10.1186/1471-2105-15-275

[127] Ronen R, Gan I, Modai S, Sukacheov A, Dror G, Halperin E, Shomron N. miRNAkey: a software for microRNA deep sequencing analysis. Bioinformatics. 2010;26:2615-2616. doi:10.1093/bioinformatics/btq493

[128] Gunaratne PH, Coarfa C, Soibam B, Tandon A. miRNA data analysis: next-gen sequencing. Next-Generation MicroRNA Expression Profiling Technology: Methods and Protocols. 2012;2012:273-288.

[129] Betel D, Wilson M, Gabow A, Marks DS, Sander C. The microRNA. org resource: targets and expression. Nucleic Acids Research. 2008;36:D149-D153. doi:10.1093/nar/gkm995

[130] Dai, X. and P.X. Zhao, psRNATarget: a plant small RNA target analysis server. Nucleic Acids Research, 2011;39:W155-W159. doi:10.1093/nar/gkr319

[131] Pruitt KD, Tatusova T, Maglott DR. NCBI reference sequences (RefSeq): a curated non redundant sequence database of genomes, transcripts and proteins. Nucleic Acids Research. 2007;35:D61-D65. doi:10.1093/nar/gkl842

[132] Liu C, Bai B, Skogerbø G, Cai L, Deng W, Zhang Y, Bu D, Zhao Y, Chen R. NONCODE: an integrated knowledge database of non-coding RNAs. Nucleic Acids Research. 2005;33:D112-D115. doi:10.1093/nar/gki041

[133] Amaral PP, Clark MB, Gascoigne DK, Dinger ME, Mattick JS. lncRNAdb: a reference database for long noncoding RNAs. Nucleic Acids Research. 2011;39:D146-D151. doi: 10.1093/nar/gkq1138

[134] Xuan H, Zhang L, Liu X, Han G, Li J, Li X, Liu A, Liao M, Zhang S. PLNlncRbase: a resource for experimentally identified lncRNAs in plants. Gene. 2015;573:328-332. doi: 10.1016/j.gene.2015.07.069

[135] Lane AK, Niederhuth CE, Ji L, Schmitz RJ. pENCODE: a plant encyclopedia of DNA elements. Annual Review of Genetics. 2013;48:49-70. doi:10.1146/annurev-genet-120213-092443

[136] Barrett T, Troup DB, Wilhite SE, Ledoux P, Rudnev D, Evangelista C, Kim IF, Soboleva A, Tomashevsky M, Edgar R. NCBI GEO: mining tens of millions of expression profiles —database and tools update. Nucleic Acids Research. 2007;35:D760-D765. doi:10.1093/nar/gkl887

[137] Geisen S, Barturen G, Alganza ÁM, Hackenberg M, Oliver JL. NGSmethDB: an updated genome resource for high quality, single-cytosine resolution methylomes. Nucleic Acids Research. 2013:gkt1202. doi:10.1093/nar/gkt1202

[138] Heinz S, Benner C, Spann N, Bertolino E, Lin YC, Laslo P, Cheng JX, Murre C, Singh H, Glass CK. Simple combinations of lineage-determining transcription factors prime cis-regulatory elements required for macrophage and B cell identities. Molecular Cell. 2010;38:576-589. doi:10.1016/j.molcel.2010.05.004

[139] Hetzl J, Foerster AM, Raidl G, Scheid OM. CyMATE: a new tool for methylation analysis of plant genomic DNA after bisulphite sequencing. The Plant Journal. 2007;51:526-536. doi:10.1111/j.1365-313X.2007.03152.x

[140] Bewick AJ, Hofmeister BT, Lee K, Zhang X, Hall DW, Schmitz RJ. FASTmC: a suite of predictive models for non-reference-based estimations of DNA methylation. G3: Genes, Genomes, Genetics. 2015:g3-115. doi:10.1534/g3.115.025668

[141] Song Q, Decato B, Hong EE, Zhou M, Fang F, Qu J, Garvin T, Kessler M, Zhou J, Smith AD. A reference methylome database and analysis pipeline to facilitate integrative and comparative epigenomics. PloS One. 2013;8:e81148. doi:10.1371/journal.pone.0081148

[142] Stroud H, Do T, Du J, Zhong X, Feng S, Johnson L, Patel DJ, Jacobsen SE. Non-CG methylation patterns shape the epigenetic landscape in *Arabidopsis*. Nature Structural & Molecular Biology. 2014;21:64-72. doi:10.1038/nsmb.2735

[143] Saze H, Tsugane K, Kanno T, Nishimura T. DNA methylation in plants: relationship to small RNAs and histone modifications, and functions in transposon inactivation. Plant and Cell Physiology. 2012;53:766-784. doi:10.1093/pcp/pcs008

[144] Cervera MT, Ruiz-Garcia L, Martinez-Zapater J. Analysis of DNA methylation in *Arabidopsis thaliana* based on methylation-sensitive AFLP markers. Molecular Genetics and Genomics. 2002;268:543-552. doi:10.1007/s00438-002-0772-4

[145] Salmon A, Clotault J, Jenczewski E, Chable V, Manzanares-Dauleux MJ. *Brassica oleracea* displays a high level of DNA methylation polymorphism. Plant Science. 2008;174:61-70. doi:10.1016/j.plantsci.2007.09.012

[146] Keyte AL, Percifield R, Liu B, Wendel JF. Infraspecific DNA methylation polymorphism in cotton (*Gossypium hirsutum* L.). Journal of Heredity. 2006;97:444-450. doi:10.1093/jhered/esl023

[147] Jin B, Robertson KD. DNA methyltransferases, DNA damage repair, and cancer. In: Adam R. Karpf, editor. Epigenetic Alterations in Oncogenesis. 754. New York: Springer; 2013. p. 3-29. doi:10.1007/978-1-4419-9967-2_1

[148] Kimatu JN, Bao L. Epigenetic polymorphisms could contribute to the genomic conflicts and gene flow barriers resulting to plant hybrid necrosis. African Journal of Biotechnology. 2015;9:8125-8133

[149] He SP, Sun JL, Zhang C, Du XM. Identification of exotic genetic components and DNA methylation pattern analysis of three cotton introgression lines from *Gossypium bickii*. Molecular Biology. 2011;45:204-210. doi:10.1134/S002689331102018x

[150] Song Q, Guan X, Chen ZJ. Dynamic roles for small RNAs and DNA methylation during ovule and fiber development in allotetraploid cotton. PLoS Genetics. 2015;11:e1005724. doi:10.1371/journal.pgen.1005724

[151] Soltani A, Kumar A, Mergoum M, Pirseyedi SM, Hegstad JB, Mazaheri M, Kianian SF. Novel nuclear-cytoplasmic interaction in wheat (Triticum aestivum) induces vigorous plants. Functional & Integrative Genomics. 2016;16:171-182. doi:10.1007/s10142-016-0475-2

[152] Sharma SK, Yamamoto M, Mukai Y. Molecular cytogenetic approaches in exploration of important chromosomal landmarks in plants. In: Rajpal VR, Rama SR, Raina, SN, editors. Molecular Breeding for Sustainable Crop Improvement. 11. New York: Springer; 2016. p. 127-148

[153] Page JT, Gingle AR, Udall JA. PolyCat: a resource for genome categorization of sequencing reads from allopolyploid organisms. G3: Genes| Genomes| Genetics. 2013;3:517-525. doi:10.1534/g3.112.005298

4

A Bioinformatics Method for the Production of Antibody-Drug Conjugates Through Site-Specific Cysteine Conjugation

Arianna Filntisi, Dimitrios Vlachakis and George K. Matsopoulos

Abstract

Antibody-drug conjugates (ADCs) have emerged as a promising class of targeted anticancer therapy, and it is distinguished from traditional chemotherapeutic approaches by its potential to kill cancer cells with limited side effects. Site-specific conjugation is one of the current challenges in ADC development because it allows for controlled conjugation and production of homogeneous ADCs. This chapter describes a computational method for the generation of antibody-drug conjugates as PDB files through site-specific cysteine conjugation, given the PDB files of a drug, a linker, and an antibody. The drug and linker are reconfigured using the rotation and translation functions of an affine transformation, which is brought in appropriate positions for the bonds to occur between the three molecules. The hydrogen and disulfide bonds are employed to connect the linker and drug as well as the linker with the antibody, respectively. Examples of conjugates produced with the presented method have been demonstrated.

Keywords: bioinformatics, cancer, targeted therapy, antibody-drug conjugates, cysteines

1. Introduction

Antibodies are large proteins produced by the immune system against an invader substance called antigen. As proteins, they consist of one or more chains of amino acid residues, two types of which are cysteines and lysines. Antibodies can also be manufactured and used as standalone therapeutic agents for a number of diseases including cancer. The largest part of the amino acid sequence of an antibody is conserved in all antibodies, with the exception of a small part that is called a hypervariable (HV) region. Due to the specialized and unique amino acid sequence of the HV region, an antibody has the ability to recognize and bind a specific antigen with higher affinity than different antigens—a property called specificity. The specificity and long half-lives of antibodies are valuable features to their therapeutic effect. However, the performance of manufactured antibodies as standalone therapeutic agents is impeded by their limited potency. On the contrary, traditional chemotherapeutic drugs are more potent but have short half-lives and no specificity, targeting all cells indiscriminately. A class of targeted anticancer therapy (TAT), that utilizes the best qualities of antibodies and cytotoxins by combining them in one molecule, is antibody-drug conjugates (ADCs). ADCs are therapeutic agents formed by three elements: an antibody whose target is a cancer-specific molecule, a cytotoxic agent, also called the drug, and a linker molecule that connects them. An ADC ideally delivers the conjugated cytotoxic molecule directly into cancer cells, limiting damage to healthy tissues [1–8].

The typical route of an ADC molecule upon its insertion to the bloodstream can be described briefly with the following steps. At first, the ADC circulates in the plasma remaining stable, while the drug remains inactive. When the target antigen is encountered, the antibody binds to it. Typically, the ADC-antigen complex undergoes antigen-mediated or antigen-independent internalization by the cancer cell. Subsequently, the ADC is degraded in its parts through a process that depends on the type of linker used. ADCs containing a non-cleavable linker usually undergo lysosomal digestion. The antibody is degraded to its amino acids and the final active drug complex is the drug with the linker and a single amino acid. On the other hand, the ADCs, containing a cleavable linker, are usually degraded through hydrolysis or enzymatic cleavage. At that stage, the drug is activated and kills the cell, in a way that depends on the type of the cytotoxic agent used. For example, auristatins and maytansinoids disrupt the microtubule assembly of the cell, while calicheamicins and duocarmycins target its DNA structure. An alternative ADC strategy is targeting the endothelial cells of the tumor vessels in order to deprive the cancer cells from blood [6, 9–12].

Conjugation usually occurs at solvent-accessible reactive amino acids of the preserved regions of the antibody, leaving the hypervariable regions of the antibody available for antigen binding. The reactive thiol groups of cysteine residues, being made available for conjugation after the reduction of the inter-chain disulfide bonds, as well as the amino groups of lysine residues, are typical conjugation sites. However, these methods by default produce ADCs with variable drug to antibody ratio (DAR) and conjugation sites, and therefore unstable pharmacokinetic properties. For example, in cysteine conjugation, ADCs with a DAR of eight have a significantly shorter circulation time compared to the unconjugated antibodies. On the other hand, ADCs with a DAR of four have a longer circulation time than ADCs with the double DAR, but the same therapeutic effect as them. More homogeneous ADCs can be produced by site-specific conjugation methods, one of which is cysteine engineering. THIOMABs are engineered antibodies in which cysteines have been introduced to the amino acid sequence by substituting original residues. The ideal sites for residue substitution can be identified with phage display techniques. Eight residues on the light chain (LC-V205C, LC-S168C, LC-A153C,

LC-S127C, LC-S121C, LC-S114C, LC-V110C, LC-V15C) and five residues on the heavy chain (HC-T116C, HC-S115C, HC-A114C, HC-S113C, HC-S112C) have been investigated as potential cysteine insertion sites (Kabat numbering). The above amino acids were substituted with cysteines, generating THIOMABs, which were subsequently conjugated with biotin-maleimide. Considering the conjugation of two molecules per antibody as 100% conjugation efficiency, most THIOMABs demonstrated more than 90% conjugation efficiency, proving to be suitable for site-specific conjugation of thiol reactive probes. THIOMAB-drug conjugates (TDCs) are the product of the conjugation of a drug to a THIOMAB [11, 13–28].

The process of drug discovery and development is rather strenuous, taking up to 15 years to produce a new drug from the early *in vitro* discovery stages to the time it is available as a treatment option. However, drug discovery can be facilitated by bioinformatic techniques. The three-dimensional structure of a molecule can be described computationally in various formats, a prevalent one being the Protein Data Bank (PDB) format [29–31]. Computational drug design and molecular mechanics methods provide a means to model molecules and assess their features. Even though there is a variety of general computer-aided drug design tools, the more specific field of ADC computational design is less evolved. JSDraw–Antibody-Drug Conjugates is an editor tool that can be used for ADCs, although it is drawing-oriented and uses 2D coordinates. In addition, a mathematical model was recently developed to describe mechanistically the pharmacokinetic behavior and preclinical efficacy of THIOMAB-drug conjugates [32–41].

An attempt to contribute to the study of ADC computational modeling was made in [42] describing the computational construction of ADCs through lysine conjugation. The three input PDB files of the antibody, the linker, and the drug were processed and merged to a PDB file that represented an ADC molecule. The linker and drug molecules were reconfigured and aligned with the selected lysine amino acid, while hydrogen bonds linked the three molecules [30, 31].

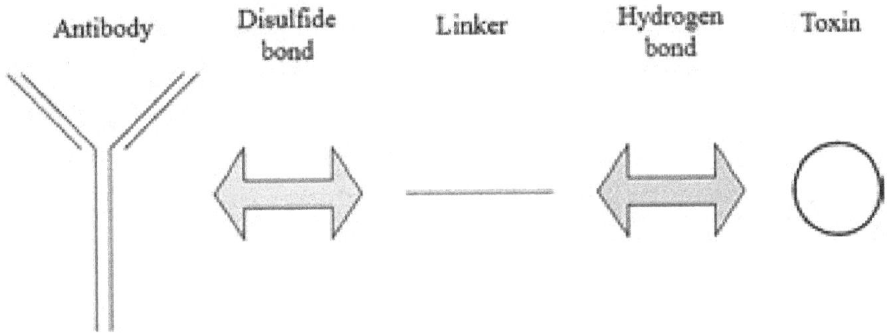

Figure 1. Scheme of the bonds formed between the drug, the linker, and the antibody molecules, according to the method described in this paper.

The present chapter aims to extend the above-mentioned technique to a more site-specific conjugation method, connecting the linker-drug complex to engineered cysteines instead of lysines. The molecule $C_{12}S$ has been used as a non-cleavable linker. The linker and drug form a hydrogen bond (H bond), while the linker-drug complex and the engineered cysteine form a disulfide bond (SS bond) (**Figure 1**). The configurational change of the linker and drug is accomplished using the rotation and translation functions of an affine transformation. Molecular modeling software was used for functions such as the visualization of molecules, the addition of hydrogen atoms to the antibody PDB files, the insertion of cysteine amino acids to the antibody sequence, as well as the generation of the linker PDB file.

2. Methods

The main aspects of the cysteine conjugation method have been analyzed here. Firstly, the steps of the method are presented in detail. The Affine Transformation subsection contains the definition of the affine transformation, as well as details concerning its application to this project. The Rotation and Translation subsections focus on the application of the rotation and translation functions to the reconfiguration of the molecules. The Axes and Distances subsections contain the specifications made in relevance to the axes of the molecules involved and the distances between them. Finally, the implementation of the method is described.

2.1. Method steps

First, the antibody, the cytotoxic agent, and the linker that will form the conjugate are determined (step S1). The antibody PDB files we have used were obtained from the RCSB database, while the drug PDB files from the NCI database. The linker PDB file was created with the molecular modeling software UCSF Chimera [36]. Given an antibody PDB file, it is possible to produce the PDB file of an engineered antibody (THIOMAB) by replacing selected antibody residues with cysteines using molecular modeling software (step S2). The antibody was selected to have solvent-accessible residues not in the hypervariable region of the antibody, that are suitable to be replaced with cysteines and to be conjugated. The drug was selected so as to have at least one hydrogen (H) atom bonded covalently to an electronegative (EN) atom, in order to be able to form a hydrogen bond with the linker (**Figure 1**). The linker molecule was designed to be able to form a disulfide bond with a cysteine. For that purpose, a sulfur (S) atom was included. Also, the linker was designed to be long enough to allow for the drug to be linked to the cysteine without colliding with the nearby residues. For that purpose, 12 carbon (C) atoms were incorporated following a linear, instead of circular, layout. Finally, the linker should theoretically be able to release the drug from the antibody inside the cancer cell. For the molecule $C_{12}S$ we used, there are three possible ways for that to happen. Breakage of the hydrogen bond between the drug and linker could occur. Also, the disulfide bond between the linker and drug could break due to disulfide exchange, a phenomenon associated with cleavable linkers. In addition, lysosomal digestion of the antibody could occur, in case the disulfide bond between the cysteine and the linker does not break, a phenomenon associated with non-cleavable linkers. In that case, the final active drug complex would contain a single residue (the cysteine connected to the linker), the linker, as well as the drug.

Even though the PDB format has more than 40 types of PDB records, the three types of records that are crucial for the representation of three-dimensional molecular structure include the PDB ATOM records, which contain information about the atoms of standard amino acids and nucleotides. The PDB HETATM records contain information about additional "non-standard", non-polymer chemical structures of the molecule, and the PDB CONECT records specify the connectivity between the atoms of the molecule [29].

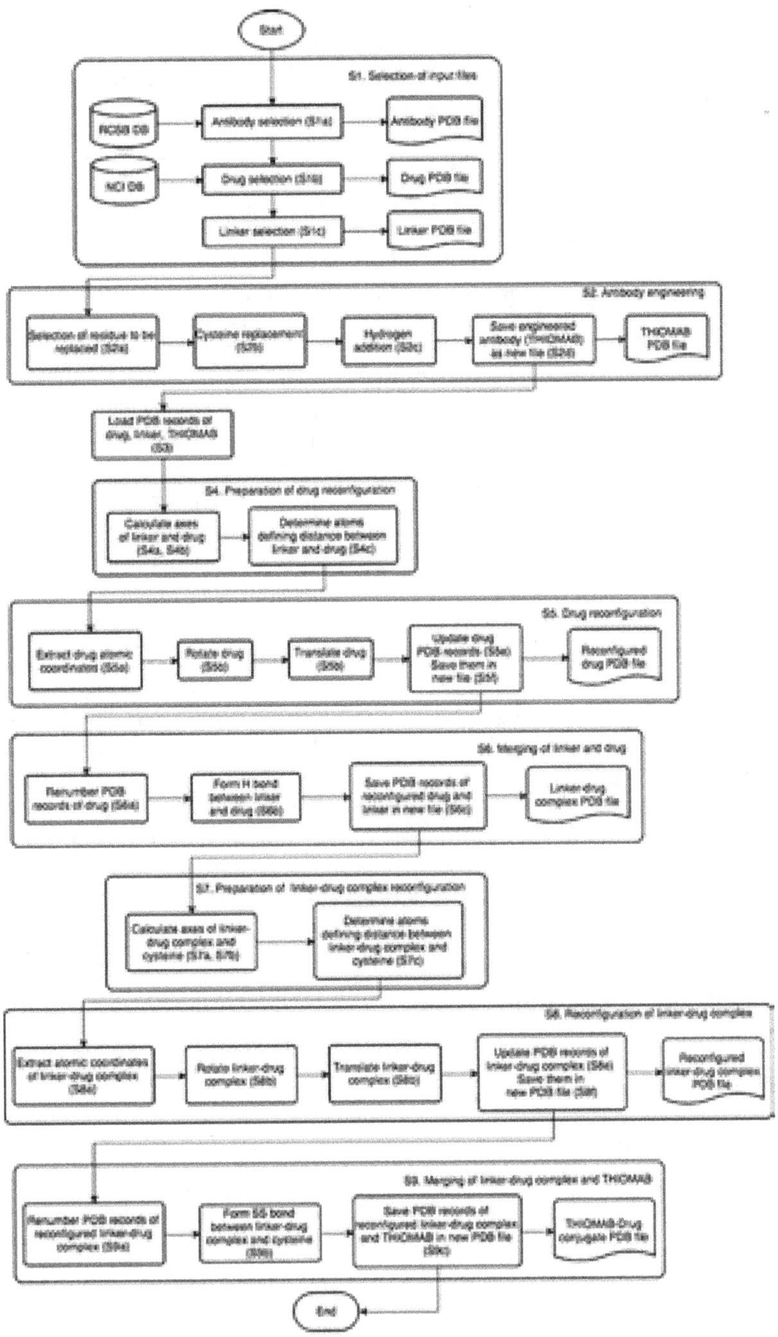

Figure 2. Flowchart of the cysteine conjugation method described in this paper. S1. Determine the three molecules (antibody, drug, linker) that will build the antibody-drug conjugate. S1a. Select the antibody PDB file from the RCSB database (e.g., ab.pdb) according to the criteria described in Section 2.1. S1b. Select the anticancer drug PDB file from the

NCI database (e.g., drug.pdb). S1c. Determine the linker PDB file (e.g., linker.pdb). S2. Engineer the antibody (ab.pdb) to create a THIOMAB (e.g., thiomab.pdb), using molecular modeling software. S2a. Select one (or more) native residue(s) of the antibody (ab.pdb) to be replaced by a cysteine residue. The residue should be solvent-accessible, on the surface of the antibody, in order to be accessible to the linker-drug complex, as well as suitable for cysteine replacement and site-specific conjugation. S2b. Replace the selected residue(s) with a cysteine residue. S2c. Add hydrogen atoms to the antibody, since the antibody PDB files from the RCSB database are provided without them. S2d. Save the engineered antibody (THIOMAB) to a new file (thiomab.pdb). S3. Load the necessary PDB records from the PDB files of the engineered antibody (thiomab.pdb), the linker (linker.pdb) and the drug (drug.pdb). Those include (i) the PDB ATOM, HETATM, and CONECT records of the engineered antibody; (ii) the PDB HETATM and CONECT records of the linker; (iii) the PDB HETATM and CONECT records of the drug (the drug and linker files do not contain PDB ATOM records). S4. Prepare the reconfiguration of the drug. S4a. Calculate the axis of the drug as specified in Eqs. (16–18). Briefly, the drug axis has been specified as the line connecting the drug atoms participating in the hydrogen bond with the linker, which are a hydrogen (H) atom and the electronegative (EN) atom bonded covalently to it. S4b. Calculate the axis of the linker as specified in Eqs. (19–21). Briefly, the linker axis has been specified as the line connecting the two most distant non-hydrogen (NH) atoms of the linker, since that line is quite representative of the shape of the linker. Given the chemical formula of the linker used ($C_{12}S$), those were always the sulfur (S) atom and its most distant carbon (C) atom. S4c. Find the atoms that will define the distance between the drug and linker according to Eqs. (30–32). Briefly, the distance between the drug and linker can be described as the distance between the atoms participating in the hydrogen bond between the two molecules, which are a hydrogen drug atom and an electronegative linker atom. S5. Reconfigure the drug. S5a. Obtain the coordinates of the drug atoms from the drug PDB HETATM records (drug.pdb). S5b. Rotate the drug using the affine transformation (Eqs. (1–15)). Apply the affine transformation to the drug for $a = 1$, $T_x = T_y = T_z = 0$ and the values of the variables r_x, r_y, r_z in [−180°, +180°] that satisfy a specific condition, as defined in Eq. (28). Briefly, the condition can be specified as the minimization of the absolute value of the angle between the axes of the drug and linker. S5c. Update (optional) the PDB drug records with the modified coordinates from step S5b. S5d. Translate the drug using the affine transformation (Eqs. (1–15)). Apply the affine transformation to the drug for $a = 1$, $r_x = r_y = r_z = 0$ and the values of the variables T_x, T_y, T_z in [−D, +D] that satisfy a specific condition, defined in Eq. (36). Briefly, the condition can be specified as the minimization of the absolute value of the difference between the distance of the two molecules (defined in step S4c) and the official length of the hydrogen bond. S5e. Update the PDB records of the drug with the modified coordinates. S5f. Save updated PDB drug records in a new PDB file (e.g., drug_rt.pdb). S6. Merge the PDB files of the linker and the reconfigured drug to produce the PDB file of the linker-drug complex. S6a. Renumber the sequence numbers of the PDB records of the reconfigured drug, since they will be placed after the PDB records of the linker in step S6c. S6b. Form computationally a hydrogen bond between the drug and linker by updating the PDB CONECT records of the drug and linker. S6c. Save the necessary data to a new file that will represent the linker-drug complex (e.g., linkerdrug.pdb). Those data include the (i) PDB HETATM records of the linker (linker.pdb; from step S1c); (ii) renumbered PDB HETATM records of the reconfigured drug (from step S6a); (iii) updated PDB CONECT records of the linker and drug, including the one that represents the new hydrogen bond (from step S6b). S6d. Load the records of the linker-drug complex (from step S6c) into appropriate structures, to use in the steps S7–S9. S7. Prepare the reconfiguration of the linker-drug complex. S7a. Calculate the axis of the linker-drug complex as specified in Eqs. (22–24). Briefly, the axis of the linker-drug complex has been specified as the line connecting the two most distant non-hydrogen (NH) atoms of the linker-drug complex. S7b. Calculate the axis of the selected engineered cysteine as specified in Eqs. (25–27). Briefly, the axis of the cysteine residue has been specified as the line connecting the alpha carbon (Cα) atom and the sulfur (S) atom of the cysteine side chain. S7c. Find the atoms that will determine the distance between the linker-drug complex and the cysteine residue, as specified in Eqs. (33–35). Briefly, the distance between the linker-drug complex and the cysteine has been specified as the distance between the atoms participating in the disulfide bond between the two molecules. S8. Reconfigure the linker-drug complex. S8a. Get the atomic coordinates of the linker-drug complex from its PDB HETATM records (linkerdrug.pdb). S8b. Rotate the linker-drug complex using the affine transformation (Eqs. (1–15)). Apply the affine transformation to the linker-drug complex for $a = 1$, $T_x = T_y = T_z$ and the values of the variables r_x, r_y, r_z in [−180°, +180°] that satisfy a specific condition, as defined in Eq. (29). Briefly, the condition is the minimization of the absolute value of the angle between the axes of the linker-drug complex and the cysteine residue. S8c. Update (optional) the PDB records of the linker-drug complex with the modified coordinates (from step S8b). S8d. Translate the linker-drug complex using the affine transformation (Eqs. (1–15)). Apply the affine transformation to the linker-drug complex for $a = 1$, $r_x = r_y = r_z = 0$, and the values of the variables T_x, T_y, T_z that satisfy a specific condition, as defined in Eq. (37). Briefly, the condition is the minimization of the absolute value of the difference between the distance of the two molecules and the official length of the disulfide bond. S8e. Update the PDB records of the linker-drug complex with the modified coordinates. S8f. Save the updated PDB records of the linker-drug complex in a new PDB file (e.g., linkerdrug_rt.pdb). S9. Merge the PDB files of the THIOMAB and the reconfigured linker-drug complex to produce the PDB file of the THIOMAB-drug conjugate. S9a. Renumber the PDB records of the linker-drug complex (from step S8), since they will be placed after the antibody records in step S9c.

S9b. Form computationally a disulfide bond between the cysteine and the linker-drug complex, by updating the PDB CONECT records of the antibody and linker. S9c. Save the necessary data to a new file that will represent the final THIOMAB-drug conjugate (e.g., tdc.pdb). Those data include (i) the PDB ATOM and HETATM records of the antibody (thiomab.pdb) (from step S2); (ii) the renumbered PDB HETATM records of the reconfigured linker-drug complex (from step S9a); (iii) the updated PDB CONECT records of the antibody and the linker-drug complex, including the one that contains the disulfide bond (from steps S9a, S9b).

Once the input files of the engineered antibody, linker, and drug are available, the necessary PDB records are loaded into instances of suitable data structures (step S3). The remaining process can be divided in two basic stages, which are the synthesis of a linker-drug complex molecule from the linker and drug (steps S4–S6), and the synthesis of an antibody-drug conjugate from the linker-drug complex and engineered antibody (steps S7–S9). More specifically, while the linker remains fixed, the drug undergoes rotation and translation, brought in a specific position near the linker (step S5). Following that, the two molecules are linked through a hydrogen bond, and their PDB files are combined to form a linker-drug complex (step S6). While the antibody remains fixed, the linker-drug complex undergoes rotation and translation (step S8). The two molecules are linked through a disulfide bond and their PDB files are combined to produce the final THIOMAB-drug conjugate (step S9). Since a drug molecule can contain more than one pair of atoms that can form a hydrogen bond with the linker, it is possible to either select the drug conjugation site, or produce more than one final TDCs. The computational method proposed in this chapter is described more analytically with the following steps (S1–S9), depicted in **Figure 2**.

2.2. Affine transformation

The modification of atomic coordinates serves the purpose of altering the relative positions between the different molecules. However, it should not disrupt the initial lines and distances between the atoms of a single molecule, since that would interfere with the validity of its chemical structure. To that end, an affine transformation was employed for the reconfiguration of a molecule [42–45]. According to [45], the transformation of a rigid body in three-dimensional space can be described with the following equations:

$$x' = R_{00} * x + R_{01} * y + R_{02} * z + R_{03} \tag{1}$$

$$y' = R_{10} * x + R_{11} * y + R_{12} * z + R_{13} \tag{2}$$

$$z' = R_{20} * x + R_{21} * y + R_{22} * z + R_{23} \tag{3}$$

$$R_{00} = a * \cos r_z * \cos r_y \tag{4}$$

$$R_{01} = a * \left(\cos r_z * \sin r_y * \sin r_x - \sin r_z * \cos r_x\right) \tag{5}$$

$$R_{02} = a * \left(\cos r_z * \sin r_y * \cos r_x + \sin r_z * \sin r_x\right) \quad (6)$$

$$R_{03} = T_x \quad (7)$$

$$R_{10} = a * \sin r_z * \cos r_y \quad (8)$$

$$R_{11} = a * \left(\sin r_z * \sin r_y * \sin r_x + \cos r_z * \cos r_x\right) \quad (9)$$

$$R_{12} = a * \left(\sin r_z * \sin r_y * \cos r_x - \cos r_z * \cos r_x\right) \quad (10)$$

$$R_{13} = T_y \quad (11)$$

$$R_{20} = -a * \sin r_y \quad (12)$$

$$R_{21} = a * \cos r_y * \sin r_x \quad (13)$$

$$R_{22} = a * \cos r_y * \cos r_x \quad (14)$$

$$R_{23} - T_z \quad (15)$$

where x, y, z are the initial coordinates of a point of the rigid body, x', y', z' are the coordinates of the same point after the application of the transformation, α is the scaling factor, r_x, r_y, r_z are the rotation angles of the body around the x, y, z axis, respectively, and T_x, T_y, T_z are the distances according to which each point will be translated in the x, y, z direction, respectively. From Eqs. (1–15), we can deduce that if we apply the values $a = 1$, $r_x = r_y = r_z = T_x = T_y = T_z = 0$, it will be $x' = x$, $y' = y$, $z' = z$.

The term point used above refers not to a set of constant coordinates but to an element of the body with variable coordinates. In the context of molecular reconfiguration, each molecule was considered as a rigid body. Each atom of a molecule was seen as a single point, since an atom is assigned one set of three-dimensional Cartesian coordinates according to the PDB format. The Cartesian coordinates of an atom of a molecule correspond to the variables x,y,z, x',y',z' of Eqs. (1–15). The scaling factor α was always set to the value of 1, because scaling the molecules would result in the disruption of their structure.

2.3. Rotation

According to the method presented in this chapter, in order to perform the rotation of a molecule A in reference to a fixed molecule B, the affine transformation is applied to A without using the translation and scaling functions, for the parameter values $a = 1$, $T_x = T_y = T_z = 0$. The values of r_x, r_y, r_z that satisfy a certain condition (defined in Eqs. (28, 29)) are searched for in the interval [−180°, +180°]. Finally, the affine transformation is applied to molecule A for the found r_x, r_y, r_z values. According to **Figure 2**, rotation occurs in two occasions in our method. First, the drug adopts the role of molecule A and the linker adopts the role of molecule B (step S5b). Later, the linker-drug complex adopts the role of molecule A and the cysteine takes the role of B (step S8b).

2.4. Translation

According to the method presented in this chapter, in order to perform the translation of a molecule A in reference to a fixed molecule B, the affine transformation is applied to A without using the rotation and scaling functions, for the parameter values $a = 1$, $r_x = r_y = r_z = 0$. The values of T_x, T_y, T_z that satisfy a certain condition (Eqs. (36, 37)), are searched in the interval [−D, +D]. Usually, the value $D = 200$ was sufficient to complete the search successfully; otherwise, a larger D value was applied. Finally, the affine transformation is applied to molecule A for the found T_x, T_y, T_z values. According to **Figure 2**, translation occurs in two occasions in our method. First, the drug adopts the role of the molecule A undergoing translation, and the linker adopts the role of the fixed molecule B (step S5d). Later, the linker-drug complex adopts the role of the molecule A undergoing translation, and the cysteine adopts the role of the fixed molecule B (step S8d).

2.5. Axes

As mentioned in Section 2.3, the rotation of a molecule A (i.e., the drug or linker-drug complex) in relation to its corresponding fixed molecule B (i.e., the linker or antibody, respectively) is performed for $a = 1$, $T_x = T_y = T_z = 0$ and the values of the variables r_x, r_y, r_z that satisfy a certain condition. That condition is the minimization of the absolute value of the angle between the axes of the two molecules (Eqs. (28, 29)).

In the context of this project, the axis of each molecule has been defined by taking into account the desired final layout of the three molecules. In the lysine [34] and cysteine conjugation methods, the same kind of bond connects the linker with the drug. Therefore, in both cases, the drug axis is defined by the drug atoms participating in the hydrogen bond with the linker, which are a hydrogen (H) atom and the electronegative (EN) atom covalently bonded to it [Eqs. (16–18); **Figure 2**, step S4a]. The linker axis has been specified so as to connect the two most distant non-hydrogen linker atoms because that line is quite representative of the linker shape [Eqs. (19–21); **Figure 2**, step S4b]. Given the chemical structure $C_{12}S$ of the linker used in cysteine conjugation, those two atoms are always the sulfur (S) atom and its most distant carbon (C) atom. On a similar note, the axis of the linker-drug complex connects the two most distant non-hydrogen atoms of the linker-drug complex [Eqs. (22–24); **Figure 2**, step S7a]. The cysteine axis has been defined as the line going through the alpha carbon (C_α) and sulfur atoms (S) of the cysteine side chain [Eqs. (25–27); **Figure 2**, step S7b]. This line was chosen for its direction towards the exterior of the antibody, which is the desired direction for the linker-drug complex. In the lysine conjugation of [34], the linker $C_{15}N$ was used, therefore the linker axis was defined by the nitrogen (N) atom with its most distant carbon (C) atom. Also, the lysine axis went through the alpha carbon (C_α) and nitrogen (N) atoms of the lysine side chain.

$$p_{d1} = (x,y,z)_{H\ atom\ of\ drug} = (x_{Hd}, y_{Hd}, z_{Hd}) \tag{16}$$

$$p_{d2} = (x,y,z)_{EN\ atom\ of\ drug} = (x_{ENd}, y_{ENd}, z_{ENd}) \tag{17}$$

$$\begin{aligned} L_d :(x,y,z) = p_{d1} + t_d * (p_{d1} - p_{d2}) = (x_{Hd}, y_{Hd}, z_{Hd}) + \\ t_d * (x_{Hd} - x_{ENd}, y_{Hd} - y_{ENd}, z_{Hd} - z_{ENd}),\ -\infty < t_d < +\infty \end{aligned} \tag{18}$$

$$p_{l1} = (x,y,z)_{NH\ atom\ of\ linker} = (x_{NHl}, y_{NHl}, z_{NHl}) \tag{19}$$

$$p_{l2} = (x,y,z)_{S\ atom\ of\ linker} = (x_{Sl}, y_{Sl}, z_{Sl}) \tag{20}$$

$$\begin{aligned} L_l :(x,y,z) = p_{l1} + t_l * (p_{l1} - p_{l2}) = (x_{NHl}, y_{NHl}, z_{NHl}) + \\ t_l * (x_{NHl} - x_{Sl}, y_{NHl} - y_{Sl}, z_{NHl} - z_{Sl}), -\infty < t_l < +\infty \end{aligned} \tag{21}$$

$$p_{ld1} = (x,y,z)_{NH\ atom\ of\ linker-drug} = (x_{NHld}, y_{NHld}, z_{NHld}) \tag{22}$$

$$p_{ld2} = (x,y,z)_{S\ atom\ of\ linker-drug} = (x_{Sld}, y_{Sld}, z_{Sld}) \tag{23}$$

$$\begin{aligned} L_{ld} :(x,y,z) = p_{ld1} + t_{ld} * (p_{ld1} - p_{ld2}) = (x_{NHld}, y_{NHld}, z_{NHld}) + \\ t_l * (x_{NHld} - x_{Sld}, y_{NHld} - y_{Sld}, z_{NHld} - z_{Sld}), -\infty < t_{ld} < +\infty \end{aligned} \tag{24}$$

$$p_{cys1} = (x,y,z)_{S\ atom\ of\ Cysteine} = (x_{Sc}, y_{Sc}, z_{Sc}) \tag{25}$$

$$p_{cys2} = (x, y, z)_{Ca\ atom\ of\ Cysteine} = (x_{Cc}, y_{Cc}, z_{Cc}) \tag{26}$$

$$L_{cys} : (x, y, z) = p_{cys1} + t_{cys} * (p_{cys1} - p_{cys2}) = (x_{Sc}, y_{Sc}, z_{Sc}) + t_{cys} * (x_{Sc} - x_{Cc}, y_{Sc} - y_{Cc}, z_{Sc} - z_{Cc}), -\infty < t_{cys} < +\infty \tag{27}$$

$$r_x, r_y, r_z : \angle(L_d, L_l) = 0° \tag{28}$$

$$r_x, r_y, r_z : \angle(L_{ld}, L_{cys}) = 0° \tag{29}$$

2.6. Distances

As mentioned in Section 2.4, the translation of a molecule A (i.e., the drug or the linker-drug complex) in relation to its corresponding fixed molecule B (i.e., the linker or the antibody, respectively) is performed for $a = 1$, $r_x = r_y\ r_z = 0$, and the values of the variables T_x, T_y, T_z that satisfy a certain condition. That condition is the minimization of the absolute value of the difference between the distance of the two molecules and a fixed value l (Eqs. (36, 37)). The definitions of the distance between two molecules and the fixed value are given below.

Taking into consideration the desired connectivity between the three molecules, in the context of this project, the distance between two molecules was defined as the distance between the atoms that will participate in the intended intermolecular bond (hydrogen or disulfide bond, **Figure 1**). As a consequence, the fixed value l was defined as the official length of that intermolecular bond (l_{Hb} = 1.5–2.5 Å for the Hydrogen bond, l_{SSb} = 2.05 Å for the disulfide bond).

In more detail, in both lysine and cysteine conjugation, the distance between the drug and linker was defined by the drug hydrogen (H) atom and the linker electronegative (EN) atom participating in the H bond between the linker and drug [Eqs. (30–32); **Figure 2**, step S4c]. In regard to the distance between the linker-drug complex and the conjugation residue, in lysine conjugation, it was defined by the linker nitrogen (N) atom and the lysine hydrogen (H) atom bonded to the side chain nitrogen atom. In cysteine conjugation, however, it has been defined as the distance between the sulfur (S) linker atom and the sulfur atom of the cysteine side chain [Eqs. (33–35); **Figure 2**, step S7c]. The translation of the drug in relation to the linker is performed for the values of the variables T_x, T_y, T_z for which the distance between the two molecules takes the official length l_{Hb} = 1.5–2.5 Å of the hydrogen bond (Eq. (36)). The translation of the linker-drug complex in relation to the cysteine is performed for the values of the variables T_x, T_y, T_z for which the distance between the two molecules takes the official length l_{SSb} = 2.05 Å of the disulfide bond (Eq. (37)):

$$p_d = (x, y, z)_{H\ atom\ of\ drug} = (x_{Hd}, y_{Hd}, z_{Hd}) = p_{d1} \tag{30}$$

$$p_{l1} = (x, y, z)_{EN\ atom\ of\ linker} = (x_{ENl}, y_{ENl}, z_{ENl}) = (x_{Cl}, y_{Cl}, z_{Cl}) \tag{31}$$

$$D_{d-l} = p_d - p_{l1} \tag{32}$$

$$p_{l2} = (x, y, z)_{S\ atom\ of\ linker} = (x_{Sl}, y_{Sl}, z_{Sl}) \tag{33}$$

$$p_{cys} = (x, y, z)_{S\ atom\ of\ Cysteine} = (x_{Scys}, y_{Scys}, z_{Scys}) \tag{34}$$

$$D_{cys-l} = p_{cys} - p_{l2} \tag{35}$$

$$T_x, T_y, T_z : |D_{d-l} - l_{Hb}| = 0 \tag{36}$$

$$T_x', T_y', T_z' : |D_{cys-l} - l_{SSb}| = 0 \tag{37}$$

where the variables T_x, T_y, T_z of Eq. (36) refer to the translation of the drug (**Figure 2**, step S5d), while the variables of T_x', T_y', T_z' Eq. (37) refer to the translation of the linker-drug complex (**Figure 2**, step S8d).

2.7. Implementation

The steps S1 and S2 (see the **Figure 2**) that involve the selection and preparation of the input files of the engineered antibody, the linker, the drug, as well as the selection of the conjugation cysteine residue, are carried out manually. The steps S3–S9 have been developed as a C++ program, and therefore can be executed automatically. In order to make the necessary data management, a number of classes and functions were created in regard to the PDB records used, the three-dimensional points and lines, as well as the affine transformation.

3. Results

The method described in the previous section has been applied to various antibody and drug combinations, taking about 10–13 seconds for each TDC to be generated. In **Figure 3**, the basic stages of the production of a THIOMAB-drug conjugate are depicted. First, the drug and linker are formed in two independent PDB files. The axis of each molecule is depicted as a white line, and the initial angle between them is 98.2° (**Figure 3a**). The drug is rotated, and the angle between the two axes becomes 0.08° (**Figure 3b**). The initial distance between the atoms that will form the hydrogen bond (the C atom of the linker and the H atom of the drug, bonded covalently to an O drug atom) is 28.02 Å. The drug is translated and the distance between the two atoms becomes 1.51 Å (**Figure 3c**). The linker has remained fixed. The two molecules are merged into a linker-drug complex, whose new axis is depicted as a white line (**Figure 3d**). The residue valine of the antibody with PDB id 4GAG, with sequence number 206 on the light chain (LC-VAL206), is replaced by a cysteine, whose axis is calculated (**Figure 3e**). The initial angle between the axis of the linker-drug complex and the axis of the cysteine is 105.6°, and the distance between the sulfur atoms of the cysteine and the linker is 54.75 Å. The linker-drug complex is rotated and translated, so that the angle becomes 0.083° and the distance becomes 2.07 Å. Finally, the linker-drug complex is connected with the cysteine (**Figure 3f**).

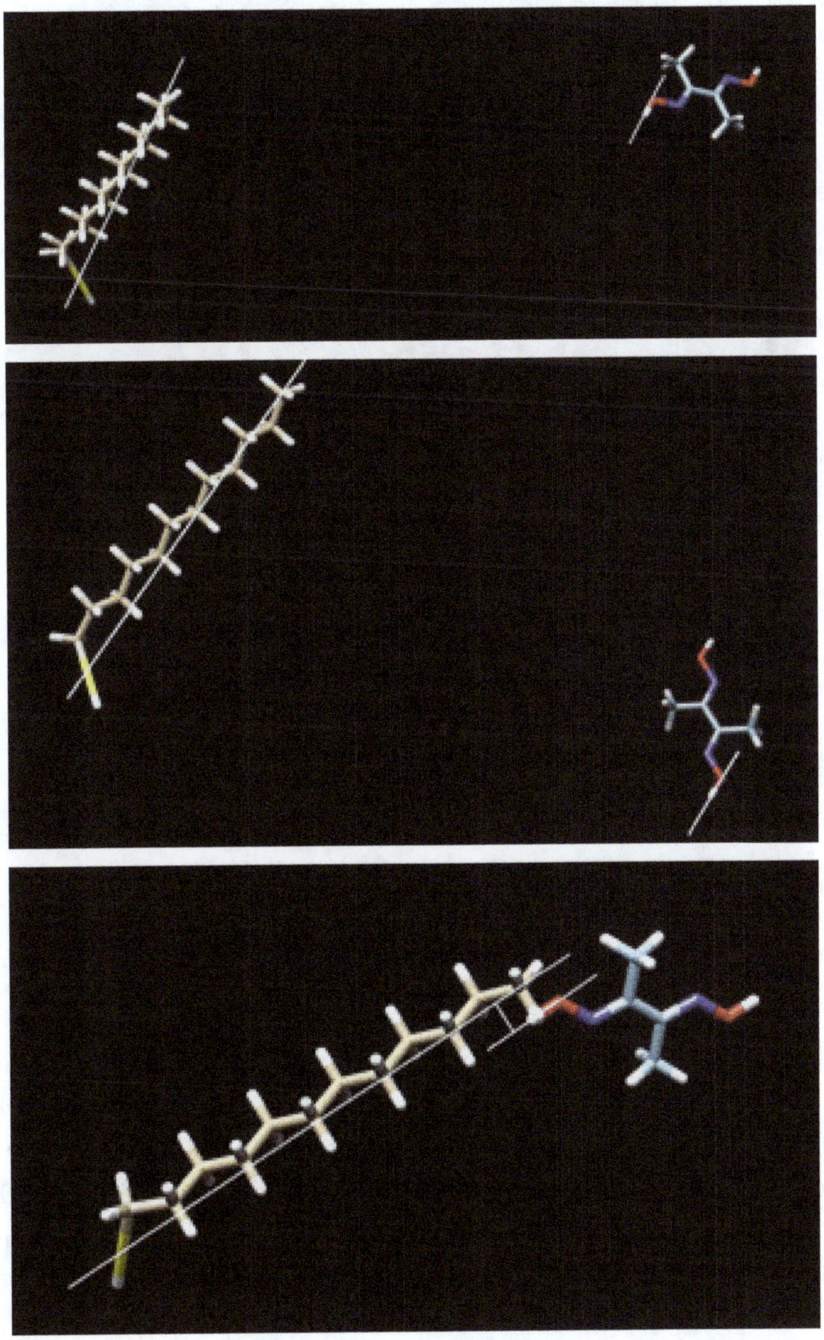

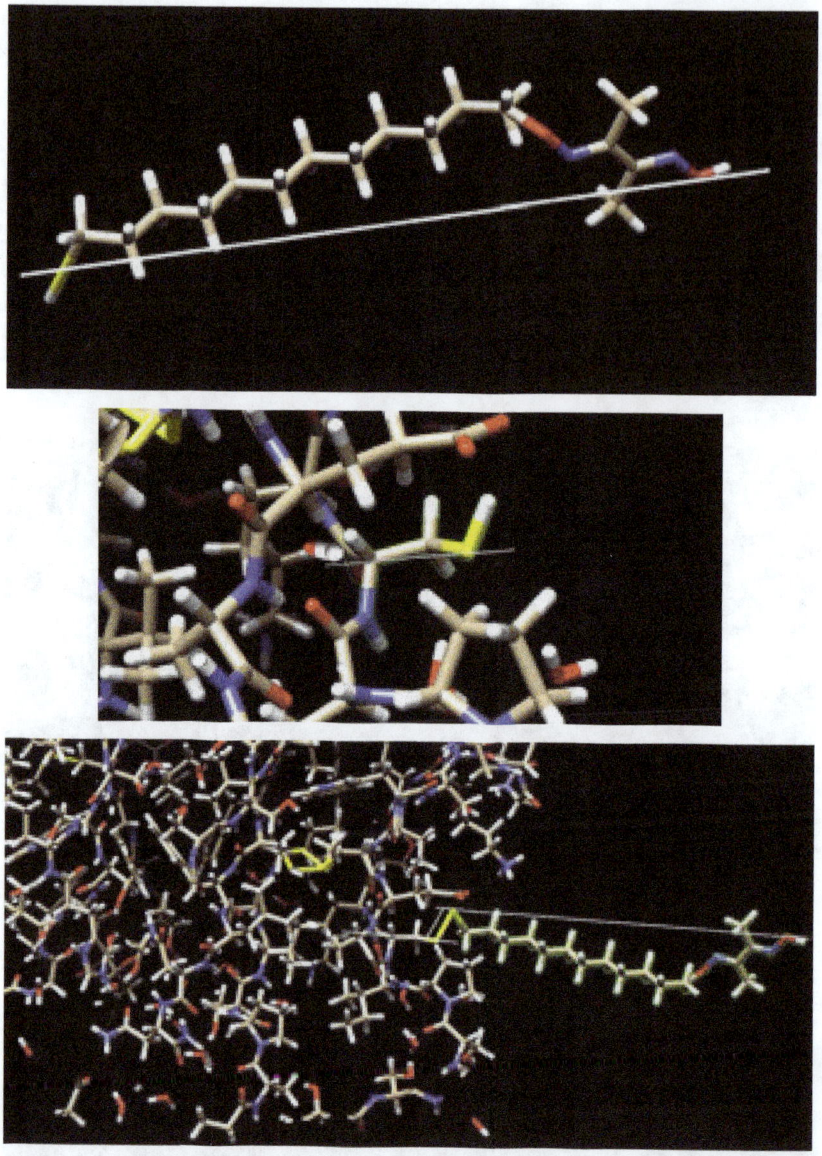

Figure 3. The basic stages of the computational production of a TDC. (a) Left: The linker molecule $C_{12}S$. Right: The drug with NCI sequence number 9, before its reconfiguration. Angle between the two axes: 98.2°. (b) The linker and the rotated drug. Angle between the two axes: 0.08°. Distance between the H atom of the drug and the C atom of the linker: 28.02 Å. (c) The linker and the translated drug. Distance between the H atom of the drug and the C atom of the linker: 1.51 Å. (d) The linker-drug complex. (e) The engineered cysteine that has replaced the residue LC-VAL206 on the antibody with PDB id 4GAG. Initial angle between the axes of the cysteine and the linker-drug complex: 105.6°. Initial distance between the two S atoms: 54.75 Å. (f) The linker-drug complex conjugated to the engineered cysteine. Angle between the axes of the cysteine and the linker-drug complex: 0.084°. Distance between the two S atoms: 2.07 Å.

More examples of produced TDCs are demonstrated in **Figure 4**. The TDC composed by the antibody with PDB id 4GAJ, the linker molecule $C_{12}S$ and the drug with sequence number 5 (nci_5.pdb) is depicted in **Figure 4(a)**. A hydrogen bond connects the drug with the linker, and a disulfide bond connects the engineered cysteine on the antibody with the linker-drug complex. The cysteine has replaced the solvent-accessible amino acid alanine with residue sequence number 114, at the heavy chain of the antibody (HC-ALA114). The same TDC is depicted from a closer distance at the right of the figure. In **Figure 4b**, a similar TDC composed by the antibody with PDB id 4GAJ, the linker molecule C12S and the drug with sequence number 7 (nci_7.pdb) is depicted. The above technique can be used for the conjugation of TDCs with more than one drug molecules per antibody. **Figure 4c** depicts a TDC with a DAR of 2, composed by the antibody with PDB id 4GAG, to which two drug molecules with sequence number 1 (nci_1.pdb) have been conjugated. The engineered cysteines that serve as conjugation sites have replaced the solvent-accessible residues alanine with sequence number 114 at the heavy chain and valine with sequence number 206 at the light chain (LC-VAL206). The conjugation areas of the same TDC are demonstrated from a closer distance at the right of the figure. The linker and the drugs used in the examples are depicted in **Figure 5**.

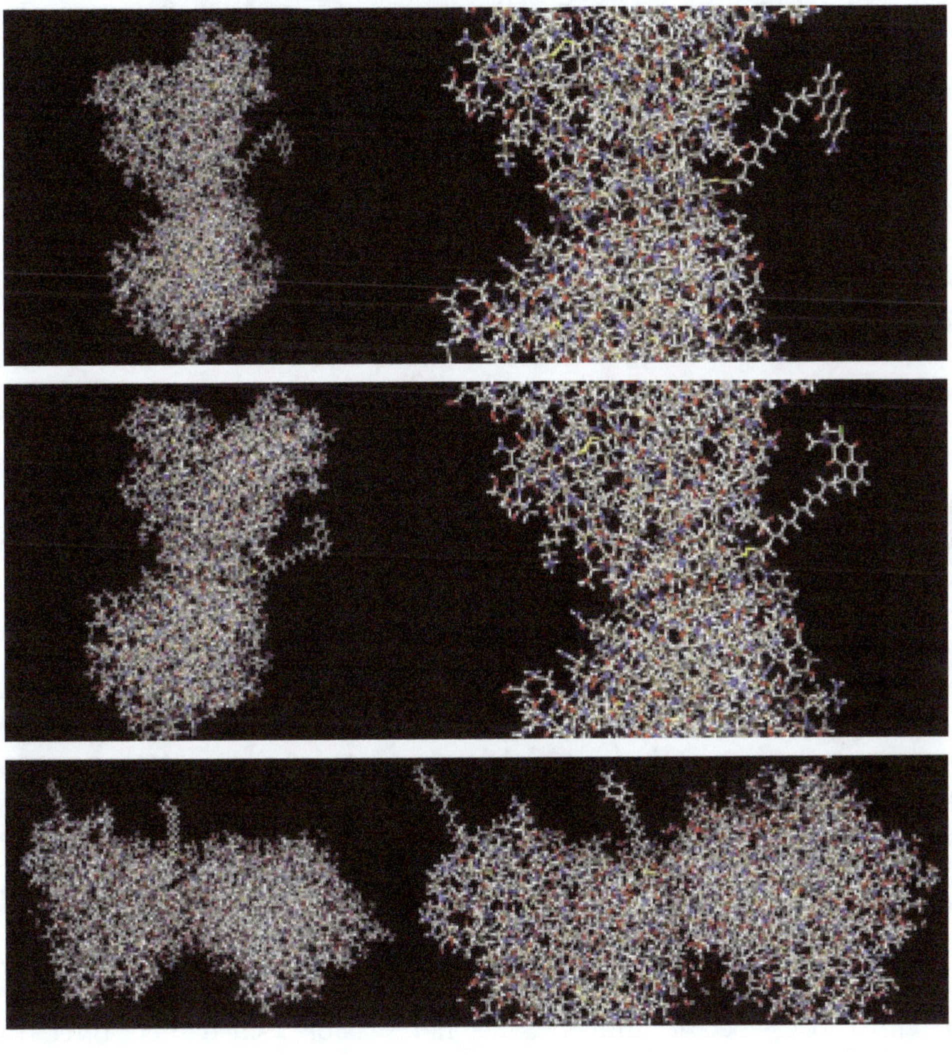

Figure 4. (a) Left: TDC composed by the antibody with PDB id 4GAJ, the linker molecule $C_{12}S$, and the drug with sequence number 5. The drug has been conjugated through the linker to the engineered cysteine that has replaced the residue HC-ALA114. Right: The same TDC from a closer distance. (b) Left: TDC composed by the antibody with PDB id 4GAJ, the linker molecule $C_{12}S$, and the drug with sequence number 7. The drug has been conjugated through the linker to the engineered cysteine that has replaced the residue HC-ALA114. Right: The same TDC from a closer distance. (c) Left: TDC composed by the antibody with PDB id 4GAG, two linker molecules $C_{12}S$, and two drug molecules with sequence number 1. The antibody conjugation sites are engineered cysteines that have replaced the residues HC-ALA114, and LC-VAL206. Right: The same TDC from a closer distance.

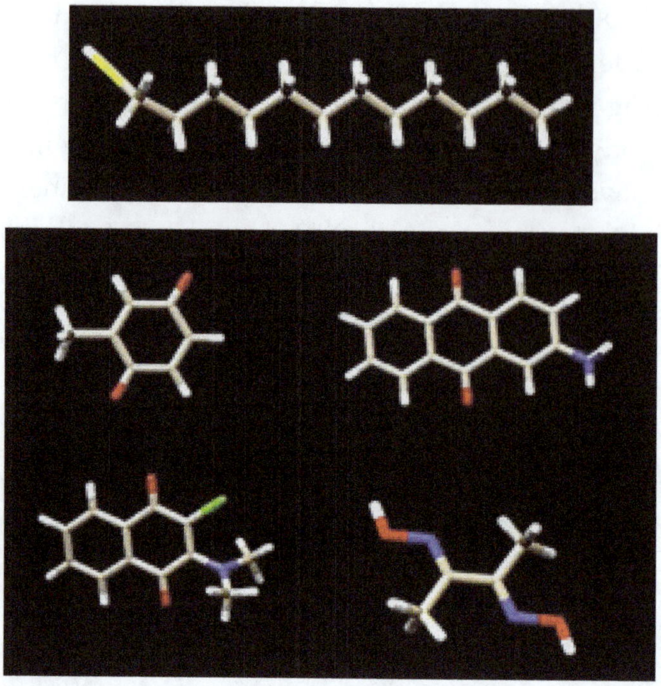

Figure 5. (a) The linker molecule $C_{12}S$. (b) The cytotoxic drugs used in **Figures 3** and **4**. Top left: the drug with sequence number 1 (nci_1.pdb). Top right: the drug with sequence number 5 (nci_5.pdb). Bottom left: the drug with sequence number 7 (nci_7.pdb). Bottom right: the drug with sequence number 9 (nci_9.pdb).

4. Discussion

This chapter has presented a computational method for site-specific formulation of ADCs, by conjugating the cytotoxin and the linker to engineered cysteines on the surface of the antibody. A hydrogen bond is employed to connect the linker with the drug, forming a linker-drug conjugate, which is then attached through a disulfide bond to the selected engineered cysteine. The change of the atomic coordinates of a molecule was implemented with an affine transformation. The generation of the linker molecule $C_{12}S$, as well as the cysteine replacement, was achieved through molecular modeling software [36, 42].

The present chapter proposes a general way to implement computationally ADC molecules using data from established databases. The common antibodies used do not target cancer-specific molecules, as opposed to antibodies that are parts of real ADCs. The employment of antibodies with cancer-specific antigens was not feasible, since their three-dimensional structures are not available. However, considering that the conjugation occurs in the preserved regions of the antibody, the same techniques could be applied using the PDB file of any full antibody. In addition, conventional anticancer drugs were used, since the three-dimensional structures of the more potent drugs used in ADCs currently are not available. It is important to note that the results of the presented method are not proposed as real anticancer drugs. The presented method is an attempt to simulate ADCs using programming techniques and available data.

Site-specific conjugation is a substantial goal in ADC development, since it gives the ability to control the pharmacokinetic and therapeutic properties of the produced ADCs. Aside from the replacement of residues with cysteines, mentioned in Section 1, a similar concept is the substitution of cysteines with serines in order to reduce the number of inter-chain cysteines available for conjugation. Another technique is the introduction of non-native amino acids in the antibody sequence, such as selenocysteine, acetylphenylalanine, para-acetylphenylalanine, and para-azidophenylalanine, utilizing their ability to provide orthogonal conjugation chemistries that are not available from original residues. Enzymes have been used to help form bonds between the antibody, the linker and drug, a method called enzymatic conjugation. An additional strategy is the conjugation at the glycosylation sites of the antibody, leaving the amino acid sequence of the antibody and the cell culture condition intact. For example, sialic acid moieties can be introduced enzymatically to the antibody, which can then be oxidized to introduce aldehyde groups. The resulting antibodies can then form stable oxime bonds with drugs containing aldehyde-reactive aminooxy groups [22, 46–59].

A number of actions could be taken to improve and expand the work described in this chapter. Some of our future goals is the implementation of more site-specific conjugation ways, the use of antibody forms alternative to the full monoclonal antibody, such as scFvs, diabodies, and minibodies, as well as other non-cleavable and cleavable linker molecules. In addition, the automatic recognition of all amino acids available for conjugation given a specific antibody would be interesting, as well as the estimation of the best drugs and linkers given a conjugation method. Finally, the conditions and equations used could be modified in order to take into consideration other factors.

Acknowledgements

Funding was received through an IKY Fellowship of Excellence for postgraduate studies in Greece—Siemens Program. The authors confirm that the funder had no influence over the study design, content of the chapter, or selection of this book.

Author details

Arianna Filntisi[1,2*], Dimitrios Vlachakis[2] and George K. Matsopoulos[1]

*Address all correspondence to: arianna.filntisi@gmail.com

1 School of Electrical and Computer Engineering, National Technical University of Athens, Athens, Greece

2 Computational Biology and Medicine Group, Biomedical Research Foundation of the Academy of Athens, Athens, Greece

References

[1] National Cancer Institute at the National Institutes of Health. 2014. Available from: http://www.cancer.gov/ [Accessed: 2014–03]

[2] American Cancer Society. 2014. Available from: http://www.cancer.org/ [Accessed: 2014–03]

[3] Cancer Research Institute. 2014. Available from: http://www.cancerresearch.org/ [Accessed: 2014–02]

[4] Li J, Chen F, Cona MM, Feng Y, Himmelreich U, Oyen R, et al. A review on various targeted anticancer therapies. Target Oncology. 2012;7(1):69–85. DOI: 10.1007/s11523-012-0212-2

[5] Adler MJ, Dimitrov DS. Therapeutic antibodies against cancer. Hematol Oncol Clin North Am.2012;26(3):447–481. DOI: 10.1016/j.hoc.2012.02.013

[6] Sapra P, Shor B. Monoclonal antibody-based therapies in cancer: advances and challenges. Pharmacol Ther. 2013;138(3):452–469. DOI: 10.1016/j.pharmthera.2013.03.004

[7] Goldmacher VS, Chittenden T, Chari RVJ, Kovtun YV, Lambert JM. Antibody-drug conjugates for targeted cancer therapy. In: Desai MC, editor. Annual Reports in Medicinal Chemistry, Volume 47. Academic Press; 2012. p. 349–366.

[8] Casi G, Neri D. Antibody-drug conjugates: basic concepts, examples and future perspectives. J Control Release. 2012;161(2):422–428. DOI: 10.1016/j.jconrel.2012.01.026

[9] Kovtun Y V, Goldmacher VS. Cell killing by antibody-drug conjugates. Cancer Lett. 2007;255(2):232–240. DOI: http://dx.doi.org/10.1016/j.canlet.2007.04.010

[10] Gerber H-P, Senter PD, Grewal IS. Antibody drug-conjugates targeting the tumor vasculature: Current and future developments. MAbs. 2009;1(3):247–253. DOI: 10.4161/mabs.1.3.8515

[11] Wu AM, Senter PD. Arming antibodies: prospects and challenges for immunoconjugates. Biotechnol. 2005;23(9):1137–1146. DOI: 10.1038/nbt1141

[12] Perez HL, Cardarelli PM, Deshpande S, Gangwar S, Schroeder GM, Vite GD, et al. Antibody-drug conjugates: current status and future directions. Drug Discov Today. 2014;19(7):869–881. DOI: 10.1016/j.drudis.2013.11.004

[13] Alley SC, Anderson KE. Analytical and bioanalytical technologies for characterizing antibody-drug conjugates. Curr Opin Chem Biol. 2013;17(3):406–411. DOI: 10.1016/j.cbpa.2013.03.022

[14] Iyer U, Kadambi VJ. Antibody drug conjugates-Trojan horses in the war on cancer. J Pharmacol Toxicol Methods. 2011;64(3):207–212. DOI: 10.1016/j.vascn.2011.07.005.

[15] Carter PJ. Potent antibody therapeutics by design. Nat Rev Immunol. 2006;6(5):343–357. DOI: 10.1038/nri1837

[16] Junutula JR, Flagella KM, Graham RA, Parsons KL, Ha E, Raab H, et al. Engineered thio-trastuzumab-DM1 conjugate with an improved therapeutic index to target human epidermal growth factor receptor 2-positive breast cancer. Clin Cancer Res. 2010;16(19): 4769–4778. DOI: 10.1158/1078-0432.CCR-10-0987

[17] Dere R, Yi J-H, Lei C, Saad OM, Huang C, Li Y, et al. PK assays for antibody-drug conjugates: case study with ado-trastuzumab emtansine. Bioanalysis. 2013;5(9):1025–1040. DOI: 10.4155/bio.13.72

[18] Hamblett KJ, Senter PD, Chace DF, Sun MMC, Lenox J, Cerveny CG, et al. Effects of drug loading on the antitumor activity of a monoclonal antibody drug conjugate. Clin Cancer Res. 2004;10(20):7063–7070. DOI: 10.1158/1078-0432.CCR-04-0789

[19] Flemming A. Antibody engineering: fine-tuning antibody-drug conjugates. Nat Rev Drug Discov. 2014;13(3):178. DOI: 10.1038/nrd4266

[20] Tian F, Lu Y, Manibusan A, Sellers A, Tran H, Sun Y, et al. A general approach to sitespecific antibody drug conjugates. Proc Natl Acad Sci U S A. 2014;111(5):1766–1771. DOI: 10.1073/pnas.1321237111

[21] Junutula JR, Raab H, Clark S, Bhakta S, Leipold DD, Weir S, et al. Site-specific conjugation of a cytotoxic drug to an antibody improves the therapeutic index. Nat Biotechnol. 2008;26(8):925–932. DOI: 10.1038/nbt.1480

[22] Senter PD. Potent antibody drug conjugates for cancer therapy. Curr Opin Chem Biol. 2009;13(3):235–244. DOI: 10.1016/j.cbpa.2009.03.023

[23] Lyon RP, Meyer DL, Setter JR, Senter PD. Conjugation of anticancer drugs through endogenous monoclonal antibody cysteine residues. Methods Enzymol. 2012;502:123–138. DOI: 0.1016/B978-0-12-416039-2.00006-9

[24] Junutula JR, Bhakta S, Raab H, Ervin KE, Eigenbrot C, Vandlen R, et al. Rapid identification of reactive cysteine residues for site-specific labeling of antibody-Fabs. J Immunol Methods. 2008;332(1–2):41–52. DOI: 10.1016/j.jim.2007.12.011

[25] Li X, Patterson JT, Sarkar M, Pedzisa L, Kodadek T, Roush WR, et al. Site-specific dual antibody conjugation via engineered cysteine and selenocysteine residues. Bioconjug Chem. 2015;26(11):2243–2248. DOI: 10.1021/acs.bioconjchem.5b00244

[26] Shen B-Q, Xu K, Liu L, Raab H, Bhakta S, Kenrick M, et al. Conjugation site modulates the in vivo stability and therapeutic activity of antibody-drug conjugates. Nat Biotechnol. 2012;30(2):184–189. DOI: 10.1038/nbt.2108

[27] Strop P, Liu S-H, Dorywalska M, Delaria K, Dushin RG, Tran T-T, et al. Location matters: site of conjugation modulates stability and pharmacokinetics of antibody drug conjugates. Chem Biol. 2013;20(2):161–167. DOI: 10.1016/j.chembiol.2013.01.010

[28] Boylan NJ, Zhou W, Proos RJ, Tolbert TJ, Wolfe JL, Laurence JS. Conjugation site heterogeneity causes variable electrostatic properties in Fc conjugates. Bioconjug Chem. 2013;24(6):1008–1016. DOI: 10.1021/bc4000564

[29] Worldwide PDB (wwPDB) organization. 2003 [Updated: 2016]. Available from: http://www.wwpdb.org/ [Accessed: 2015]

[30] RCSB Protein Data Bank. 2000 [Updated: 2016]. Available from: http://www.rcsb.org [Accessed: 2016]

[31] Berman HM, Westbrook J, Feng Z, Gilliland G, Bhat TN, Weissig H, Shindyalov IN, Bourne PE. The Protein Data Bank. Nucl. Acids Res. 2000;28(1):235–242. DOI: 10.1093/nar/28.1.235

[32] Chemical Computing Group Inc. Molecular Operating Environment (MOE). 2013. [Accessed: 2014]

[33] Filntisi A, Papangelopoulos N, Bencurova E, Kasampalidis I, Matsopoulos G, Vlachakis D, et al. State-of-the-Art Neural Networks Applications in Biology. Int J Syst Biol Biomed Technol. 2013;2(4):63–85. DOI: 10.4018/ijsbbt.2013100105

[34] Filntisi A, Vlachakis D, Matsopoulos G, Kossida S. 3D structural bioinformatics of proteins and antibodies: state of the art perspectives and challenges. Int J Syst Biol Biomed Technol. 2013;2(3):67–74. DOI: 10.4018/ijsbbt.2013070105

[35] Vlachakis D, Tsagrasoulis D, Megalooikonomou V, Kossida S. Introducing Drugster: a comprehensive and fully integrated drug design, lead and structure optimization toolkit. Bioinformatics. 2013;29(1):126–128. DOI: DOI: 10.1093/bioinformatics/bts637

[36] Pettersen EF, Goddard TD, Huang CC, Couch GS, Greenblatt DM, Meng EC, et al. UCSF Chimera-a visualization system for exploratory research and analysis. J Comput Chem. 2004;25(13):1605–1612. DOI: 10.1002/jcc.20084

[37] Toshimoto K, Wakayama N, Kusama M, Maeda K, Sugiyama Y, Akiyama Y. In silico prediction of major drug clearance pathways by support vector machines with feature selected descriptors. Drug Metab Dispos. 2014;42(11):1811–1819. DOI: 10.1124/dmd.114.057893

[38] Mitra A, Kesisoglou F, Dogterom P. Application of absorption modeling to predict bioequivalence outcome of two batches of etoricoxib tablets. AAPS PharmSciTech. 2015;16(1):76–84. DOI: 10.1208/s12249-014-0194-8

[39] Vlachakis D, Kossida S. Antibody drug conjugate bioinformatics: drug delivery through the letterbox. Comput Math Methods Med. 2013;2013(2013):282398. DOI:http://dx.doi.org/10.1155/2013/282398

[40] Scilligence. JSDraw - Antibody-Drug Conjugate [Internet]. 2014 [Updated: 2015]. Available from: http://www.elncloud.com/jsdrawapp/jsdraw/ADC.htm [Accessed: 2016]

[41] Sukumaran S, Gadkar K, Zhang C, Bhakta S, Liu L, Xu K, et al. Mechanism-based pharmacokinetic/pharmacodynamic model for THIOMABTM drug conjugates. Pharm Res. 2015;32(6):1884–1893. DOI: 10.1007/s11095-014-1582-1

[42] Filntisi A, Vlachakis D, Matsopoulos GK, Kossida S. Computational construction of antibody-drug conjugates using surface lysines as the antibody conjugation site and a non-cleavable linker. Cancer Inform. 2014;13:179–186. DOI: 10.4137/CIN.S19222

[43] Wolfram Mathworld. http://mathworld.wolfram.com/AffineTransformation.html [Internet]. 1999 [Updated: 2006]. Available from: http://mathworld.wolfram.com/AffineTransformation.html [Accessed: 2014]

[44] Michigan Technical University. Geometric Transformations [Internet]. 1998 [Updated: 2011]. Available from: http://www.cs.mtu.edu/~shene/COURSES/cs3621/NOTES/geometry/geo-tran.html [Accessed: 2015]

[45] Steven M. Lavalle, Cambridge University Press. Planning Algorithms [Internet]. 2006 [Updated: 2012]. Available from: http://planning.cs.uiuc.edu/node104.html [Accessed: 2015]

[46] Panowski S, Bhakta S, Raab H, Polakis P, Junutula JR. Site-specific antibody drug conjugates for cancer therapy. MAbs. 2014;6(1):34–45. DOI: 10.4161/mabs.27022

[47] McDonagh CF, Turcott E, Westendorf L, Webster JB, Alley SC, Kim K, et al. Engineered antibody-drug conjugates with defined sites and stoichiometries of drug attachment. Protein Eng Des Sel. 2006;19(7):299–307. DOI: 10.1093/protein/gzl013

[48] Wakankar A, Chen Y, Gokarn Y, Jacobson FS. Analytical methods for physicochemical characterization of antibody drug conjugates. MAbs. 2011;3(2):161–172. DOI: 10.4161/mabs.3.2.14960

[49] Rabuka D, Rush JS, de Hart GW, Wu P, Bertozzi CR. Site-specific chemical protein conjugation using genetically encoded aldehyde tags. Nat Protoc. 2012;7(6):1052–1067. DOI: 0.1038/nprot.2012.045

[50] Zhou Q, Stefano JE, Manning C, Kyazike J, Chen B, Gianolio DA, et al. Site-specific antibody-drug conjugation through glycoengineering. Bioconjug Chem. 2014;25(3): 510–520. DOI: 10.1021/bc400505q

[51] Zhu Z, Ramakrishnan B, Li J, Wang Y, Feng Y, Prabakaran P, et al. Site-specific antibody-drug conjugation through an engineered glycotransferase and a chemically reactive sugar. MAbs. 2014;6(5):1190–1200. DOI: 10.4161/mabs.29889

[52] Cai X, Janda KD. A chemoenzymatic approach toward the preparation of site-specific antibody-drug conjugates. Tetrahedron Lett. 2015;56(23):3172–3175. DOI: 10.1016/j.tetlet.2014.12.025

[53] Kiick KL, Saxon E, Tirrell DA, Bertozzi CR. Incorporation of azides into recombinant proteins for chemoselective modification by the Staudinger ligation. Proc Natl Acad Sci U S A. 2002;99(1):19–24. DOI: 10.1073/pnas.012583299

[54] Hofer T, Skeffington LR, Chapman CM, Rader C. Molecularly defined antibody conjugation through a selenocysteine interface. Biochemistry. 2009;48(50):12047–12057. DOI: 10.1021/bi901744t

[55] Hofer T, Thomas JD, Burke TR, Rader C. An engineered selenocysteine defines a unique class of antibody derivatives. Proc Natl Acad Sci U S A. 2008;105(34):12451–12456. DOI: 10.1073/pnas.0800800105

[56] Li X, Yang J, Rader C. Antibody conjugation via one and two C-terminal selenocysteines. Methods. 2014;65(1):133–138. DOI: 10.1016/j.ymeth.2013.05.023

[57] Zimmerman ES, Heibeck TH, Gill A, Li X, Murray CJ, Madlansacay MR, et al. Production of site-specific antibody-drug conjugates using optimized non-natural amino acids in a cell-free expression system. Bioconjug Chem. 2014;25(2):351–361. DOI: 10.1021/bc400490z

[58] Axup JY, Bajjuri KM, Ritland M, Hutchins BM, Kim CH, Kazane SA, et al. Synthesis of site-specific antibody-drug conjugates using unnatural amino acids. Proc Natl Acad Sci U S A. 2012;109(40):16101–16106. DOI: 10.1073/pnas.1211023109

[59] Sochaj AM, Świderska KW, Otlewski J. Current methods for the synthesis of homogeneous antibody-drug conjugates. Biotechnol Adv. 2015;33(6):775–784. DOI: 10.1016/j.biotechadv.2015.05.001

5

The Study of Hepatitis B Virus Using Bioinformatics

Trevor Graham Bell and Anna Kramvis

Abstract

Hepatitis refers to the inflammation of the liver. A major cause of hepatitis is the hepatotropic virus, hepatitis B virus (HBV). Annually, more than 786,000 people die as a result of the clinical manifestations of HBV infection, which include cirrhosis and hepatocellular carcinoma. Sequence heterogeneity is a feature of HBV, because the viral-encoded polymerase lacks proof-reading ability. HBV has been classified into nine genotypes, A to I, with a putative 10th genotype, "J," isolated from a single individual. Comparative analysis of HBV strains from various geographic regions of the world and from different eras can shed light on the origin, evolution, transmission and response to anti-HBV preventative, and treatment measures. Bioinformatics tools and databases have been used to better understand HBV mutations and how they develop, especially in response to antiviral therapy and vaccination. Despite its small genome size of ~3.2 kb, HBV presents several bioinformatic challenges, which include the circular genome, the overlapping open reading frames, and the different genome lengths of the genotypes. Thus, bioinformatics tools and databases have been developed to facilitate the study of HBV.

Keywords: alignments, computation, databases, genotypes, phylogenetics

1. Introduction

Primarily, bioinformatics is the use of computational science to study biological and clinical data using statistics, mathematics, and information theory. This field is developing and evolving; thus, the definition cannot be precise. Moreover, the field is broad, ranging from the study of DNA and proteins, to structural biology, drug design and comparative genomics, transcriptomics, proteomics, and metagenomics. The optimization of computational technology is paramount in order to handle, store, manage, and analyze the large volumes of data generated in the last decade. The data include molecular sequencing data of host and pathogen genomes and their associations to demographic and clinical records, laboratory test results, as well as information on treatment. Moreover, bioinformatics can aid in the investigation of virus–host genome and environmental interactions and in the identification of both host and viral biomarkers. This analysis can lead to a better understanding of clinical manifestation of disease and effective design of preventative and treatment measures [1].

In the first section, we describe the unique genomics and molecular biology of hepatitis B virus (HBV). Using illustrative examples, we showed how bioinformatics analyses can facilitate the understanding of the origin, evolution, transmission, and response to antiviral agents of HBV. Next, we described the bioinformatics challenges posed by HBV and present the public databases and tools currently available for the study of HBV.

2. Hepatitis B virus

2.1. Hepatitis

Hepatitis refers to the inflammation of the liver. A major cause of hepatitis is the hepatotropic virus, HBV. HBV infection is a public health problem of worldwide importance. Globally, 2 billion people have been exposed to this virus at some stage of their lives, and 240 million are chronic carriers of the virus [2].

This infection can lead to a spectrum of clinical consequences. In the majority of cases, the infection is subclinical and transient, whereas in 25% of cases, it can cause self-limited acute hepatitis and in 1% of these progress to acute liver failure. The virus can persist in 90% of neonates and 5–10% of adults, leading to chronic infection that can progress to either chronic hepatitis or an asymptomatic carrier state. Both of these states can ultimately develop liver cancer or hepatocellular carcinoma (HCC), with or without the intermediate cirrhotic stage. Annually, more than 786,000 people die as a result of these clinical manifestations of HBV infection [3].

2.2. Prevalence

The prevalence of HBV in a community can be estimated by the proportion of the population, who are hepatitis B surface antigen (HBsAg)-positive carriers. HBV prevalence varies widely in the world [3]. The prevalence is low (<1%) in northern Europe, Australia, New Zealand, Canada, and the United States of America. Northern Asia, the Indian subcontinent, parts of Africa, Eastern and south-eastern Europe, and parts of Latin America are areas of intermediate prevalence (1–5%). The high prevalence areas (5–20%) include East and Southeast Asia, the Pacific Islands, and sub-Saharan Africa.

2.3. Classification and structure

HBV, the prototype member of the family *Hepadnaviridae*, belongs to the genus *Orthohepadnavirus*. With a diameter of 42 nm and a DNA genome of ~3.2 kilobases (kb), it is the smallest

DNA virus infecting man. The genome is circular and partially double stranded. One DNA strand is complete, except for a small nick (the minus strand), and the other is short and incomplete (the plus strand). The minus strand contains four overlapping open reading frames (ORFs; **Figure 1**) [4] that represent: (1) the *preS/S* gene that codes for the envelope proteins, large, middle, and small HBsAgs; (2) the *P* gene for DNA polymerase/reverse transcriptase (POL); (3) the *X* gene for the X protein, a key regulator during the natural infection process, which has transcriptional trans-activation activity and is required to initiate and maintain HBV replication [5]; and (4) the *precore/core* gene that codes for the HBcAg or core protein that forms the capsid and for an additional protein known as HBeAg, which is not incorporated into the virus itself but is expressed on the liver cells and secreted into the serum. **Figure 2** illustrates the structure of the hepatitis B virion.

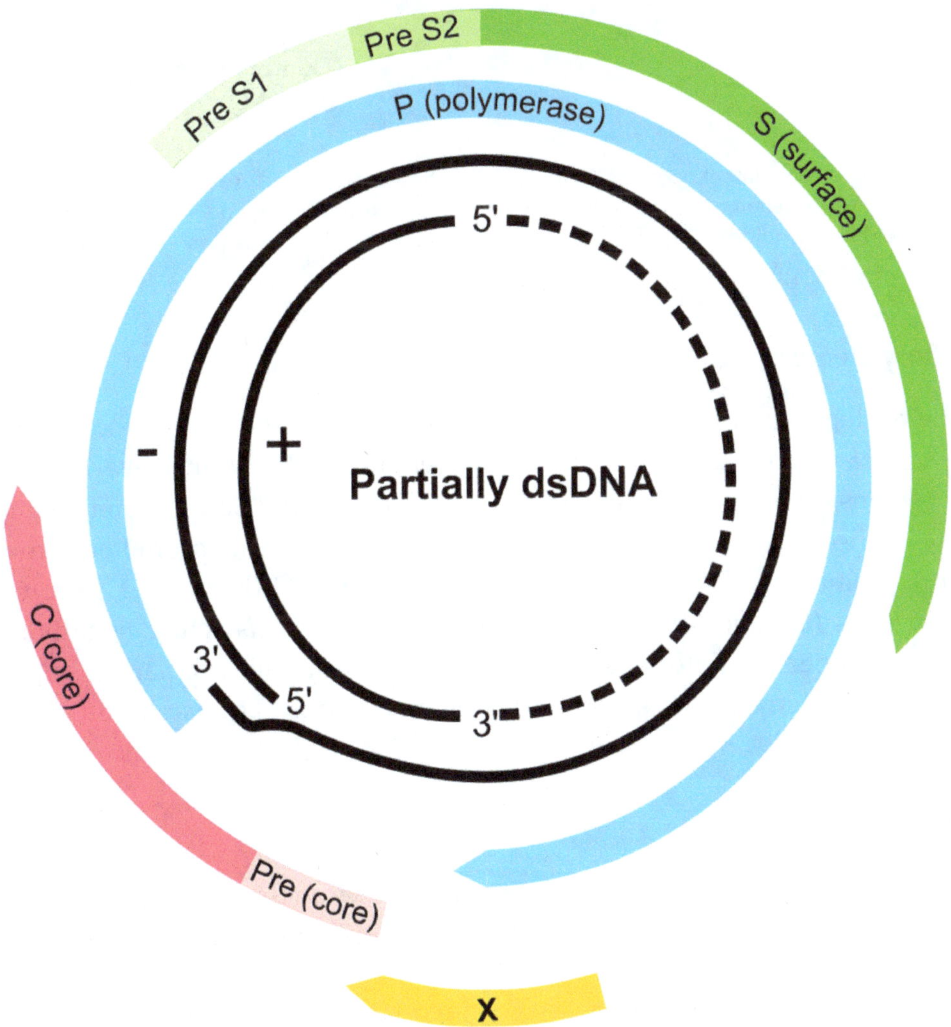

Figure 1. The genome of hepatitis B virus (HBV). The partially double-stranded DNA (dsDNA) with the complete minus (–) strand and the incomplete (+) strand. The four open reading frames (ORFs) are shown: *precore/core (preC/C)* that encodes the e antigen (HBeAg) and core protein (HBcAg); *P* for polymerase (reverse transcriptase), *PreS1/PreS2/S* for surface proteins [three forms of HBsAg, small (S), middle (M), and large (L)] and *X* for a transcriptional trans-activator protein.

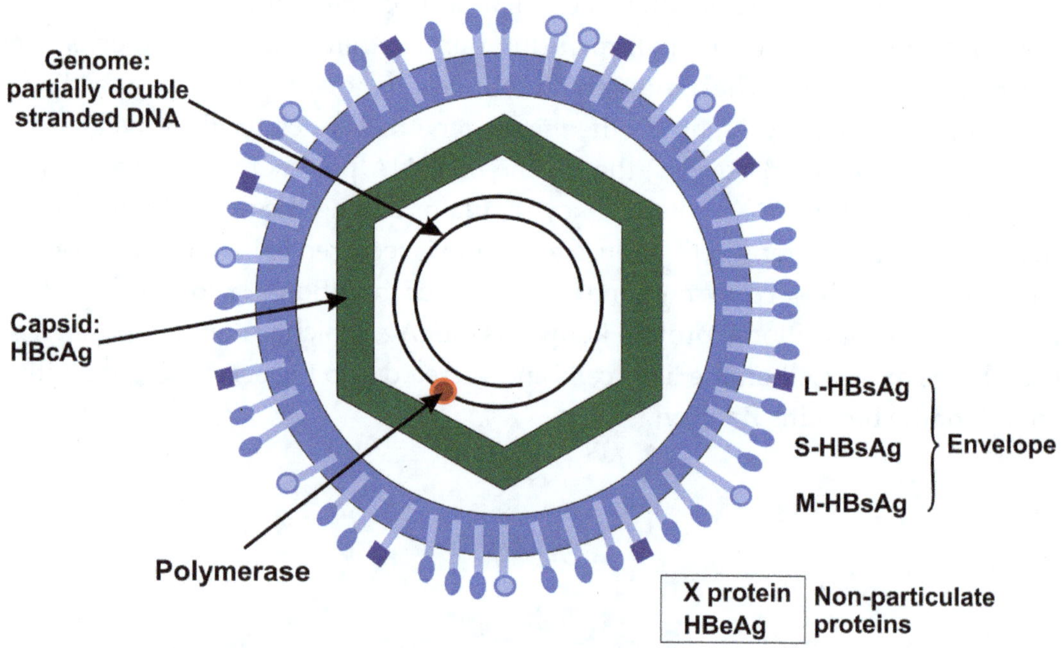

Figure 2. Schematic representation of hepatitis B virus (HBV), showing the structure of the virion, composed of a partially double-stranded DNA genome, enclosed by a capsid, comprised of HBcAg and surrounded by a lipid envelope containing large (L)-HBsAg, middle (M)-HBsAg, and small (S)-HBsAg. The virus also expresses two non-particulate proteins X protein and HBeAg.

2.4. Regulatory elements of HBV

Every single nucleotide of the HBV genome is necessary for the translation of a protein and may also be part of one of the regulatory elements of HBV, which overlap with protein expressing regions. The regulatory elements include the S1 and S2 promoters, which overlap both the preS region and polymerase ORFs; the preC/pregenomic promoter, which includes the basic core promoter (BCP) and overlaps the X and preC ORF; and the X promoter. There are two enhancers (enhancer I and enhancer II) as well as *cis*-acting negative regulatory elements (URR: upper regulatory region, CURS: core upstream regulatory sequence, NRE: negative regulatory element). These regulatory elements control transcription (reviewed in [6, 7]).

2.5. Replication of HBV

HBV and other members of the family *Hepadnaviridae* have an unusual replication cycle. These DNA viruses replicate by reverse transcription of a RNA intermediate known as the pregenomic RNA (pgRNA) [8]. Entry into the cell is via the sodium taurocholate cotransporting polypeptide (NTCP), a multiple transmembrane transporter predominantly expressed in the liver [9]. After entry, the virion is uncoated and the core particle is actively transported to the nucleus [10], where the partially double strand relaxed circular DNA molecule is released. The single-stranded gap is closed by the viral polymerase to yield a covalently closed circular molecule of DNA (cccDNA) [11], which is the template for transcription by the host RNA polymerase II [12]. The mRNAs are transported into the cytoplasm where they are translated into the seven viral proteins. In addition to being translated into the polymerase and the core protein, the pgRNA is packaged into immature core particles by the process known as encapsidation. In order to be encapsidated, the 5′ end of the pgRNA has to be folded into a particular secondary structure known as the encapsidation signal (ε) [13].

The encapsidation signal (ε) is a bipartite stem-loop structure, consisting of an upper and lower stem, the bulge, and an apical loop. Besides encapsidation, ε has a number of other functions (reviewed in [13]) and references therein. It acts in template restriction so that not any piece of RNA is encapsidated, and it also plays a role in the activation of the viral polymerase, so that there is no indiscriminate reverse transcription. It is also involved in the initiation of reverse transcription. The polymerase or reverse transcriptase acts as a primer of RNA-directed DNA synthesis by the binding of the polymerase to the bulge of ε. The first three nucleotides of the negative stand of DNA are synthesized at the bulge and are transferred to an acceptor site on the 3′ end of the pgRNA, where DNA synthesis proceeds toward the 5′ end of the pgRNA [14], giving rise to the immature virion. The virus matures by acquiring its glycoprotein envelope, containing HBsAg, in the endoplasmic reticulum and is exported by vesicular transport from the cell [15].

2.6. Genotypes and subgenotypes of HBV

Sequence heterogeneity is a feature of HBV, because the viral-encoded polymerase lacks proof-reading ability as mentioned above [16]. Using phylogenetic analysis of the complete genome of HBV and an intergroup divergence of greater than 7.5%, HBV has been classified into nine genotypes, A to I [17, 18, 19], with a putative 10th genotype, "J," isolated from a single individual [20]. With between ~4 and ~8% intergroup nucleotide difference across the complete genome and good bootstrap support, genotypes A–D, F, H, and I are classified further into at least 35 subgenotypes [21]. The genotypes differ in genome length, the size of ORFs and the proteins translated [17], as well as the development of various mutations [22]. Generally, the genotypes, and in some cases the subgenotypes, have a distinct geographic distribution (**Table 1**).

Genotype	Length	Differentiating features	Subgenotypes	Geographic distribution	Serological subtype	Transmission route
A	3221	6-nucleotide insert at carboxyl end of core gene	A1	Africa[#]	*adw*2/*ayw*2	Horizontal: parenteral or sexual
			A2	Europe/North America	*adw*2	
			Quasi-subgenotype A3[(A3,A4,A5)§]	Africa, Haiti	*ayw*1	
			A4[(A6)§]	Africa	*ayw*1 *adw*4*	

Genotype	Length	Differentiating features	Subgenotypes	Geographic distribution	Serological subtype	Transmission route
B	3215		B1	Japan	*adw*2	
			B2	China	*adw*2	
			Quasi-subgenotype B3[(B3,B5,B7–B9,B6(China)§]	Indonesia Philippines/China	*adw*2	Perinatal
			B4	Vietnam/Cambodia/ France	*ayw*1/*adw*2	
			B5[(B6)§]	Eskimos/Inuits	*adw*2 *adr*fntab1_3	
C	3215		C1	Thailand/Myanmar/ Vietnam	*adr**	
			Quasi-subgenotype C2[(C2,C14, undefined sequences)§]	Japan/China/Korea	*adr*	
			C3	New Caledonia/ Polynesia	*adr*	Perinatal
			C4	Australian Aborigines	*ayw*2/*ayw*3	
			C5	Philippines/ Indonesia	*adw*2	
			C6–C12	Indonesia/ Philippines	*adr*	
			C13–C15	Indonesia	*adr*	
			C16	Indonesia	*ayr**	
D	3182	33-nucleotide deletion at the amino terminus of the preS1 region	D1	Middle East, Central Asia	*ayw*2	
			D2	Europe/Japan/Lebanon	*ayw*3	
			D3	Worldwide	*ayw*2/*ayw*3	
			D4	Australian aborigines, Micronesians, Papua New Guineans, Arctic Denes	*ayw*2	Horizontal: parenteral with intravenous drug use being a risk factor
			D5	India	*ayw*3/*ayw*2	

Genotype	Length	Differentiating features	Subgenotypes	Geographic distribution	Serological subtype	Transmission route
			D6	Tunisia/Nigeria	*ayw*2 *ayw*4*/*adw*3*	
E	3212	3-nucleotide deletion at the amino terminus of the preS1 region	–	Western/Central Africa	*ayw*4	Horizontal
F	3215		F1	Argentina/Costa Rica/El Salvador, Alaska	*adw*4	
			F2	Nicaragua/ Venezuela/Brazil	*adw*4	Horizontal
			F3	Venezuela/Colombia	*adw*4	
			F4	Argentina	*adw*4 *adw*2*/*ayw*4*	

¶Summarizes data compiled from Kramvis [21] and references cited therein.
§Earlier subgenotype designation.
*Rare serological subtype for that genotype.
#And in regions outside Africa where there was historical forced migration as a result of the slave trade [23].
¥Vietnamese residing in Canada [24].

Table 1. Comparison of the virological and clinical characteristics of the genotypes and subgenotypes of HBV¶.

2.7. Genotyping and subgenotyping methods

HBV genotypes, and in some cases subgenotypes and various mutations, can influence the clinical course of disease [22] as well as response to antiviral therapy [25] and can be used to show transmission [26] and to trace human migrations [23]. Thus, HBV genotyping is becoming increasingly relevant in the clinical setting and may contribute to future personalized treatment [27] and may be important in epidemiological and transmission studies. Bioinformatics has played a major role in the development of various tools that can be used for identifying genotypes/subgenotypes and detecting various mutations. Therefore, a number of methods have been developed [28, 29].

Although analysis of the HBV *S* gene sequence is sufficient to classify HBV into genotypes [30], the complete genome sequence provides additional information with respect to phylogenetic relatedness [31, 32], including the identification of recombinants. Furthermore, even though complete genome analysis is the gold standard for genotyping, it does not allow for rapid and direct analysis on a large scale basis [17] and requires expertise and thus capacity development in computer processing coupled with phylogenetic analyses. In order to expedite and facilitate genotyping, a number of methods have been developed [17, 28, 29]. Each one has its advantages and disadvantages [17, 28, 29], which should be taken into account, when selecting the genotyping method appropriate for a particular study or application.

2.8. Phylogenetic analyses of HBV

Although, as already mentioned, the error-prone polymerase of HBV leads to sequence heterogeneity [16], the degree, at which this can occur, is constrained by the partially overlapping ORFs and the presence of secondary RNA structures, such as ε, coded by non-overlapping regions [33, 34]. The HBV genome has been estimated to evolve with an error rate of ~10^{-3}–10^{-6} nucleotide substitutions/site/year [35–41], although this rate is not constant within the different regions of the HBV genome [41]. The progress of computers and information technology has played an important role in the development of phylogenetic analysis as a powerful tool in the analysis of the molecular evolution of viruses.

As exemplified in the next sections, comparative analysis of HBV strains from various geographic regions of the world and from different eras can shed light on the origin, evolution, transmission, and response to anti-HBV preventative and treatment measures.

2.9. Origin

The origin and age of the family *Hepadnaviridae* remains controversial. However, until the issues with the estimation of the substitution rate of HBV [41] are overcome, the debate on the origin of HBV will continue ([17, 41] and references cited therein). Nonetheless, bioinformatics, coupled with growing number of hepadnaviral sequences in the databases, with accurate sampling times, and advances in phylogenetic and coalescent methodology [42], is beginning to shed light on this issue. For example, according to Suh and colleagues [43], analysis of the endogenous sequences in the zebra finch provides direct evidence that the compact genomic organization of hepadnaviruses has not changed during the last 482 million years of hepadnaviral evolution. Furthermore, phylogenetic analyses and distribution of HBV relics suggest that birds potentially are the ancestral hosts of the family *Hepadnaviridae* and that mammalian hepatitis B viruses probably emerged after a bird–mammal host switch [43].

2.10. Evolution

Genetic variation is important in viral evolution. The sequence heterogeneity displayed by HBV because of the lack of proof-reading ability of the polymerase is limited by functional constraints [33], leading to non-random variation [44]. Moreover, mutations can be affected by host–virus interaction and selective pressure, imposed endogenously by the immune system and exogenously by vaccination and antiviral treatment [17]. Phenotypic resistance to antiviral drugs occurs because of mutations in the reverse transcriptase of POL, whereas mutations in the *BCP/preC* and *preS* regions have been implicated as risk factors for the development of HCC. Mutations in the *S* region coding for HBsAg can lead to both vaccine and detection escape of HBV. At any time, the virus population can be composed of a number of different mutants referred to as "quasispecies" [45]. Direct sequencing and more recently next generation sequencing (NGS), parallel with bioinformatics, provide us with powerful tools to study the evolution of the various HBV mutations. NGS or ultra-deep sequencing generates large volumes of data, which can only be analyzed using bioinformatics tools and provides large coverage that can detect minor quasispecies populations of HBV [46–51] that may be important in understanding HBV pathogenicity and response to treatment. In order to minimize the number of artifactual calls of single-nucleotide variations in NGS, it is important that the correct reference sequences are used [51, 52].

By designing a circular construct, Homs and co-workers [53] were able to use NGS to study evolution of both the precore and polymerase regions. They demonstrated the presence of precore mutants in HBeAg-positive phase, wild-type precore in the HBeAg-negative phase as well as lamivudine resistance strains in treatment naïve patients. This demonstrates that viral strains occurring at low frequencies can act as reservoirs or memory genomes, which are selected and evolve in response to both intrinsic (host immune response) and extrinsic (drug administration) factors.

2.11. Transmission and tracing human migrations

Sequencing and bioinformatics have played an important role in demonstrating transmission routes, for which previous evidence could only be anecdotal. For example, molecular characterization of HBV together with phylogenetic analysis was used to demonstrate inter-spousal transmission of HBV even after long marriages, in two Japanese patients, who developed acute liver failure [54]. Similarly, the first known case of transfusion-transmitted HBV infection by blood screened using individual donor nucleic acid testing was confirmed by the 99.7% sequence homology between the complete genome sequences of the donor and the recipient HBV strains [26]. When migration events were estimated by ancestral state reconstruction using the criterion of parsimony, it was shown that Africa was the most probable source of dispersal of subgenotype A1 of HBV globally and its dispersal to Asia and Latin America occurred as a result of the slave and trade routes [23, 55].

2.12. Treatment response and resistance to treatment

According to international chronic hepatitis B treatment guidelines, the most desirable endpoint of treatment is HBsAg loss. Following HBsAg loss, patients have better clinical outcomes, including decreased risk of developing cirrhosis and HCC, and death [56]. However, the currently available treatments, which include either nucleos(t)ide analogues (NAs) for direct inhibition of the viral polymerase or pegylated interferon (PegIFN) for immune-mediated HBV control, generally achieve HBV DNA suppression and HBeAg loss only, which are not enduring. In an attempt to identify viral factors associated with HBsAg loss, Charuworn et al. [57] demonstrated that viral diversity could differentiate those patients, who would lose HBsAg when treated with tenofovir disoproxil fumarate. Lower diversity was seen in the protein-encoding regions of HBV from patients who lost HBsAg compared to those who did not. On the other hand, higher diversity in regulatory elements of HBV was found to be a predictor of HBsAg loss [57]. These findings need to be confirmed by studies incorporating larger numbers of patients, as well as genotypes other than A and D.

The high mutation rate of HBV means that it can evolve to develop resistance against NAs that target the viral DNA polymerase. Drug-resistant mutants develop under drug pressure in order for HBV to survive in the presence of the NA. The development of drug resistance mutations can be affected by HBV DNA levels at baseline, rate of viral suppression, length of NA treatment, and prior exposure to NA treatment [58]. Sequential treatment with different NAs, following drug failure, can lead to the development of multidrug resistance, which cannot be treated using currently available drugs [59]. The most frequent lamivudine drug resistance mutants are rtM204V/I, which are also selected by the L-pyrimidine analogues, emitricitabine, clevudine, and telbivudine but are susceptible to the purine analogues adefovir and tenofovir [59]. rtA181V develops following lamivudine treatment but is sensitive to other NAs, whereas rtN236T is resistant to adefovir only. In deciding on treatment options, the detection of genotypic resistance, which is defined as the detection of viral mutations conferring drug resistance, is a priority in clinics. Direct and NGS of the polymerase region of the HBV genome can detect both well-defined and novel mutations.

Bioinformatics tools and databases have been used to better understand HBV mutations and how they develop, especially in response to antiviral therapy and vaccination. Although laboratory methods have been used to study mutations, they are both labor intensive and expensive and limited in the degree of complexity they can investigate. As a more economical alternative, bioinformatics and computer simulation can use available biological data, such as the protein sequence and structural information, to investigate interactions by virus, host, and the environment [60]. Thus, Shen et al. [60] showed that most mutations develop in the hydrophobic regions of HBsAg and POL and that the amino acids that are more likely to be mutated are serine and threonine [60]. Understanding how amino acids mutations develop in HBV proteins can facilitate the rational design of both vaccines and drugs [60], for the prevention and treatment of HBV infection, respectively. By the use of bioinformatics to compare viral and host genomic patterns, together with clinical information, to data from databases can lead to enhanced and individualized antiviral therapy.

3. Bioinformatics tools and databases

3.1. Bioinformatics challenges of HBV

Despite its small genome size of ~3.2 kb, HBV presents several bioinformatic challenges:

1. The genome is circular, with position 1 conventionally taken to be the first "T" nucleotide in the *Eco*R1 restriction site ("GAA<u>T</u>TC"). Historically, position 1 was the start of the "Core" region, which is position 1901 in the current numbering system. Therefore, a number of sequences deposited earlier in the public databases are numbered using this outdated system and thus require processing before they can be used in alignments, together with more recently submitted sequences.

2. Four overlapping reading frames are encoded in the circular genome, whereas nucleotides or amino acids are sequenced and processed linearly. Extracting nucleotide or amino acid

sequences for the *S* and *POL* ORFs, which span the *EcoRI* site, from full-length or subgenomic fragments, requires additional processing.

3. The differences in genome lengths between the nine HBV genotypes (ranging from 3182 to 3248 base pairs in length) mean that direct comparison of loci between genotypes is not always possible using the current numbering system. These differences in genome lengths result in genotype alignments containing several regions of gaps, ranging from 3 to 33 nucleotides in length. A possible solution is the implementation of a standardized "universal numbering system" for all HBV genotypes, which we are currently developing.

4. Sequence variability is a feature of HBV. It is, therefore, essential to check all sequences carefully, to distinguish between artifacts and true variation (mutations). Variation within a population at a locus may result in two overlapping peaks on a chromatogram. Superinfections or co-infections with different strains may result in mixed populations, which appear as multiple or misaligned peaks on sequencing chromatograms. Disambiguating these is essential for robust downstream analyses.

3.2. Public sequence databases

The first public sequence database, "GenBank," was established in 1982, having arisen from the earlier Los Alamos database, established in 1979 [61, 62]. Since then, the number of nucleotides in GenBank has doubled approximately every 18 months [63]. The International Nucleotide Sequence Database Collaboration (INSDC) is a collection of three publicly available nucleotide (DNA or RNA) sequence databases, which synchronize data daily [64]. The collection consists of the DNA DataBank of Japan (DDBJ, located in Japan), the European Molecular Biology Laboratory (EMBL, located in the United Kingdom) and GenBank (located in the United States of America). The latest release of the database (release 211.0, from 15 December, 2015; [65]) contains 189,232,925 loci and 203,939,111,071 bases, from 189,232,925 sequences, totaling approximately 742 gigabytes. In addition to the INSDC, many other databases exist, including genome databases, protein sequence, structure and interaction databases, microarray databases, and meta-databases. A list of biological databases on Wikipedia includes over 200 entries [66].

When searching for "hepatitis b virus" across all fields, the GenBank database [63], accessed on 27th January 2016, contained 105,745 sequences. When searching for "hepatitis b virus" in the "organism" field only, 84,119 sequences were found, with the oldest sequence submitted in the early 1980s. Refining this search to include only sequences of 200 nucleotides or longer, and excluding words such as "recombinant," "clone," and "patent," resulted in 68,762 sequences. When this same query was previously executed on 29 November 2015, 67,893 sequences were returned. Therefore, in the 59 days between the two queries, 869 new sequences (of at least 200 nucleotides in length, and not containing the words mentioned previously) were uploaded to GenBank. On average, this equates to almost 15 new HBV sequences added to GenBank per day.

Making use of these sequences in downstream applications, such as multiple sequence alignments or phylogenetic analyses, is often challenging, as it is difficult to query for sufficient sequences, of the correct genotype, or subgenotype, and covering the required genomic region. In order to overcome this limitation, we have developed a bioinformatics solution, whereby all sequences matching a query are downloaded, curated, and aligned. The algorithm developed allows for the generation of a multiple sequence alignment for each genotype, which contains all the available sequences matching the query and in their correct position and orientation [67].

3.3. Bioinformatics tools for HBV

Workflow	Tool name and description
Chromatograms	**Quality score analyzer**
	• *Plots chromatogram quality scores*
	Automatic contig generator tool
	• *Generates a contig from a forward and reverse chromatogram*
Alignment	**Automatic alignment clean-up tool**
	• *Eliminates "gap-columns" and disambiguate ambiguous bases*
	Mind the gap
	• *Splits FASTA file based on gap threshold per column*
Analysis	**Babylon**
	• *Extracts HBV protein sequences (ORFs)*
	Wild-type 2 × 2
	• *Calculates 2 × 2 wild-type/mutant contingency tables*
	Divergence calculator*
	• *Intra- and Inter-group divergence with custom groups*
	Rafael*
	• *Generates random subsets from an input FASTA file*
Serotyping	**HBV serotyper tool**
	• *Determines HBV serotype*
Phylogenetics	**Pipeline: TreeMail**
	• *Generates a phylogenetic tree*
GenBank Submission	**PadSeq**
	• *Places two HBV sequence fragments on a backbone template*

¶Table modified from Bell and Kramvis [68].
*Described for the first time here.

Table 2. List of the online tools developed and the workflow process at which each would be used¶.

A standard molecular biology laboratory workflow includes DNA extraction, polymerase chain reaction (PCR) amplification, direct DNA sequencing, viewing and checking of chromatograms, preparation of curated sequences, multiple sequence alignment, sequence analysis, serotyping, genotyping, phylogenetic analysis, and preparation of sequences for submission to the GenBank public sequence database [68]. Each of these steps presents data processing challenges, many of which have been addressed by the development of a suite of online tools (**Table 2**) [68].

Name	URL	Usage	References
Geno2Pheno [hbv]	http://www.genafor.org/g2p_hbv/index.php	Drug resistance mutations, escape mutant analysis	[69]
HBV Blast Search	http://www.bioafrica.net/blast/hbvblast.html	Genotyping, drug resistance database	[70]
HBV STAR	http://www.vgb.ucl.ac.uk/starn.shtml	Genotyping	[71]
HBVRegDB	http://lancelot.otago.ac.nz/HBVRegDB/	Annotation, alignments, information about conserved regions	[72]
HBVdb	http://hbvdb.ibcp.fr/	Genotyping, annotation, drug resistance database	[73]
HBVseq	http://hivdb.stanford.edu/HBV/HBVseq/ development/HBVseq.html		
Hepatitis virus database	http://s2as02.genes.nig.ac.jp/	Genotyping, sequence alignment and map viewing	
Hepatitis virus database	http://www.ibibiobase.com/projects/hepatitis/ index.htm		
NCBI genotyping tool	http://www.ncbi.nlm.nih.gov/projects/ genotyping/	Genotyping	[74]
Oxford HBV subtyping tool	http://www.bioafrica.net/rega-genotype/html/ subtypinghbv.html	Genotyping	[70]
RegaDB	http://regaweb.med.kuleuven.be/software/regadb		
SeqHepB	http://www.seqhepb.com/	Sequence analysis, genotyping, detection of clinically important mutations	[75]

¶Table modified from [67].

Table 3. Currently available HBV websites and databases¶.

Any operating system platform from any location with an internet connection can be used to access stand-alone, web-based tools. There is no requirement to install and learn new bioinformatics software, as these tools can be used when required. A system for processing ultra-deep pyrosequencing (amplicon resequencing) data has also been developed [51]. In addition, a number of HBV-specific websites and databases are currently available, a selection of which are represented in **Table 3**.

3.4. New bioinformatics tools for HBV

Here, we present two newly developed tools for the bioinformatic analysis of HBV.

3.4.1. Divergence calculator [http://hvdr.bioinf.wits.ac.za/divergence/]

One method of classifying HBV sequences into genotype or subgenotype is to examine nucleotide sequence divergence between sequences. This divergence calculation is performed by totaling the number of nucleotides, which differ, between two aligned sequences and computing the percentage difference. The divergence calculator (**Figure 3**) performs various divergence calculations on groups of sequences from nucleotide or amino acid multiple sequence alignments in FASTA format. A minimum of one group containing two sequences, or two groups containing one sequence each, must be specified.

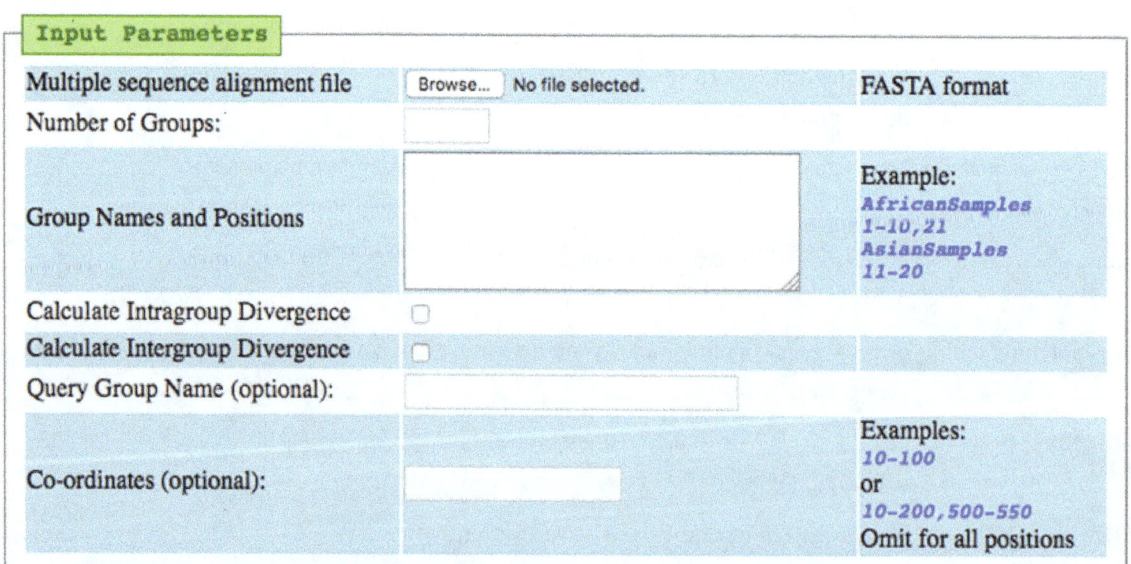

Figure 3. The input screen of the divergence calculator in which sequences are extracted and allocated to groups and other parameters specified.

As an example, consider an alignment of 10 genotype A sequences (group 1) and 10 genotype D sequences (group 2). Intra-group divergence, for each group, is calculated by comparing each sequence in group 1 with each other sequence in group 1 and then calculating the median, mean, and standard deviation of the divergences. This is then repeated for group 2. The inter-group divergence compares each sequence in group 1 with each sequence in group 2, and then calculates the median, mean, and standard deviation. If more than two groups are specified, the calculations iterate over all groups in turn.

If the optional "query" group is specified, the tool compares each sequence in the query group with each sequence in the other group or groups, but outputs statistics for each sequence in the query group individually. This method would typically be used with a set of unknown query sequences and one or more groups of reference sequences. A comprehensive list of descriptive statistics is included on the output page for each analysis.

3.4.2. *Random FASTA extraction and allocation (RAFAEL) [http://hvdr.bioinf.wits.ac.za/rafael/]*

In some analyses, particularly when constructing phylogenetic trees, it may be desirable to extract one or more random subsets of sequences from a master or reference alignment. The "RAFAEL" tool was designed to perform this task. This tool takes an input file in FASTA format, which does not have to be aligned and generates one or more subsets of the file, each containing a random selection of the specified number of sequences. The number of sequences may be specified as a count, or as a percentage of the number of sequences in the input file. There are guaranteed to be no duplicate sequences within each subset. However, duplicates may exist in multiple subsets, as subsets are not unique.

3.5. Open-source software

In addition to biological databases, a large variety of biological analysis software, which is generally genome agnostic, is available. As with software in any field, the licensing terms and commercial costs of these packages vary widely. Packages, which may be free of cost, may not necessarily be open-source, for example.

The Free Software Foundation (FSF) [76, 77] defines free software as software which "respects the users' freedom" in the sense that "users have the freedom to run, copy, distribute, study, change, and improve the software". As such, "free" is "a matter of liberty, not price". Free software, therefore, does not necessarily have to be made available at no cost or be a non-commercial project. Furthermore, software, which is provided at no cost, may not be "free" in the sense described above.

The term "open-source" is often used when referring to "free" software. However, the two terms are not synonymous, although there is some overlap. Open-source software may, or may not, be free software, depending on the restrictions placed on users by the software. If the user is not free to distribute, change, and improve the software, even if it is open source, then it cannot be considered to be free software. Most software, for which a license is purchased, is not free, or open source. The user does not have the freedom to distribute the software, or to use it on any computer chosen.

3.6. Recommended software

A list of recommended freely available download software is presented in **Table 4**. Comprehensive lists of open-source bioinformatics software can be found elsewhere [78].

Software name	Software description	Website (http://)	Lin*	Mac*	Win*	References
Unipro UGene	Integrated bioinformatics suite	ugene.net	Yes	Yes	Yes	[79]
MEGA 6	Integrated bioinformatics suite; command-line version available	megasoftware.net	Emu	Yes	Yes	[80]
BioEdit	Multiple sequence alignment viewer and editor	www.mbio.ncsu.edu/bioedit/bioedit.html	No	No	Yes	
SeaView	Multiple sequence alignment viewer and editor, and molecular phylogenetics; opens GeneDoc "MSF" files	doua.prabi.fr/software/seaview	Yes	Yes	Yes	[81]
AliView	Multiple sequence alignment viewer and editor	www.ormbunkar.se/aliview	Yes	Yes	Yes	[82]
GeneDoc	Multiple sequence alignment viewer and editor, and shading utility	iubio.bio.indiana.edu/soft/molbio/ibmpc/genedoc-readme.html	No	No	Yes	
PHYLIP	Programs for interring phylogenies; website includes comprehensive list of other phylogenetics programs	evolution.genetics.washington.edu/phylip.html	Yes	Yes	Yes	[83]
MrBayes	Bayesian inference and model choice using Markov Chain Monte Carlo	mrbayes.sourceforge.net	Com	Yes	Yes	[84]
BEAST	Bayesian analysis of molecular sequences using Markov Chain Monte Carlo	beast.bio.ed.ac.uk	Yes	Yes	Yes	[85]
FigTree	Graphical viewer and editor of phylogenetic trees	tree.bio.ed.ac.uk/software/figtree/	Yes	Yes	Yes	
Archaeopteryx	Graphical viewer and editor of phylogenetic trees	sites.google.com/site/cmzmasek/home/software/archaeopteryx	Yes	No	Yes	[86]
EMBOSS	A suite of command-line tools for molecular biology	emboss.sourceforge.net/	Yes	Yes	Emu	
JalView	Multiple sequence alignment viewer and editor	www.jalview.org	Yes	Yes	Yes	[87]
Finch TV	DNA sequence trace (chromatogram) viewer	www.geospiza.com/Products/finchtv.shtml	Yes	Yes	Yes	

Software name	Software description	Website (http://)	Lin*	Mac*	Win*	References
Chromas	DNA sequence trace (chromatogram) viewer	technelysium.com.au/?page_id=13	No	No	Yes	

*"GUI" = graphical user interface, "CL" = command line interface, "OSS" = open-source software, "Lin" = GNU/Linux, "Mac" = Apple MacIntosh, "Win" = Microsoft Windows, "Emu" = emulator or virtual machine recommended by authors, "Com" = compilation from source code required.

Table 4. Bioinformatics software available free of charge for various computer operating system platforms.

4. Conclusion

The unique genome structure and molecular biology of HBV pose a number of challenges, and thus, the development of bioinformatic tools has facilitated a more comprehensive and detailed analysis and understanding of the origin, evolution, transmission, and response to antiviral agents of HBV and its interaction with the host. There are a wide range of free and commercially available tools, which have been developed for different applications. The availability and applications of high-throughput sequencing techniques and the advancement of "-omics" will continue to provide additional challenges, which will need to be addressed by further computational solutions.

Acknowledgements

Trevor Bell is the recipient of a National Research Foundation (NRF) Scarce Skills Post-Doctoral Fellowship (GUN#86215) and Anna Kramvis received funding from the National Research Foundation (GUN#65530, GUN#93516).

Author details

Trevor Graham Bell* and Anna Kramvis

*Address all correspondence to: TrevorGrahamBell@gmail.com

Hepatitis Virus Diversity Research Unit, Faculty of Health Sciences, University of the Witwatersrand, Johannesburg, South Africa

References

[1] Qi H, Wu NC, Du Y, Wu TT, Sun R. High-resolution genetic profile of viral genomes: why it matters. Curr Opin Virol. 2015;14:62–70. PubMed PMID: 26364133.

[2] Hepatitis B Fact sheet no. 204 [Internet]. 2015. Available from: http://www.who.int/mediacentre/factsheets/fs204/en/[Accessed: 2016-01-16]

[3] Ott JJ, Stevens GA, Groeger J, Wiersma ST. Global epidemiology of hepatitis B virus infection: new estimates of age-specific HBsAg seroprevalence and endemicity. Vaccine. 2012;30(12):2212–9. PubMed PMID: 22273662. Epub 2012/01/26. eng.

[4] Tiollais P, Pourcel C, Dejean A. The hepatitis B virus. Nature. 1985;317(6037):489–95. PubMed PMID: 2995835.

[5] Lucifora J, Arzberger S, Durantel D, Belloni L, Strubin M, Levrero M, et al. Hepatitis B virus X protein is essential to initiate and maintain virus replication after infection. J Hepatol. 2011;55(5):996–1003. PubMed PMID: 21376091.

[6] Kramvis A, Kew MC. The core promoter of hepatitis B virus. J Viral Hepat. 1999;6(6): 415–27. PubMed PMID: 10607259. Epub 1999/12/22. eng.

[7] Moolla N, Kew M, Arbuthnot P. Regulatory elements of hepatitis B virus transcription. J Viral Hepat. 2002;9(5):323–31. PubMed PMID: 12225325. Epub 2002/09/13. eng.

[8] Summers J, Mason WS. Replication of the genome of a hepatitis B-like virus by reverse transcription of an RNA intermediate. Cell. 1982;29(2):403–15. PubMed PMID: 6180831. Epub 1982/06/01. eng.

[9] Yan H, Zhong G, Xu G, He W, Jing Z, Gao Z, et al. Sodium taurocholate cotransporting polypeptide is a functional receptor for human hepatitis B and D virus. eLife. 2012;1:e00049. PubMed PMID: 23150796. Pubmed Central PMCID: 3485615.

[10] Rabe B, Glebe D, Kann M. Lipid-mediated introduction of hepatitis B virus capsids into nonsusceptible cells allows highly efficient replication and facilitates the study of early infection events. J Virol. 2006;80(11):5465–73. PubMed PMID: 16699026. Pubmed Central PMCID: 1472160.

[11] Kock J, Rosler C, Zhang JJ, Blum HE, Nassal M, Thoma C. Generation of covalently closed circular DNA of hepatitis B viruses via intracellular recycling is regulated in a virus specific manner. Plos Pathogens. 2010;6(9):e1001082. PubMed PMID: 20824087. Pubmed Central PMCID: 2932716.

[12] Beck J, Nassal M. Hepatitis B virus replication. World J Gastroenterol. 2007;13(1):48–64. PubMed PMID: 17206754. Pubmed Central PMCID: 4065876.

[13] Kramvis A, Kew MC. Structure and function of the encapsidation signal of hepadnaviridae. J Viral Hepat. 1998;5(6):357–67. PubMed PMID: 9857345. Epub 1998/12/19. eng.

[14] Kramvis A, Kew M, Francois G. Hepatitis B virus genotypes. Vaccine. 2005;23(19):2409–23. PubMed PMID: 15752827. Epub 2005/03/09. eng.

[15] Wang GH, Seeger C. Novel mechanism for reverse transcription in hepatitis B viruses. J Virol. 1993;67(11):6507–12. PubMed PMID: 7692081. Pubmed Central PMCID: 238087.

[16] Glebe D, Bremer CM. The molecular virology of hepatitis B virus. Semin Liver Dis. 2013;33(2):103–12. PubMed PMID: 23749666.

[17] Steinhauer DA, Holland JJ. Direct method for quantitation of extreme polymerase error frequencies at selected single base sites in viral RNA. J Virol. 1986;57(1):219–28. PubMed PMID: 3001347. Pubmed Central PMCID: 252718. Epub 1986/01/01. eng.

[18] Yu H, Yuan Q, Ge SX, Wang HY, Zhang YL, Chen QR, et al. Molecular and phylogenetic analyses suggest an additional hepatitis B virus genotype "I". Plos One. 2010;5(2):e9297. PubMed PMID: 20174575. Pubmed Central PMCID: 2824819. Epub 2010/02/23. eng.

[19] Norder H, Courouce AM, Coursaget P, Echevarria JM, Lee SD, Mushahwar IK, et al. Genetic diversity of hepatitis B virus strains derived worldwide: genotypes, subgenotypes, and HBsAg subtypes. Intervirology. 2004;47(6):289–309. PubMed PMID: 15564741. Epub 2004/11/27. eng.

[20] Tatematsu K, Tanaka Y, Kurbanov F, Sugauchi F, Mano S, Maeshiro T, et al. A genetic variant of hepatitis B virus divergent from known human and ape genotypes isolated from a Japanese patient and provisionally assigned to new genotype J. J Virol. 2009;83(20):10538–47. PubMed PMID: 19640977. Pubmed Central PMCID: 2753143. Epub 2009/07/31. eng.

[21] Kramvis A. Genotypes and genetic variability of hepatitis B virus. Intervirology. 2014;57(3–4):141–50. PubMed PMID: 25034481. Epub 2014/07/19. eng.

[22] Kramvis A, Kew MC. Relationship of genotypes of hepatitis B virus to mutations, disease progression and response to antiviral therapy. J Viral Hepat. 2005;12(5):456–64. PubMed PMID: 16108759. Epub 2005/08/20. eng.

[23] Kramvis A, Paraskevis D. Subgenotype A1 of HBV—tracing human migrations in and out of Africa. Antivir Ther. 2013;18(3 Pt B):513–21. PubMed PMID: 23792935. Epub 2013/06/26. eng.

[24] Osiowy C, Kaita K, Solar K, Mendoza K. Molecular characterization of hepatitis B virus and a 9-year clinical profile in a patient infected with genotype I. J Med Virol. 2010;82(6): 942–8. PubMed PMID: 20419807. Epub 2010/04/27. eng.

[25] Wiegand J, Hasenclever D, Tillmann HL. Should treatment of hepatitis B depend on hepatitis B virus genotypes? A hypothesis generated from an explorative analysis of published evidence. Antivir Ther. 2008;13(2):211–20. PubMed PMID: 18505172.

[26] Vermeulen M, Dickens C, Lelie N, Walker E, Coleman C, Keyter M, et al. Hepatitis B virus transmission by blood transfusion during 4 years of individual-donation nucleic acid testing in South Africa: estimated and observed window period risk. Transfusion. 2012;52:880–92. PubMed PMID: 21981386. Epub 2011/10/11. Eng.

[27] Martinot-Peignoux M, Marcellin P. Virological and serological tools to optimize the management of patients with chronic hepatitis B. Liver Int. 2016;36 Suppl 1:78–84. PubMed PMID: 26725902.

[28] Bartholomeusz A, Schaefer S. Hepatitis B virus genotypes: comparison of genotyping methods. Rev Med Virol. 2004;14(1):3–16. PubMed PMID: 14716688.

[29] Guirgis BS, Abbas RO, Azzazy HM. Hepatitis B virus genotyping: current methods and clinical implications. Int J Infect Dis IJID. 2010;14(11):e941–53. PubMed PMID: 20674432.

[30] Kramvis A, Arakawa K, Yu MC, Nogueira R, Stram DO, Kew MC. Relationship of serological subtype, basic core promoter and precore mutations to genotypes/subgenotypes of hepatitis B virus. J Med Virol. 2008;80(1):27–46. PubMed PMID: 18041043. Epub 2007/11/28. eng.

[31] Hu X, Margolis HS, Purcell RH, Ebert J, Robertson BH. Identification of hepatitis B virus indigenous to chimpanzees. Proc Natl Acad Sci USA. 2000;97(4):1661–4. PubMed PMID: 10677515. Pubmed Central PMCID: 26492. Epub 2000/03/04. eng.

[32] Norder H, Courouce AM, Magnius LO. Complete genomes, phylogenetic relatedness, and structural proteins of six strains of the hepatitis B virus, four of which represent two new genotypes. Virology. 1994;198(2):489–503. PubMed PMID: 8291231. Epub 1994/02/01. eng.

[33] Mizokami M, Orito E, Ohba K, Ikeo K, Lau JY, Gojobori T. Constrained evolution with respect to gene overlap of hepatitis B virus. J Mol Evol. 1997;44 Suppl 1:S83–90. PubMed PMID: 9071016. Epub 1997/01/01. eng.

[34] Torres C, Fernandez MD, Flichman DM, Campos RH, Mbayed VA. Influence of overlapping genes on the evolution of human hepatitis B virus. Virology. 2013;441(1): 40–8. PubMed PMID: 23541083. Epub 2013/04/02. eng.

[35] Tedder RS, Bissett SL, Myers R, Ijaz S. The 'Red Queen' dilemma–running to stay in the same place: reflections on the evolutionary vector of HBV in humans. Antivir Ther. 2013;18(3 Pt B):489–96. PubMed PMID: 23792884. Epub 2013/06/26. eng.

[36] Andernach IE, Hunewald OE, Muller CP. Bayesian Inference of the Evolution of HBV/E. Plos One. 2013;8(11):e81690. PubMed PMID: 24312336. Pubmed Central PMCID: 3843692. Epub 2013/12/07. eng.

[37] Fares MA, Holmes EC. A revised evolutionary history of hepatitis B virus (HBV). J Mol Evol. 2002;54(6):807–14. PubMed PMID: 12029362. Epub 2002/05/25. eng.

[38] Orito E, Mizokami M, Ina Y, Moriyama EN, Kameshima N, Yamamoto M, et al. Host-independent evolution and a genetic classification of the hepadnavirus family based on nucleotide sequences. Proc Natl Acad Sci USA. 1989;86(18):7059–62. PubMed PMID: 2780562. Pubmed Central PMCID: 297993. Epub 1989/09/01. eng.

[39] Gunther S, Sommer G, Von Breunig F, Iwanska A, Kalinina T, Sterneck M, et al. Amplification of full-length hepatitis B virus genomes from samples from patients with low levels of viremia: frequency and functional consequences of PCR-introduced

mutations. J Clin Microbiol. 1998;36(2):531–8. PubMed PMID: 9466771. Pubmed Central PMCID: 104572. Epub 1998/02/18. eng.

[40] Paraskevis D, Magiorkinis G, Magiorkinis E, Ho SY, Belshaw R, Allain JP, et al. Dating the origin and dispersal of hepatitis B virus infection in humans and primates. Hepatology. 2013;57(3):908–16. PubMed PMID: 22987324. Epub 2012/09/19. eng.

[41] Bouckaert R, Alvarado-Mora MV, Pinho JR. Evolutionary rates and HBV: issues of rate estimation with Bayesian molecular methods. Antivir Ther. 2013;18(3 Pt B):497–503. PubMed PMID: 23792904. Epub 2013/06/26. eng.

[42] Alvarado Mora MV, Romano CM, Gomes-Gouvea MS, Gutierrez MF, Botelho L, Carrilho FJ, et al. Molecular characterization of the Hepatitis B virus genotypes in Colombia: a Bayesian inference on the genotype F. Infect Genet Evol. 2011;11(1):103–8. PubMed PMID: 20951841. Epub 2010/10/19. eng.

[43] Suh A, Brosius J, Schmitz J, Kriegs JO. The genome of a Mesozoic paleovirus reveals the evolution of hepatitis B viruses. Nat Commun. 2013;4:1791. PubMed PMID: 23653203.

[44] Yang Z, Lauder IJ, Lin HJ. Molecular evolution of the hepatitis B virus genome. J Mol Evol. 1995;41(5):587–96. PubMed PMID: 7490773.

[45] Domingo E, Sabo D, Taniguchi T, Weissmann C. Nucleotide sequence heterogeneity of an RNA phage population. Cell. 1978;13(4):735–44. PubMed PMID: 657273.

[46] Buti M, Tabernero D, Mas A, Homs M, Prieto M, Rodriguez-Frias F, et al. Hepatitis B virus quasispecies evolution after liver transplantation in patients under long-term lamivudine prophylaxis with or without hepatitis B immune globulin. Transpl Infect Dis. 2015;17(2):208–20. PubMed PMID: 25641570.

[47] Homs M, Buti M, Tabernero D, Quer J, Sanchez A, Corral N, et al. Quasispecies dynamics in main core epitopes of hepatitis B virus by ultra-deep-pyrosequencing. World J Gastroenterol. 2012;18(42):6096–105. PubMed PMID: 23155338. Pubmed Central PMCID: 3496886.

[48] Homs M, Caballero A, Gregori J, Tabernero D, Quer J, Nieto L, et al. Clinical application of estimating hepatitis B virus quasispecies complexity by massive sequencing: correlation between natural evolution and on-treatment evolution. Plos One. 2014;9(11):e112306. PubMed PMID: 25393280. Pubmed Central PMCID: 4231103.

[49] Ramirez C, Gregori J, Buti M, Tabernero D, Camos S, Casillas R, et al. A comparative study of ultra-deep pyrosequencing and cloning to quantitatively analyze the viral quasispecies using hepatitis B virus infection as a model. Antivir Res. 2013;98(2):273–83. PubMed PMID: 23523552.

[50] Rodriguez-Frias F, Buti M, Tabernero D, Homs M. Quasispecies structure, cornerstone of hepatitis B virus infection: mass sequencing approach. World J Gastroenterol.

2013;19(41):6995–7023. PubMed PMID: 24222943. Pubmed Central PMCID: 3819535. Epub 2013/11/14. Eng.

[51] Yousif M, Bell TG, Mudawi H, Glebe D, Kramvis A. Analysis of ultra-deep pyrosequencing and cloning based sequencing of the basic core promoter/precore/core region of hepatitis B virus using newly developed bioinformatics tools. Plos One. 2014;9(4):e95377. PubMed PMID: 24740330. Epub 2014/04/18. eng.

[52] Liu WC, Lin CP, Cheng CP, Ho CH, Lan KL, Cheng JH, et al. Aligning to the sample-specific reference sequence to optimize the accuracy of next-generation sequencing analysis for hepatitis B virus. Hepatol Int. 2016;10(1):147–57. PubMed PMID: 26208819. Pubmed Central PMCID: 4722079.

[53] Homs M, Buti M, Quer J, Jardi R, Schaper M, Tabernero D, et al. Ultra-deep pyrosequencing analysis of the hepatitis B virus preCore region and main catalytic motif of the viral polymerase in the same viral genome. Nucleic Acids Res. 2011;39(19):8457–71. PubMed PMID: 21742757. Pubmed Central PMCID: 3201856. Epub 2011/07/12. eng.

[54] Okamoto D, Nakayama H, Ikeda T, Ikeya S, Nagashima S, Takahashi M, et al. Molecular analysis of the interspousal transmission of hepatitis B virus in two Japanese patients who acquired fulminant hepatitis B after 50 and 49 years of marriage. J Med Virol. 2014;86(11):1851–60. PubMed PMID: 25132075.

[55] Lago BV, Mello FC, Kramvis A, Niel C, Gomes SA. Hepatitis B virus subgenotype A1: evolutionary relationships between Brazilian, African and Asian Isolates. Plos One. 2014;9(8):e105317. PubMed PMID: 25122004. Pubmed Central PMCID: 4133366. Epub 2014/08/15. eng.

[56] Liu J, Yang HI, Lee MH, Lu SN, Jen CL, Batrla-Utermann R, et al. Spontaneous seroclearance of hepatitis B seromarkers and subsequent risk of hepatocellular carcinoma. Gut. 2014;63(10):1648–57. PubMed PMID: 24225939.

[57] Charuworn P, Hengen PN, Aguilar Schall R, Dinh P, Ge D, Corsa A, et al. Baseline interpatient hepatitis B viral diversity differentiates HBsAg outcomes in patients treated with tenofovir disoproxil fumarate. J Hepatol. 2015;62(5):1033–9. PubMed PMID: 25514556.

[58] Gish RG, Given BD, Lai CL, Locarnini SA, Lau JY, Lewis DL, et al. Chronic hepatitis B: virology, natural history, current management and a glimpse at future opportunities. Antivir Res. 2015;121:47–58. PubMed PMID: 26092643.

[59] Zoulim F, Durantel D, Deny P. Management and prevention of drug resistance in chronic hepatitis B. Liver Int. 2009;29 Suppl 1:108–15. PubMed PMID: 19207973. Epub 2009/02/12. eng.

[60] Shen K, Shen L, Wang J, Jiang Z, Shen B. Understanding amino acid mutations in hepatitis B virus proteins for rational design of vaccines and drugs. Adv Protein Chem Struct Biol. 2015;99:131–53. PubMed PMID: 26067819.

[61] Kanehisa M, Fickett JW, Goad WB. A relational database system for the maintenance and verification of the Los Alamos sequence library. Nucleic Acids Res. 1984;12(1 Pt 1): 149–58. PubMed PMID: 6694899. Pubmed Central PMCID: 320992.

[62] Bilofsky HS, Burks C, Fickett JW, Goad WB, Lewitter FI, Rindone WP, et al. The GenBank genetic sequence databank. Nucleic Acids Res. 1986;14(1):1–4. PubMed PMID: 3945546. Pubmed Central PMCID: 339347.

[63] Benson DA, Clark K, Karsch-Mizrachi I, Lipman DJ, Ostell J, Sayers EW. GenBank. Nucleic Acids Res. 2015;43(Database issue):D30–5. PubMed PMID: 25414350. Pubmed Central PMCID: 4383990.

[64] Karsch-Mizrachi I, Nakamura Y, Cochrane G, International Nucleotide Sequence Database C. The international nucleotide sequence database collaboration. Nucleic Acids Res. 2012;40(Database issue):D33–7. PubMed PMID: 22080546. Pubmed Central PMCID: 3244996.

[65] Genetic Sequence Data Bank [Internet]. 2016. Available from: ftp://ftp.ncbi.nih.gov/genbank/gbrel.txt [Accessed: 2016-01-27]

[66] List of biological databases [Internet]. 2016. Available from: https://en.wikipedia.org/wiki/List_of_biological_databases [Accessed: 2016-01-27]

[67] Bell, T.G., Yousif, Y., Kramvis, A. Bioinformatic curation and alignment of genotyped hepatitis B virus (HBV) sequence data from the GenBank public database. Submitted.

[68] Bell TG, Kramvis A. Bioinformatics tools for small genomes, such as hepatitis B virus. Viruses. 2015;7(2):781–97. PubMed PMID: 25690798. Epub 2015/02/19. eng.

[69] Beerenwinkel N, Daumer M, Oette M, Korn K, Hoffmann D, Kaiser R, et al. Geno2pheno: Estimating phenotypic drug resistance from HIV-1 genotypes. Nucleic Acids Res. 2003;31(13):3850–5. PubMed PMID: 12824435. Pubmed Central PMCID: 168981.

[70] de Oliveira T, Deforche K, Cassol S, Salminen M, Paraskevis D, Seebregts C, et al. An automated genotyping system for analysis of HIV-1 and other microbial sequences. Bioinformatics. 2005;21(19):3797–800. PubMed PMID: 16076886.

[71] Myers R, Clark C, Khan A, Kellam P, Tedder R. Genotyping Hepatitis B virus from whole- and sub-genomic fragments using position-specific scoring matrices in HBV STAR. J Gen Virol. 2006;87(Pt 6):1459–64. PubMed PMID: 16690910.

[72] Panjaworayan N, Roessner SK, Firth AE, Brown CM. HBVRegDB: annotation, comparison, detection and visualization of regulatory elements in hepatitis B virus sequences. Virol J. 2007;4:136. PubMed PMID: 18086305. Pubmed Central PMCID: 2235840. Epub 2007/12/19. eng.

[73] Hayer J, Jadeau F, Deleage G, Kay A, Zoulim F, Combet C. HBVdb: a knowledge database for hepatitis B virus. Nucleic Acids Res. 2013;41(Database issue):D566–70. PubMed PMID: 23125365. Pubmed Central PMCID: 3531116.

[74] Rozanov M, Plikat U, Chappey C, Kochergin A, Tatusova T. A web-based genotyping resource for viral sequences. Nucleic Acids Res. 2004;32(Web Server issue):W654–9. PubMed PMID: 15215470. Pubmed Central PMCID: 441557.

[75] Yuen LK, Ayres A, Littlejohn M, Colledge D, Edgely A, Maskill WJ, et al. SeqHepB: a sequence analysis program and relational database system for chronic hepatitis B. Antivir Res. 2007;75(1):64–74. PubMed PMID: 17215050.

[76] Free Software Foundation [Internet]. 2016. Available from: http://www.fsf.org/ [Accessed: 2016-01-27]

[77] What is free software? [Internet]. 2016. Available from: http://www.gnu.org/philosophy/free-sw.html [Accessed: 2016-01-27]

[78] List of open-source bioinformatics software [Internet]. 2016. Available from: https://en.wikipedia.org/wiki/List_of_open-source_bioinformatics_software [Accessed: 2016-01-28]

[79] Okonechnikov K, Golosova O, Fursov M, team U. Unipro UGENE: a unified bioinformatics toolkit. Bioinformatics. 2012;28(8):1166–7. PubMed PMID: 22368248.

[80] Tamura K, Stecher G, Peterson D, Filipski A, Kumar S. MEGA6: Molecular Evolutionary Genetics Analysis version 6.0. Mol Biol Evol. 2013;30(12):2725–9. PubMed PMID: 24132122. Pubmed Central PMCID: 3840312.

[81] Gouy M, Guindon S, Gascuel O. SeaView version 4: a multiplatform graphical user interface for sequence alignment and phylogenetic tree building. Mol Biol Evol. 2010;27(2):221–4. PubMed PMID: 19854763.

[82] Larsson A. AliView: a fast and lightweight alignment viewer and editor for large datasets. Bioinformatics. 2014;30(22):3276–8. PubMed PMID: 25095880. Pubmed Central PMCID: 4221126.

[83] Phylogeny Programs [Internet]. 2016. Available from: http://evolution.genetics.washington.edu/phylip/software.html [Accessed: 2016-03-07]

[84] Huelsenbeck JP, Ronquist F. MRBAYES: Bayesian inference of phylogenetic trees. Bioinformatics. 2001;17(8):754–5. PubMed PMID: 11524383.

[85] Drummond AJ, Suchard MA, Xie D, Rambaut A. Bayesian phylogenetics with BEAUti and the BEAST 1.7. Mol Biol Evol. 2012;29(8):1969–73. PubMed PMID: 22367748. Pubmed Central PMCID: 3408070.

[86] Han MV, Zmasek CM. phyloXML: XML for evolutionary biology and comparative genomics. BMC Bioinform. 2009;10:356. PubMed PMID: 19860910. Pubmed Central PMCID: 2774328.

[87] Waterhouse AM, Procter JB, Martin DM, Clamp M, Barton GJ. Jalview Version 2—a multiple sequence alignment editor and analysis workbench. Bioinformatics. 2009;25(9):1189–91. PubMed PMID: 19151095. Pubmed Central PMCID: 2672624.

6

Databases and Algorithms in Allergen Informatics

Kiran Kadam, Sangeeta Sawant, V.K. Jayaraman and Urmila Kulkarni-Kale

Abstract

Allergic diseases are considered as one of the major health problems worldwide due to their increasing prevalence. Advancements in genomic, proteomic, and analytical techniques have resulted in considerable progress in the field of allergology, which has led to accumulation of huge amount of data. Allergen bioinformatics comprises allergen-related data resources and computational methods/tools, which deal with an efficient archival, management, and analysis of allergological data. Significant work has been done in the area of allergen bioinformatics that has proven pivotal for the development and progress of this field. In this chapter, we describe the current status of databases and algorithms, encompassing the field of allergen bioinformatics by examining work carried out thus far with respect to features such as allergens and allergenicity, allergen databases, algorithms/tools for allergen/allergenicity prediction, allergen epitope prediction, and allergenic cross-reactivity assessment. This chapter illustrates concepts and algorithms in allergen bioinformatics, as well as it outlines the key areas for potential development in allergology field.

Keywords: allergen databases, allergen epitope prediction, allergenicity, allergenic proteins, bioinformatics

1. Introduction

The immune system represents a very complex system comprising numerous biological molecules and processes, which combine to form body's defense against infectious agents and other threats. Immunity is basically divided into two types such as innate immunity and adaptive immunity [1]. Innate immunity also referred to as natural, native, or nonspecific immunity acts as a first line of defense against common harmful agents. Innate immune response provides immediate protection and involves number of components such as monocytes, macrophages, neutrophils, cytokines, complement, and epithelial barriers. Adaptive or acquired immunity comprises highly specific immune responses that are elicited against particular pathogens or antigens. These immune responses are either cell mediated or antibody mediated (humoral) and executed by specialized lymphocytes or immunoglobulins, respectively. On certain occasions, the immune system produces immune responses that are harmful for the host organism. Autoimmunity denotes one such case wherein the body elicits immune responses against its own cells and tissues (self-antigens) which lead to development of autoimmune diseases. In some cases, immune system produces inappropriate immune responses known as hypersensitivity, which has deleterious effects on the host organism. Hypersensitivity reactions are categorized into four groups based on the type of immune response and the effector mechanism involved. These are (i) immediate hypersensitivity (type I), (ii) antibody-mediated hypersensitivity (type II), (iii) immune complex–mediated hypersensitivity (type III), and (iv) cell-mediated hypersensitivity (type IV) [2].

Allergic reactions are type I hypersensitivity reactions, which are characterized by induction of specific class of antibodies known as immunoglobulin E (IgE). These reactions are elicited against specific type of antigens commonly referred to as allergens. An allergic reaction involves specialized cells and specific molecules of the immune system [3]. IgE antibodies induced by allergens upon allergic sensitization bind to effector cells such as basophils and mast cells via specific Fc receptors present on the surfaces of those cells. Subsequent exposure to the allergen causes cross-linking of membrane-bound IgE on these effector cells, which leads to their degranulation and release of pharmacologically active agents such as histamine. These pharmacological mediators are responsible for clinical manifestations of allergic reactions in the affected individuals. Immunogenicity in general refers to the potential of an antigen to elicit an immune response, while in case of allergens, allergenicity is considered as a reflection of its allergenic potential. Allergenicity indicates the capability of an allergen to induce clinical symptoms of allergy as well as to induce and bind to IgE antibodies [4]. The prevalence of allergic reactions has increased significantly in the last few years, especially in the developing countries [5]. This has resulted in considerable increase in disease burden as well as economic issues due to costs associated with these diseases. Therefore, the study of allergic diseases has gained tremendous importance as they represent one of the major health problems in urban and rural regions.

The field of allergy research has rapidly progressed in the last few years [6]. Recent advances in genomic, proteomic, and analytical methods have given rise to large amounts of data relevant to allergens. This data can be correlated with pathology of various allergic diseases based on experimental, clinical, and epidemiological data for allergic reactions. The continuous growth of data calls for the efficient archival, management, and analysis of data. This has led to the development of the field of allergen informatics, which comprises allergen specific databases/resources and computational methods/tools. Allergen informatics constitutes an important branch of immunoinformatics [7]. In this chapter, a review of the existing status of allergen informatics with respect to important aspects such as allergens and allergenicity, allergen databases, algorithms/tools for allergen/allergenicity prediction, allergen epitope prediction, and allergenic cross-reactivity assessment has been presented (**Figure 1**).

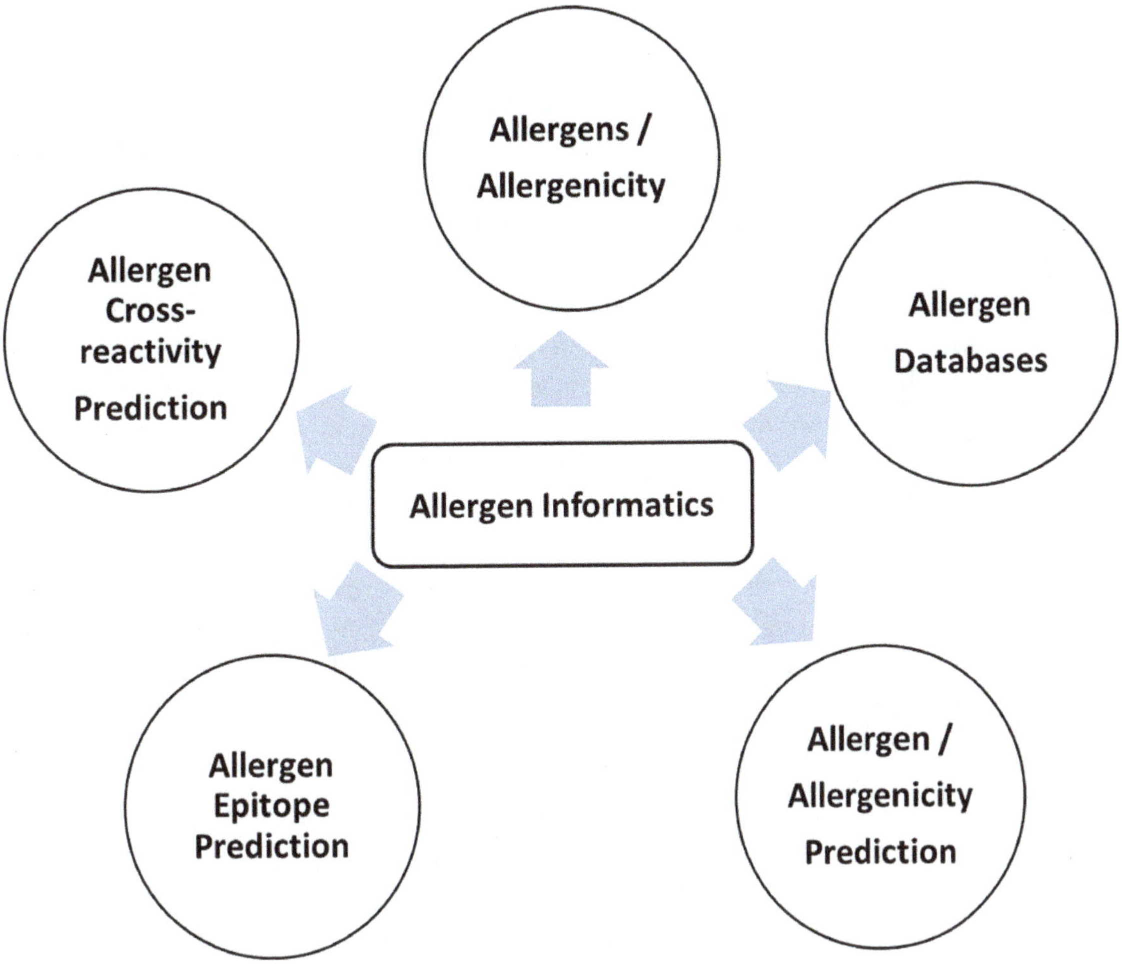

Figure 1. Topics covered for allergen informatics in the current review.

2. Allergens and allergenicity

Allergens represent the most critical component of an allergic reaction, although IgE antibody, Fc receptors, mast cells, and basophils as well as pharmacological mediators such as histamine and heparin also play very significant roles. Allergens are ubiquitous substances, which arise from a variety of sources such as foods, plants, animals, or environment. An allergen can either be a chemical substance (e.g., penicillin) or a protein (e.g., albumin, profilin, etc.). Majority of the allergens are proteins or glycoproteins that possess high water solubility. Several bio-chemical and structural features of allergens such as stability, hydrophobicity, and ligand-binding domains are known to contribute to their allergenicity [8]. However, common molecular and structural features of allergens that are responsible for allergenicity have not yet been conclusively discovered.

Allergens are provided with a unique, unambiguous, and systematic nomenclature which has been developed and maintained by the World Health Organization (WHO) and International Union of Immunological Societies' (IUIS) "Allergen Nomenclature Sub-committee" [9, 10]. The nomenclature is based on the Linnean system, and an allergen, which satisfies certain biochemical and immunological criteria, is included in the WHO/IUIS nomenclature. An allergen name consists of an abbreviation of the scientific name of the allergen source organism. First 3–4 letters denote the genus name, while the subsequent 1–2 letters represent species, followed by an Arabic numeral that denotes the order of its identification. For instance, Der p 1 represents the first allergen to be characterized from the house dust mite *Dermatophagoides pteronyssinus*. An allergen may possess isoallergens or isoforms/variants, which are considered as multiple molecular forms of the same allergen. The WHO/IUIS nomenclature defines isoallergen as an allergen belonging to a single species, with a similar molecular size and identical biological function, and possessing ≥67% amino acid sequence identity while a variant or isoform corresponds to allergen sequences that differ by only a limited number of amino acid substitutions [11]. It is very important to archive and study the data on isoallergens and isoforms/variants in a differentiated manner as it has been shown that variations in allergens significantly affect their allergenicity and cross-reactivity as well as influence recognition of epitopes by T cells and IgE [12]. An allergen can be considered as a major or minor allergen based on the measure of its allergenicity. Major allergens are the ones to which >50% of patients with an allergy to its source are sensitized, while minor allergens are recognized by a limited number of patients [13].

Allergens display important features such as epitopes and cross-reactivity that are very critical with respect to understanding of allergic reactions and developing newer approaches for diagnosis and treatment of allergic diseases. Epitope or antigenic determinant refers to the immunologically active region of the allergen. An epitope can be an IgE-binding epitope or a T-cell epitope depending on whether it interacts with an IgE or a T-lymphocyte. An IgE epitope can be either sequential (linear) that consists of contiguous stretch of amino acids or conformational (discontinuous), which comprises amino acids present at different loci in an antigen. An antibody is said to be cross-reactive when it recognizes and binds to multiple antigens.

2.1. IgE-binding epitopes

IgE-binding epitopes refer to the IgE recognition sites in allergens that are involved in specific interaction of allergens and IgE antibody. Inferences drawn from allergen–antibody complexes and other important studies have shown that majority of IgE-binding epitopes are conformational in nature [14]. IgE epitopes possess some defining structural and immunological features such as they are more cross-reactive in nature and have higher intrinsic flexibility. These features make them distinct from other antibody epitopes and contribute significantly in the allergenicity [15, 16]. Identification and in-depth analysis of IgE-binding epitopes has the potential to contribute immensely in accurate diagnosis and allergen-specific immunotherapy of allergies, especially the food allergy [17, 18]. Large amount of data on allergen epitopes are generated by employing strategies based on the use of overlapping synthetic peptides, recombinant allergenic fragments, cocrystal structure complexes, etc. However, it is believed that insights obtained from study of allergen–antibody complexes will be the most helpful in understanding the role these epitopes play in allergic reactions [19–21].

2.2. T-cell epitopes

T-cell epitopes are the antigenic determinants of allergens that interact with T-lymphocytes via specific T-cell receptors. T-cell epitopes of allergens have shown to be very important for the modulation of allergic response and thereby contributing to symptoms associated with allergic diseases [22]. They have enormous potential in the development of allergy vaccines as well as newer strategies in allergen immunotherapy, considering their fundamental role in allergic response [23, 24]. Recent findings have indicated that T-cell epitope repertoire in allergens is diverse than IgE epitopes, and it can be very useful in specific immunotherapy in allergy [25]. An analysis carried out on available epitope data has shown that T-cell epitopes are known to occur more commonly in the airborne allergens as compared to food allergens [26].

2.3. Cross-reactivity

Cross-reactivity denotes a clinically and immunologically critical phenomenon displayed by allergens from various sources and is the cause of pollen-food syndromes, such as the one seen in case of birch and apple. Cross-reactivity is considered as a property of antibodies and it arises when an antibody or a subgroup of antibodies recognizes more than one allergen or epitope [27]. Two allergens are considered cross-reactive if they are recognized by a single antibody (or T-cell receptor). It has been stated that cross-reactivity among allergens at the level of B cells, T cells, and mast cells reflects clinical sensitivities and contributes very significantly in the regulation of allergic sensitization [28].

Cross-reactivity is predominantly an antibody defined phenomenon and IgE antibodies are shown to be more cross-reactive in nature. Affinity of the antibodies toward the allergen is known to play an important role in cross-reactivity. However, the properties of the allergenic protein are also very important and shared features on the level of both primary and tertiary structures of the cross-reactive proteins are found to be responsible for cross-reactivity [4]. Similarity at the level of sequence is an important indicator and cross-reactivity seems to require more than 70% sequence identity. In addition to this, other factors such as the host immune response against the allergen, dosage of allergen, and mode of exposure also contribute in clinical relevance of allergic cross-reactivity. Inferences drawn from studying a large number of allergens have led to the conclusion that structural similarity among proteins from diverse sources is the molecular basis of allergic cross-reactivity [29]. Considering the role it plays in the development of allergic symptoms, a detailed analysis of cross-reactivity has the potential to contribute in the development of new strategies in diagnosis and therapy of allergic diseases.

3. Allergen databases

Last few years have witnessed substantial technological advances in the field of genomics and proteomics along with tremendous improvements in analytical methods. This has led to a significant progress in the area of allergy research. As a result of this, there has been a steady and continuous increase in the number of characterized protein allergens over the last few years. Efficient storage and management of data has become very important because of such incessant accumulation of molecular and clinical data on allergens. Therefore, allergen databases represent very crucial resources for basic allergy research as they are involved in archival of available allergen knowledge.

Database	Developed by (URL)	Type of data archived	Computational tools (if any)	Updates
IUIS Allergen [36]	WHO/IUIS Allergen Nomenclature Sub-committee (http://www.allergen.org)	Sequence (isoallergens/isoforms), structure, allergenicity	–	Updated continuously
Allergome [39]	Centre for Clinical and Experimental Allergology, Italy (http://www.allergome.org)	Sequence (isoallergens/isoforms), structure, clinical, epidemiological, cross-reactivity, etc.	–	Updated continuously
Structural Database of Allergenic Proteins (SDAP) [41]	Sealy Centre for Structural Biology, University of Texas, USA (https://fermi.utmb.edu)	Sequence, structure, structural models, IgE epitopes	Yes	2013
Allergen Database For Food Safety (ADFS) [44]	National Institute of Health, Japan (http://allergen.nihs.go.jp/ADFS/)	Sequence, structure, IgE epitopes, small molecule allergens	Yes	2016
AllergenOnline [46]	Food Allergy and Resource Program (FARP) (http://www.allergenonline.org)	Sequence, allergenicity	Yes	2016
AllFam [48]	Department of Pathophysiology and Allergy Research, Medical University of Vienna, Austria (http://www.meduniwien.ac.at/allfam)	Allergen family data, cross-link to Pfam database	–	2011
AllergenPro [53]	The National Agricultural	Sequence, IgE epitopes	Yes	2015

Database	Developed by (URL)	Type of data archived	Computational tools (if any)	Updates
	Biotechnology Information Centre, Korea (http://nabic.rda.go.kr/allergen)			
AllerBase [61]	Bioinformatics Centre, Savitribai Phule Pune University, India (www.bioinfo.net.in/AllerBase/Home.html)	Sequence and structure (cross-links), IgE epitopes, IgE antibody, IgE cross-reactivity, experimental evidences of allergenicity	–	Updated continuously

Table 1. Summary of major allergen databases and their relevant features.

Many allergen-specific databases have been developed in the past few years although they differ from each other with respect to their objectives, type of data archived, accessibility of contents, and the level of annotation and applications [30]. In addition to dedicated allergen databases, primary bioinformatics databases also document significant data on allergens. Examples of these databases include GenBank/GenPept [31, 32], UniProtKB [33], and Protein Data Bank (PDB) [34], which archive sequence and structure data on allergens along with its annotation. The Summary of allergen-specific databases is provided in **Table 1**. In the following section, the existing allergen databases are described.

3.1. IUIS Allergen database

The IUIS Allergen Nomenclature Sub-Committee, under the auspices of the WHO, provides the systematic nomenclature of allergenic proteins and it has developed and maintained Allergen database [35, 36]. The database archives all of the WHO/IUIS–recognized allergens along with their isoallergens and isoforms (variants). In order to maintain a consistent allergen nomenclature for newly discovered allergens, researchers are required to submit newly described allergens to the Allergen Nomenclature Sub-Committee before submitting their manuscript to a journal for consideration for publication.

Each allergen in this database is provided with annotation that includes biochemical name, molecular weight, information on its allergenicity, reference, etc. Additionally, sequence data for allergens and isoallergens/isoforms are also stored in the database, along with cross-references to GenBank [31], GenPept [32], and UniProtKB [33], as well as to PDB [34], for nucleotide, protein sequences, and 3D structure data, respectively. Allergen database can be searched by using allergen name, biochemical name, allergen source organism, taxonomic group, etc., as search criteria. The database is updated continuously with specific names assigned to newly discovered allergens and isoallergens/variants [37]. Allergen database does not exemplify the comprehensive allergen data although it documents majority of the characterized allergens. This is because there are a large number of allergens that have been reported in literature which are not recognized by IUIS-Allergen. The database does not archive data on allergen epitopes and cross-reactivity.

3.2. Allergome database

Allergome [38] represents an extensive repository of information on allergen molecules causing IgE-mediated (allergic, atopic) diseases [39]. The database comprises comprehensive data on WHO/IUIS-approved allergens along with other non-recognized allergens. These allergenic molecules are selected and curated from the published literature and web-based resources. It also contains data on allergenic sources based on whether they possess identified molecules or not. Allergome documents information on allergen and isoallergens/isoforms along with their sequences. Cross-links to sequence and structure databases like UniProtKB [33] and PDB [34] are also provided.

Allergome can be searched by using basic and advanced search options. Basic search employs numerous search criterions such as allergen name, biochemical name, source organism, etc., while advanced search enables the user to search using specific attributes. Each allergen molecule is represented by a monograph which represents information about the three parts of allergen such as basic information, data on the native form, and its recombinant form. The most important and unique feature of Allergome platform is the presence of several support modules that deal with archival of specific aspects of allergen data. A couple of important modules are RefArray, for easy access to references stored in the Allergome, and Real Time Monitoring of IgE sensitization (ReTiME), for real-time data collection and storage of IgE sensitization data and the number of other utilities. Allergome is updated regularly and allergen data curated from literature is documented.

3.3. Structural database of allergenic proteins

Structural Database of Allergenic Proteins (SDAP) [40] is an allergen database that prominently deals with structural aspects of allergens [41]. It houses comprehensive cross-referenced sequence data on allergens, IgE-binding epitopes, 3D structures, and models of allergens. Each allergen in SDAP is provided with cross-links to primary databases such as UniProtKB [33], PDB [34], as well as to important resources such as NCBI Taxonomy Browser [42] and PubMed [42] for literature references. SDAP also has a utility as a web server that integrates various computational tools, which assist structural biology–related studies dealing with allergens and their epitopes. It employs an algorithm based on the conserved properties of amino acid side chains to detect regions associated with allergenicity in novel sequences. The database consists of number of tools that can be used to assess potential cross-reactivity of allergens and also help in screening of IgE epitopes in novel proteins. The last update of the database was carried out on February 25, 2013. SDAP does not archive complete data for allergens that are not recognized by IUIS while data on allergen cross-reactivity is also not documented.

3.4. Allergen database for food safety

Allergen Database for Food Safety (ADFS) [43] is developed as a project of the Division of Biochemistry and Immunochemistry of National Institute of Health Sciences (Japan). The aim of the database is to archive allergenic proteins and their IgE epitopes with a special emphasis on food allergens and food safety [44]. Allergens archived in ADFS are grouped into eight categories such as pollen, mite, animal, fungus, insect, food, latex, and others, and each allergen entry is provided with the primary database accession numbers of their genes and 3D structure information. The database is also equipped with homology-based sequence search tool for the evaluation of allergenicity. One of the most distinct features of ADFS is the archival of data on small molecule, nonprotein (chemical) allergens. The database does not archive data on allergen cross-reactivity.

3.5. AllergenOnline database

AllergenOnline [45] is a well curated allergen database that documents a peer reviewed allergen list, which is compiled from various resources such IUIS-Allergen, PubMed, scientific publications, and other allergen databases. The database was developed within the Food Allergy Research and Resource Program (FARRP) at the University of Nebraska [46]. For each allergen, data on source organism, common name, IUIS official nomenclature, protein length, class of allergen like food allergen, contact allergen, etc., and a link to the NCBI protein (GenPept) [32] database are provided. AllergenOnline also provides the utility for sequence-based searches for allergens, which include alignments by FASTA and an eight-amino acid short-sequence identity search. This utility can be very useful in the identification of proteins that may present a potential risk of allergenic cross-reactivity. AllergenOnline is updated every year and the last update that resulted in version 16 of the database was reported on January 27, 2016. It does not archive data on allergen epitopes as well as on allergenic cross-reactivity.

3.6. AllFam database

AllFam [47] represents a very important resource for allergens as it is involved in classification of allergens into protein families [48]. This study has shown that allergens are distributed into relatively few protein families and possess a limited number of biochemical functions. The structural classification of allergens in AllFam is performed by using family information from PFam [49] and the Structural Classification of Proteins (SCOP) database [50], while biochemical functions of allergens were extracted from the Gene Ontology annotation database [51]. The database provides the option of browsing lists of allergen families based on allergen source (plants, animals, and fungi) and route of exposure (inhalation, ingestion, etc.) while search for specific protein families can also be performed. Each allergen family in AllFam is linked to a family fact sheet that describes the biochemical properties of the family members as well as a list of key references related to this family. The last update of AllFam was reported on September 12, 2011. AllFam does not archive data on molecular features of individual allergens although cross-link to IUIS-Allergen and Allergome is provided for each documented allergen.

3.7. AllergenPro database

AllergenPro [52] is a recently developed allergen database that archives data on allergen sequences, structures, and epitopes from various sources. It is an integrated database which provides information about allergens in foods, microorganisms, fungi, animals, and plants [53]. It has been provided with a utility to search for allergens based on keywords as well as the sequence. AllergenPro is also equipped with a computational tool for the prediction of allergenicity. Prediction is based on three different approaches such as FAO/WHO guidelines (sequence)–based approach, motif-based approach, and epitope-based approach. The database was last updated on June 4, 2015. AllergenPro does not archive data on allergen cross-reactivity while the literature references for documented allergens and epitopes have also not been provided.

3.8. Archival of allergen epitope data

As mentioned earlier, epitopes denote very important feature of allergens as they play vital role in allergic diseases. Because the molecular characterization of allergens has risen immensely in recent years, the data on allergenic epitopes has also increased significantly. Therefore, it has become necessary to store and manage the epitope data for its efficient utilization.

Some of the existing allergen databases described above, such as SDAP, ADFS, and AllergenPro are involved in storage of allergen epitope data. There are few databases available that are dedicated for epitope data from all types of antigens, which also document information on allergen epitopes [54–56]. However, the allergy-associated epitope data stored in these databases may not be comprehensive. The Immune Epitope Database (IEDB) [57], which is a repository of immune epitope reactivity data, is also a major database of allergy-derived epitope data [58]. It archives extensive allergen epitope data along with biological assays associated with them, including IgE-binding as well as T-cell epitopes curated and compiled from allergy-related references. IEDB is also equipped with several strategies for efficient searching and visualization of data on allergy-related epitopes [59]. Therefore, it represents a very useful and user friendly platform to access and retrieve allergy-related epitope data for the community of allergists. In a study involving classification of all the epitope-specific literature in various immunological domains, it is stated that IEDB comprises relatively fewer references for allergy-derived epitopes as compared to Cancer and Infectious Diseases [60]. This indicates that there is considerable scope for more in-depth archival of allergen epitope data from literature. Another study on meta-analysis of the allergy-associated epitope data in IEDB has indicated that relatively lesser data is archived for allergen T-cell epitopes as compared to IgE epitopes [26].

3.8.1. AllerBase database

Observations from the study of the existing allergen databases indicated that they archive significant data on various aspects of allergen and allergenicity, although the level of completeness of data differs considerably for diverse allergen features. AllerBase [61] is a recently developed comprehensive database of allergens and allergen features which addresses some of the limitations associated with the existing allergen databases [Kadam et al. 2016, unpublished]. The database comprises extensive data on experimentally validated allergens and allergen specific features such as IgE-binding epitopes, IgE cross-reactivity, IgE antibodies, and evidences for experimental validation of allergens. AllerBase is provided with basic and advanced search utilities along with browse database option to retrieve desired allergen data. The Completeness Index, which represents availability of data for various features for each allergen and a structure visualization utility, denote important features of the database. AllerBase also provides cross-references to several immunological and allergen databases and represents a notable instance of integration of allergen data from number of resources.

4. Computational prediction of allergens/allergenicity

Allergens mainly comprise commonly occurring proteins in foods, pollens, and other biological entities in the environment. It has become necessary to assess the potential allergenicity of these proteins considering the health hazards associated with allergic reactions to them. In recent years, genetic engineering and food processing methods are routinely employed for modifying the existing proteins or introducing new ones. Analysis of allergenicity of such proteins/products along with newly introduced biopharmaceuticals is absolutely essential in order to avoid transfer of an allergenic molecule. Computational assessment or prediction of allergenicity represents the major approach to test for allergenicity, and numerous bioinformatics tools/methods have been employed successfully for this purpose [62]. The majority of these methods utilize the amino acid sequence of allergens along with its different features, while a very few approaches use structure information. **Table 2** denotes the list of computational tools/servers available for the prediction of allergens/allergenicity. In the following section, the prominent approaches used for the computational assessment or prediction of allergens/allergenicity are described briefly.

No.	Method (URL)	Approach used	Efficiency
1	SDAP (http://fermi.utmb.edu/) [41]	FAO/WHO guidelines	–
2	AllergenOnline (http://www.allergenonline.org) [46]	FAO/WHO guidelines	–
3	AllergenPro (http://nabic.rda.go.kr/allergen) [53]	FAO/WHO guidelines, sequence motifs, epitopes	–
4	Allermatch (http://allermatch.org) [66]	FAO/WHO guidelines	–
5	AllerTool (http://research.i2r.a-star.edu.sg/AllerTool/)	FAO/WHO guidelines, global representation	A_{ROC} = 0.90, SE = 86%, SP = 86%

No.	Method (URL)	Approach used	Efficiency
	[67]	of protein sequence and SVM	
6	WebAllergen (http://weballergen.bii.a-star.edu.sg/) [73]	Sequence motifs	–
7	AlgPred (http://www.imtech.res.in/raghava/algpred/) [75]	Sequence features and SVM, sequence motifs, epitopes, allergen representative peptides	Accuracy = 85%, SE = 88%, SP = 81%
8	AllerTOP (http://www.ddg-pharmfac.net/AllerTOP) [83]	Sequence based descriptors, auto and cross-covariance, machine learning	Accuracy = 85.3%, SE = 82.5%, SP = 88.1%
9	EVALLER (http://bioinformatics.bmc.uu.se/evaller.html) [86]	DFLAP algorithm and SVM	–
10	AllerHunter (http://tiger.dbs.nus.edu.sg/AllerHunter/) [89]	Iterative pairwise sequence similarity and SVM	A_{ROC} = 0.928, accuracy = 95.3%, SE = 83.4%, SP = 96.4%
11	APPEL (http://jing.cz3.nus.edu.sg/cgi-bin/APPEL) [91]	Sequence based features and SVM	MCC = 0.95, SE = 93%, SP = 99.9%
12	SORTALLER (http://sortaller.gzhmu.edu.cn/) [93]	AFFP dataset, normalized BLAST *E*-values and SVM	MCC = 0.97 SE = 98.6%, SP = 98.4%
13	PREAL (http://gmobl.sjtu.edu.cn/PREAL/index.php) [96]	Biochemical and physicochemical descriptors, sequence features, subcellular locations, mRMR, SVM	Accuracy = 93.42%
14	Allerdictor (http://allerdictor.vbi.vt.edu/) [99]	Sequences as text documents, Naive Bayes classifier and SVM	–
15	proAP (http://gmobl.sjtu.edu.cn/proAP/main.html) [100]	Integration of methods based on FAO/WHO guidelines, sequence motifs and SVM	–
16	AllergenFP (http://ddg-pharmfac.net/AllergenFP/) [103]	Auto and cross-covariance, descriptor-based fingerprints	Accuracy = 88%, MCC = 0.759

No.	Method (URL)	Approach used	Efficiency
		of residues	
17	FuzzyApp (http://fuzzyapp.bicpu.edu.in/fuzzyapp.php) [107]	Fuzzy rule based system	–

A_{ROC}: area under the receiver-operating curve; SE: sensitivity; SP: specificity; MCC: Matthews correlation coefficient.

Table 2. List of computational tools/servers for allergen/allergenicity prediction.

4.1. Sequence similarity-based approaches

One of the first studies dealing with analysis of allergenicity was put forth by Metcalfe et al. [63]. They have proposed a decision tree–based approach for allergenicity assessment of foods derived from genetically modified crops. The first computational approach for the assessment of allergenicity was provided by "Codex Alimentarius Commission" of FAO/WHO [64, 65]. It stated that a protein can be regarded as an allergen if it consists of an exact match with at least six contiguous amino acids or showed more than 35% similarity over a window of 80 amino acids when compared with a sequence of known allergen. This approach has been widely used to predict allergenicity and there are number of web servers for allergen prediction, which are based on it. Allermatch [66], AllerTool [67], and AllergenPro [53] are some of the prominent web servers which employ these FAO/WHO guidelines for allergen prediction. Additionally, some of the major allergen databases such as SDAP [41] and AllergenOnline [46] also utilize this strategy for allergenicity prediction. A recent study performed by Verma et al. [68] has shown that the sequence similarity-based approach gives substantially better results when used in combination with other bioinformatics methods. However, results obtained by certain studies indicated that approaches based on these guidelines are not highly efficient for identifying allergenic proteins and many of times they lead to false or irrelevant allergenicity estimations [69–71]. As a result of these observations, it became necessary to discover and employ other strategies for the prediction of allergenicity.

4.2. Motif-based approaches

In a study carried out by Stadler and Stadler [71], it was observed that the use of sequence motifs, which represent the secondary structures of proteins, performs significantly better than the approach based on FAO/WHO guidelines. This method employs MEME motifs of a length of 50 residues for the prediction of allergenicity by using pairwise sequence alignment with certain threshold. WebAllergen [72] is a web server for the prediction of allergenic proteins which is also based on specific detectable allergenic motifs in known allergens [73]. Furthermore, a study carried out by Kong et al. showed that an approach based on search of multiple motifs is more specific and efficient than the conventional single motif search [74]. AlgPred [75] and AllergenPro [53] are important web servers for allergen prediction in which one of the prediction approaches is based on allergen-derived motifs. A recent study that employs computational approaches for comparison of allergens and metazoan parasite proteins stated that significant sequence and structure similarity exists between parasite proteins and allergenic proteins [76]. The analysis was carried out using sequence and structural motifs in allergens and a workflow was developed for the computational analysis of parasite proteins.

4.3. Machine learning–based approaches

Recent years have witnessed tremendous increase in the application of machine learning methods for solving biological problems. Machine learning–based approaches have been widely used for predicting various aspects of protein function [77]. These methods are also employed routinely for the development of algorithms to predict allergenicity of novel proteins.

Although Support Vector Machine (SVM) is the most commonly used machine learning method for allergen prediction, other methods have also been frequently employed. One of the earliest methods was developed by Zorzet et al. [78] that utilizes a k-Nearest-Neighbor (kNN) classification algorithm for the prediction of allergenicity, while a Bayesian classifier was employed by Soeria-Atmadja et al. [79] for the same purpose. An approach based on the combination of hidden Markov model (HMM) and conserved motifs in allergen was also used to successfully predict protein allergenicity [80]. Dimitrov et al. [81] developed two artificial neural network (ANN)-based algorithms for allergenicity prediction, which utilize descriptors derived from amino acids that denote their structural and physicochemical properties. AllerTOP [82] is an online bioinformatics tool to perform the computational prediction of allergens [83]. This algorithm employs descriptors that denote the chemical properties of amino acids in allergen sequences and auto- and cross-covariance transformation along with five machine learning methods for classification. These methods are random forest, multilayer perceptron, logistic regression, decision tree, naïve Bayes, and kNN.

There are number of web-based tools/servers developed which use SVM for performing classification/prediction of allergens. AlgPred [84] is one of the earlier web servers developed for the prediction of allergenic proteins [75]. It employs SVM with amino acid and dipeptide composition as features of allergens to achieve accuracy of 85.02 and 84.00%, respectively. EVALLER [85] is another web server created for *in silico* determination of potential allergenicity with very good efficiency [86]. It performs detection based on filtered length–adjusted allergen peptides (DFLAP) algorithm and SVM. AllerTool [87] web server also applies SVM-based algorithm for the prediction of allergenicity and provides sensitivity and specificity of 86.00% [67]. AllerHunter [88] is an important web-based computational system for allergenicity assessment which uses a scheme based on iterative pairwise sequence similarity encoding along with SVM [89]. The method is very efficient with a sensitivity of 83.4% and a specificity of 96.4%.

A web-based tool APPEL [90] is developed for the prediction of allergenic proteins that employs physicochemical and structural features derived from allergen sequence in combination with SVM [91]. Zhang et al. have developed an online allergen prediction tool titled SORTALLER [92], which is based on allergen family featured peptide (AFFP) dataset and employs SVM as a classifier [93]. An algorithm developed by Mohabatkar et al. [94] for the prediction of allergenic proteins utilizes pseudoamino acid composition (PseAAC) along with SVM and provides an accuracy of 91.19%. PREAL [95] is web-based tool that performs allergen prediction by using SVM along with feature selection methods such as maximum relevance minimum redundancy (mRMR) and incremental feature selection (IFS) [96]. A combination of hydrophobicity amino acid index and discrete Fourier transform along with an SVM classifier is employed for highly efficient prediction of allergenicity in a signal-processing bioinformatics approach [97]. Allerdictor [98] is web server that specializes in large-scale allergen discovery. It models protein sequences as text documents and employs SVM in text classification for carrying out allergen prediction [99].

4.4. Other approaches

A study carried out by Wang et al. evaluated sequence-, motif-, and SVM-based approaches for the computational prediction of allergens and also performed parameter optimization to obtain better performance [100]. The resulting methods from this study are integrated and made available as a web application titled proAP [101]. AllergenFP [102] is a recently developed web server for allergenicity prediction that utilizes alignment-free descriptor-based fingerprint approach [103]. The descriptors used here are important properties of amino acid such as size, hydrophobicity, relative abundance, helix, and beta-strand forming propensities, etc. In a structure-based approach proposed by Bragin et al. [104], information derived from protein 3D structure is used for the representation of protein surface as patches designated as discontinuous peptides. It is observed that prediction of allergenic proteins based on this approach gave better accuracy. Vijayakumar and Lakshmi have developed a fuzzy inference system–based algorithm for allergenicity prediction that utilizes five different modules [105]. These modules consist of a machine learning classifier, motif search, sequence similarity, FAO/WHO evaluation scheme, etc. FuzzyApp [106], a web server based on fuzzy rule–based system, is then developed for the prediction of allergenicity [107]. Jiang et al. performed an analysis of food allergens using a computational model that simulates gastric fluid digestion [108]. This study stated that food allergens could be classified as alimentary canal-sensitized and nonalimentary canal-sensitized allergens based on the digestibility of these allergens in simulated gastric fluid.

5. Computational prediction of allergen epitopes

Epitopes represent distinctive amino acid residues on the antigens and are important determinants of an immune response. Identification of epitopes is considered a key aspect of designing highly effective multiple-subunit vaccines and developing efficient diagnostic and therapy methods against allergens. Although experimental methods have been very useful for the identification of epitopes, their usefulness is restricted because of their time- and cost-intensive nature and inability in dealing with large-scale elucidation of epitopes. Hence, computational approaches are considered to be very beneficial alternative as they are cost and time effective.

Large number of highly efficient algorithms and tools have been developed over the years for the computational prediction of epitopes. These methods deal with the prediction of both B-cell and T-cell epitopes as well as sequential (linear) and discontinuous (conformational) epitopes. Based on the information (data) utilized for performing prediction, the methodologies can be grouped as sequence-based or structure-based approaches. Many sequence-based linear epitope prediction methods for B cells have been developed and used since long time and majority of them are propensity scale and machine learning–based methods [109, 110]. Some of the major tools/servers that deal with the prediction of linear B-cell epitopes are listed in **Table 3**.

No.	Method (URL)	Approach used	Efficiency
1	ABCPred (http://www.imtech.res.in/raghava/abcpred/) [111]	Fixed length epitope patterns, recurrent ANNs	Accuracy = 65.93%, SE = 67.14%, SP = 64.71%
2	APCpred (http://ccb.bmi.ac.cn/APCpred/) [112]	Amino acid anchoring pair composition (APC) and SVM	A_{ROC} = 0.809, accuracy = 72.94%
3	BCPreds (http://ailab.cs.iastate.edu/bcpreds/) [113]	SVM classifiers with string kernels	A_{ROC} = 0.758
4	BcePred (http://www.imtech.res.in/raghava/bcepred/) [114]	Physicochemical properties of epitope residues	Accuracy = 58.7%
5	BepiPred (http://www.cbs.dtu.dk/services/BepiPred/) [115]	Parker's hydrophilicity scale and HMM	–
6	BEST (http://biomine.ece.ualberta.ca/BEST/) [116]	Antigen sequence features, SVM	A_{ROC} = 0.85
7	Bayesb (http://www.immunopred.org/bayesb/index.html) [117]	Bayes feature extraction and SVM	A_{ROC} = 0.84, accuracy = 74.5%
8	COBEpro (http://scratch.proteomics.ics.uci.edu) [118]	Antigen fragment score and SVM	A_{ROC} = 0.829
9	EPMLR (http://www.bioinfo.tsinghua.edu.cn/epitope/EPMLR/) [119]	Sequence features and multiple linear regression (MLR)	A_{ROC} = 0.728, SE = 81.8%, SP = 64.1%
10	IEDB Analysis Resource (http://tools.iedb.org/bcell/) [120]	A collection of tools based on various methods	–
11	LBtope (http://www.imtech.res.in/raghava/lbtope/) [121]	Large datasets of epitopes, KNN, SVM	Accuracy = 86%
12	SVMTriP (http://sysbio.unl.edu/SVMTriP/) [122]	Tri-peptide similarity, propensity scores and SVM	A_{ROC} = 0.702, SE = 80.1%, SP = 55.2%

A_{ROC}: area under the receiver-operating curve; SE: sensitivity; SP: specificity; MCC: Matthews correlation coefficient.

Table 3. List of major tools/servers for sequential (linear) epitope prediction.

Number of methods that utilize 3D structure of antigens for discontinuous epitope prediction have also been developed. These methods use different approaches for prediction such as solvent accessibility of surface residues [123, 124], solvent accessibility with propensity scores [125], and propensity scores with packing density of amino acids [126]. An account of major tools/servers that are involved in conformational epitope prediction is provided in **Table 4**.

No.	Method (URL)	Approach used	Efficiency
1	BEpro (formerly PEPITO) (http://pepito.proteomics.ics.uci.edu/) [127]	3D structure of antigen, amino acid propensity scores	A_{ROC} = 0.75
2	B-Pred (http://immuno.bio.uniroma2.it/bpred) [128]	3D structure or model of antigen, solvent exposure of residues	SE = 0.70
3	CBTOPE (http://www.imtech.res.in/raghava/cbtope/) [129]	Sequence features and SVM	A_{ROC} = 0.9, Accuracy = 85%
4	CEP (http://196.1.114.49/cgi-bin/cep.pl) [124]	3D structure of antigen, solvent accessibility of amino acids	Accuracy = 75%
5	DiscoTope 2.0 (http://www.cbs.dtu.dk/services/DiscoTope/) [125]	3D structure of antigen, epitope propensity scores, surface accessibility	A_{ROC} = 0.824
6	ElliPro (http://tools.immuneepitope.org/tools/ElliPro) [130]	3D structure of antigen, Thornton's method, residue clustering algorithm	A_{ROC} = 0.732
7	Epitopia (http://epitopia.tau.ac.il/) [131]	Antigen sequence or 3D structure, Naïve Bayes classifier	A_{ROC} = 0.59
8	EPSVR (http://sysbio.unl.edu/EPSVR/) [132]	3D structure of antigen, Support vector regression (SVR)	A_{ROC} = 0.597
9	EPMeta (http://sysbio.unl.edu/EPMeta/) [132]	Meta server integrating EPSVR with other methods	A_{ROC} = 0.638
10	EPCES (http://sysbio.unl.edu/EPCES/) [133]	3D structure of antigen, surface features	–
11	SEPPA 2.0 (http://badd.tongji.edu.cn/seppa/) [126]	3D structure of antigen, subcellular localization of antigen, residue propensity, etc.	A_{ROC} = 0.745

A_{ROC}: area under the receiver-operating curve; SE: sensitivity; SP: specificity; MCC: Matthews correlation coefficient.

Table 4. List of major tools/servers for conformational (discontinuous) epitope prediction.

Studies have shown that the analysis of antigen–antibody complex structures is very useful for the characterization of conformational epitopes [134]. A dedicated resource titled AgAbDb [135] that archives the interactions derived from antigen–antibody complexes is available, which can be very useful for the analysis of epitopes [20, 21]. Several algorithms have also been developed for the prediction of T-cell epitopes in antigens. These methodologies deal with the prediction of peptides that possess the ability to interact with specific major histocompatibility complex (MHC) molecules [136]. Machine learning–based approaches are very commonly employed for this purpose and are found to be very efficient [137]. The details of epitope prediction methods/tools for B cells and T cells have been reviewed elsewhere [138, 139]. Some of the important tools/servers that perform the prediction of T-cell epitopes are listed in **Table 5**. Recently, it has been shown that epitope prediction can be performed over the whole proteome by integrating multiple epitope prediction methods [149]. Antibody-specific epitope prediction has emerged as a significant alternative to the traditional antibody-independent epitope prediction methods [150].

No.	Method (URL)	Approach used	Efficiency
1	CTLPred (http://www.imtech.res.in/raghava/ctlpred/index.html) [140]	Cytotoxic T-lymphocyte epitopes, SVM, ANN	Accuracy = 75.2%
2	EpiJen (http://www.ddg-pharmfac.net/epijen/EpiJen/EpiJen.htm) [141]	Multi-step algorithm that employs integrated approach	Accuracy = 60%
3	EpiTOP (http://www.pharmfac.net/EpiTOP/) [142]	QSAR approach based on proteochemometrics	Accuracy = 89%
4	NetCTLpan (http://www.cbs.dtu.dk/services/NetCTLpan/) [143]	Integrated method employing proteasomal cleavage, TAP transport efficiency, and MHC class I binding affinity	$A_{\mathrm{ROC}} = 0.95$
5	NetMHCIIpan-3.0 (http://www.cbs.dtu.dk/services/NetMHCIIpan-3.0/) [144]	A method for all HLA class II molecules based on peptide-binding MHC environment	$A_{\mathrm{ROC}} = 0.807$
6	PREDIVAC (http://predivac.biosci.uq.edu.au/) [145]	Based on specificity-determining residues	–
7	SYFPEITHI (http://www.syfpeithi.de/bin/MHCServer.dll/EpitopePrediction.htm) [146]	Scoring system based on position of residue in the epitopes	–
8	TEPITOPEpan (http://www.biokdd.fudan.edu.cn/Service/TEPITOPEpan/) [147]	Algorithm based on HLA-DR binding pocket similarity	–
9	WAPP (http://abi.inf.uni-tuebingen.de/Services/WAPP/) [148]	Combination of methods based on proteasomal cleavage, TAP transport and MHC binding	–

A_{ROC}: area under the receiver-operating curve.

Table 5. List of major tools/servers for T-cell epitope prediction.

Epitopes represent critical components of allergens from the perspective of allergic reactions and development of new diagnosis and treatment strategies. Therefore, the computational prediction of these epitopes in allergens is of immense importance. Due to limitations associated with detailed archival of allergen epitope data and highly heterogeneous nature of the data, the number of tools available for allergen epitope prediction are far less, especially when compared with the number of tools available for allergen/allergenicity prediction. Therefore, the general epitope prediction methods listed above can be employed for epitope prediction studies in allergens. Kleter and Peijnenburg [70] developed a strategy to screen for the potential linear IgE-epitopes using sequence comparison with a minimal length of six amino acids. The approach was moderately effective and it showed that further verification of IgE binding of epitopes by experimental tests is necessary. AlgPred [84] developed by Saha and Raghava is one of the first and major tools for the computational assessment of IgE epitopes [75]. Here, a database of known IgE epitopes is created and it is used to accurately predict allergenic proteins. AllerPred is a SVM-based computational system for the assessment of overlapping continuous and discontinuous B-cell epitope binding patterns in allergenic proteins [151]. This approach is successfully used to predict allergenicity of novel proteins. Dall'Antonia et al. [152] have developed a software tool titled Surface comparison–based Prediction of Allergenic Discontinuous Epitopes (SPADE). The algorithm consists of a structure-based comparison of allergen surfaces and IgE cross-reactivity data and is able to predict IgE epitopes from three important allergen families. A recent work performed by Lollier et al. [153] on meta-analysis of IgE-binding epitopes provided some important findings regarding these epitopes. They computed the fraction of allergen amino acids that are involved in epitopes and modeled a relationship between the rising number of literature references and the amino acid fractions to assess the possibility of binary classification of epitopes and nonepitopes. A web-based tool LocAllEpi [154] is also developed for the visualization of allergen epitopes along the protein sequence and their structural features.

6. Computational prediction of allergenic cross-reactivity

Cross-reactivity plays an important role in allergic reaction from the immunological and clinical context. Therefore, the computational prediction of allergenic cross-reactivity has been considered of substantial significance. The prediction of cross-reactivity in allergens is associated with the prediction of allergenicity for the majority of the cases. This is mainly because the antigenic determinants that contribute to the cross-reactivity in allergens are also responsible for their allergenicity. As a result of this, many of the tools/algorithms that have been developed for the prediction of allergens/allergenicity also perform cross-reactivity prediction.

The criteria defined by FAO/WHO experts, which have been mentioned earlier, help to identify cross-reactivity in allergens [155]. AllerTool [87] is a web server that performs cross-reactivity prediction based on amino acid sequence and WHO/FAO guidelines [67]. It also provides a graphical representation of the published and predicted cross-reactivity patterns of allergens. Stadler and Stadler [71] developed a sequence-based approach and stated that motif-based strategy provides better results for the computational assessment of cross-reactivity than the FAO/WHO guidelines. SDAP [40], which is a specialized allergen database described before, also comprises a sequence-based tool for the identification of cross-reactivity among allergens [41]. AllerHunter [88] is a SVM-based web server that deals with efficient assessment of allergic cross-reactivity in proteins [89]. A recently developed fuzzy inference system–based algorithm for allergenicity prediction is also able to predict cross-reactivity in allergens [105].

7. Future perspectives and challenges

Allergy represents a serious problem, as allergic diseases are known to affect millions of people worldwide. Advancements in genomic, proteomic, and analytical techniques have led to the generation of large amount of data related to allergy and allergens. Archival and analysis of these data denotes a major challenge in allergen bioinformatics. Data integration is one of the key limitations for efficient and useful storage of allergy associated data. This is mainly due to the heterogeneous nature of the data, which is derived from various sources such as molecular data from experimental characterization of allergic reactions, clinical, and epidemiological data from patients/populations. Bioinformatics resources and tools have an important role to play in overcoming this problem. In the wake of ever-expanding volume of data, it is vital to focus on developing databases/resources that will integrate information from different sources as well as from literature and provide rapid access to it. Analysis of such data can be further utilized to obtain important insights to understand allergic reactions. Structural features of allergens contribute significantly to their allergenicity and therefore this knowledge can be employed for developing more efficient methods for allergen/allergenicity and allergic cross-reactivity prediction. Recent advances in epitope prediction methodologies focus on antibody-specific epitope prediction approaches [150]. Application of such approaches for predicting IgE-binding epitopes will be extremely important in the development of newer and effective strategies for diagnosis and treatment of allergic diseases. Allergen immunotherapy (AIT), which is an individualized and allergens-based treatment approach, has been considered as a prototype of precision medicine or personalized medicine [156]. Bioinformatics has the potential to play an important role in the development of novel approaches in AIT as well as contribute for further enrichment of the field of allergen informatics. This will surely aid in gaining better understanding of allergic diseases and positively influence upcoming research in the field.

Acknowledgements

The work was supported under the Senior Research Fellowship (SRF) granted to KK by the University Grants Commission (UGC), New Delhi, India. UKK and SS would like to acknowledge the Centre of Excellence grant from the Department of Biotechnology (DBT), New Delhi, India. The authors would also like to acknowledge the Bioinformatics Centre, Savitribai Phule Pune University, for providing the infrastructure and resources.

Author details

Kiran Kadam[1], Sangeeta Sawant[1], V.K. Jayaraman[2] and Urmila Kulkarni-Kale[1*]

*Address all correspondence to: urmila@bioinfo.net.in

1 Bioinformatics Centre, Savitribai Phule Pune University, Pune, India

2 Shiv Nadar University, Gautam Buddha Nagar, Uttar Pradesh, India

References

[1] Parkin J, Cohen B. An overview of the immune system. Lancet. 2001;357:1777–1789. DOI: 10.1016/S0140-6736(00)04904-7

[2] Kindt T, Goldsby R, Osborne B, Kuby J. Kuby Immunology. 6th ed. New York: W.H. Freeman; 2007. 574 p. ISBN: 9780716785903

[3] Janeway CA Jr, Travers P, Walport M, Shlomchik MJ. Immunobiology: The Immune System in Health and Disease. 5th ed. New York: Garland Science; 2001. 732 p. ISBN: 9780815336426

[4] Aalberse RC. Structural biology of allergens. J Allergy Clin Immunol. 2000;106:228–238. DOI: 10.1067/mai.2000.108434

[5] Pawankar R. Allergic diseases and asthma: a global public health concern and a call to action. World Allergy Organ J. 2014;7:12. DOI: 10.1186/1939-4551-7-12

[6] Bochner BS, Rothenberg ME, Boyce JA, Finkelman F. Advances in mechanisms of allergy and clinical immunology in 2012. J Allergy Clin Immunol. 2013;131:661–667. DOI: 10.1016/j.jaci.2012.12.676

[7] Ghosh D, Gupta-Bhattacharya S. Allergen Bioinformatics: Recent Trends and Developments. In: Xia X, editor. Selected Works in Bioinformatics. Croatia: InTech; 2011. ISBN: 978-953-307-281-4. DOI: 10.5772/20269

[8] Scheurer S, Toda M, Vieths S. What makes an allergen? Clin Exp Allergy. 2015;45:1150–1161. DOI: 10.1111/cea.12571

[9] King TP, Hoffman D, Lowenstein H, Marsh DG, Platts-Mills TA, Thomas W. Allergen nomenclature. Int Arch Allergy Immunol. 1994;105:224–233. DOI: 10.1159/000236761

[10] King TP, Hoffman D, Lowenstein H, Marsh DG, Platts-Mills TA. Thomas W. Allergen nomenclature. Allergy. 1995;50:765–774. DOI: 10.1111/j.1398-9995.1995.tb01222.x

[11] Breiteneder H, Chapman MD. Allergen Nomenclature. In: Lockey RF, Ledford DK, editors. Allergens and Allergen Immunotherapy: Subcutaneous, sublingual and oral. 5th ed. New York: CRC Press; 2014. pp. 37–49.

[12] Lagares A, Puerta L, Caraballo L. Polymorphism in allergens. Biomedica. 2002;22:51–62.

[13] Lowenstein H. Quantitative immunoelectrophoretic methods as a tool for the analysis and isolation of allergens. Prog Allergy. 1978;25:1–62. DOI: 10.1159/000401700

[14] Pomés A. Relevant B cell epitopes in allergic disease. Int Arch Allergy Immunol. 2010;152:1–11. DOI: 10.1159/000260078

[15] Bufe A. Significance of IgE-binding epitopes in allergic disease. J Allergy Clin Immunol. 2001;107:219–221. DOI: 10.1067/mai.2001.112850

[16] Aalberse RC, Crameri R. IgE-binding epitopes: a reappraisal. Allergy. 2011;66:1261–1274. DOI: 10.1111/j.1398-9995.2011.02656.x

[17] Bannon GA, Ogawa T. Evaluation of available IgE-binding epitope data and its utility in bioinformatics. Mol Nutr Food Res. 2006;50:638–644. DOI: 10.1002/mnfr.200500276

[18] Matsuo H, Yokooji T, Taogoshi T. Common food allergens and their IgE-binding epitopes. Allergol Int. 2015;64:332–343. DOI: 10.1016/j.alit.2015.06.009

[19] Meno KH. Allergen structures and epitopes. Allergy. 2011;66:19–21. DOI: 10.1111/j.1398-9995.2011.02625.x

[20] Ghate AD, Bhagwat BU, Bhosle SG, Gadepalli SM, Kulkarni-Kale UD. Characterization of antibody-binding sites on proteins: development of a knowledgebase and its applications in improving epitope prediction. Protein Pept Lett. 2007;14:531–535. DOI: 10.2174/092986607780989921

[21] Kulkarni-Kale U, Raskar-Renuse S, Natekar-Kalantre G, Saxena SA. Antigen–antibody interaction database (AgAbDb): a compendium of antigen–antibody interactions. In: De RK, Tomar N, editors. Immunoinformatics. New York: Springer; 2014. pp. 149–164. DOI: 10.1007/978-1-4939-1115-8_8

[22] Cavkaytar O, Akdis CA, Akdis M. Modulation of immune responses by immunotherapy in allergic diseases. Curr Opin Pharmacol. 2014;17:30–37. DOI: 10.1016/j.coph.2014.07.003

[23] Larché M. T cell epitope-based allergy vaccines. In: Valenta R, Coffman R, editors. Vaccines against Allergies. Berlin, Heidelberg: Springer; 2011. pp. 107–119. DOI: 10.1007/82_2011_131

[24] Prickett SR, Rolland JM, O'Hehir RE. Immunoregulatory T cell epitope peptides: the new frontier in allergy therapy. Clin Exp Allergy. 2015;45:1015–1026. DOI: 10.1111/cea.12554

[25] Frazier A, Schulten V, Hinz D, Oseroff C, Sidney J, Peters B, Sette A. Allergy-associated T cell epitope repertoires are surprisingly diverse and include non-IgE reactive antigens. World Allergy Organ J. 2014;7:26. DOI: 10.1186/1939-4551-7-26

[26] Vaughan K, Greenbaum J, Kim Y, Vita R, Chung J, Peters B, Broide D, Goodman R, Grey H, Sette A. Towards defining molecular determinants recognized by adaptive immunity in allergic disease: an inventory of the available data. J Allergy (Cairo). 2011;2010:628026. DOI: 10.1155/2010/628026

[27] Aalberse RC. Assessment of allergen cross-reactivity. Clin Mol Allergy. 2007;5:1. DOI: 10.1186/1476-7961-5-2

[28] Aalberse RC, Akkerdaas J, Van Ree R. Cross-reactivity of IgE antibodies to allergens. Allergy. 2001;56:478–490. DOI: 10.1034/j.1398-9995.2001.056006478.x

[29] Ferreira F, Hawranek T, Gruber P, Wopfner N, Mari A. Allergic cross-reactivity: from gene to the clinic. Allergy. 2004;59:243–267. DOI: 10.1046/j.1398-9995.2003.00407.x

[30] Brusic V, Millot M, Petrovsky N, Gendel SM, Gigonzac O, Stelman SJ. Allergen databases. Allergy. 2003;58:1093–1100. DOI: 10.1034/j.1398-9995.2003.00248.x

[31] Benson DA, Clark K, Karsch-Mizrachi I, Lipman DJ, Ostell J, Sayers EW. GenBank. Nucl Acids Res. 2015;43:D30–D35. DOI: 10.1093/nar/gku1216

[32] O'Leary NA, Wright MW, Brister JR, Ciufo S, Haddad D, McVeigh R, Rajput B, et al. Reference sequence (RefSeq) database at NCBI: current status, taxonomic expansion, and functional annotation. Nucl Acids Res. 2016;4:D733–D745. DOI: 10.1093/nar/gkv1189

[33] UniProt Consortium. Reorganizing the protein space at the Universal Protein Resource (UniProt). Nucl Acids Res. 2012;40:D71–D75. DOI: 10.1093/nar/gkr981

[34] Berman HM, Westbrook J, Feng Z, Gilliland G, Bhat TN, Weissig H, Shindyalov IN, Bourne PE. The protein data bank. Nucl Acids Res. 2000;28:235–242. DOI: 10.1093/nar/28.1.235

[35] WHO/IUIS Allergen database. Available from: http://www.allergen.org [Accessed: 2016-03-07].

[36] Chapman MD, Pomés A, Breiteneder H, Ferreira F. Nomenclature and structural biology of allergens. J Allergy Clin Immunol. 2007;119:414–420. DOI: 10.1016/j.jaci.2006.11.001

[37] Radauer C, Nandy A, Ferreira F, Goodman RE, Larsen JN, Lidholm J, Pomés A, Raulf-Heimsoth M, Rozynek P, Thomas WR, Breiteneder H. Update of the WHO/IUIS Allergen Nomenclature Database based on analysis of allergen sequences. Allergy. 2014;69:413–419. DOI: 10.1111/all.12348

[38] Allergome. Available from: http://www.allergome.org [Accessed: 2016-03-07].

[39] Mari A, Scala E, Palazzo P, Ridolfi S, Zennaro D, Carabella G. Bioinformatics applied to allergy: allergen databases, from collecting sequence information to data integration. The Allergome platform as a model. Cell Immunol. 2006;244:97–100. DOI: 10.1016/j.cellimm.2007.02.012

[40] SDAP – Structural Database of Allergenic Proteins. Available from: https://fermi.utmb.edu [Accessed: 2016-03-07].

[41] Ivanciuc O, Schein C, Braun W. SDAP: database and computational tools for allergenic proteins. Nucl Acids Res. 2003;31:359–362. DOI: 10.1093/nar/gkg010

[42] NCBI Resource Coordinators. Database resources of the National Center for Biotechnology Information. Nucl Acids Res. 2015;43:D6–D17. DOI: 10.1093/nar/gku1130

[43] Allergen Database for Food Safety (ADFS). Available from: http://allergen.nihs.go.jp/ADFS/ [Accessed: 2016-06-01].

[44] Nakamura R, Teshima R, Takagi K, Sawada J. Development of Allergen Database for Food Safety (ADFS): an integrated database to search allergens and predict allergenicity. Kokuritsu Iyakuhin Shokuhin Eisei Kenkyujo hokoku. 2004;123:32–36.

[45] AllergenOnline. Available from: http://www.allergenonline.org [Accessed: 2016-03-07].

[46] Goodman RE, Ebisawa M, Ferreira F, Sampson HA, van Ree R, Vieths S, et al. AllergenOnline: a peer-reviewed, curated allergen database to assess novel food proteins for potential cross-reactivity. Mol Nutr Food Res. 2016;60:1183–98. DOI: 10.1002/mnfr.201500769

[47] AllFam. Available from: http://www.meduniwien.ac.at/allfam [Accessed: 2016-03-07].

[48] Radauer C, Bublin M, Wagner S, Breiteneder H. Allergens are distributed into few protein families and possess a restricted number of biochemical functions. J Allergy Clin Immunol. 2008;121:847–852. DOI: 10.1016/j.jaci.2008.01.025

[49] Finn RD, Bateman A, Clements J, Coggill P, Eberhardt RY, Eddy SR, Heger A, Hetherington K, Holm L, Mistry J, Sonnhammer EL, Tate J, Punta M. Pfam: the protein families database. Nucl Acids Res. 2013;42:D222–D230. DOI: 10.1093/nar/gkt1223

[50] Murzin AG, Brenner SE, Hubbard T, Chothia C. SCOP: a structural classification of proteins database for the investigation of sequences and structures. J Mol Biol. 1995;247:536–540. DOI: 10.1016/S0022-2836(05)80134-2

[51] Gene Ontology Consortium. The Gene Ontology (GO) database and informatics resource. Nucl Acids Res. 2004;32:D258–D261. DOI: 10.1093/nar/gkh036

[52] AllergenPro. Available from: http://nabic.rda.go.kr/allergen [Accessed: 2016-06-01].

[53] Kim CK, Seol YJ, Lee DJ, Jeong IS, Yoon UH, Lee JY, Lee GS, Park DS. AllergenPro: an integrated database for allergenicity analysis and prediction. Bioinformation. 2014;10:378. DOI: 10.6026/97320630010378

[54] Saha S, Bhasin M, Raghava GP. Bcipep: a database of B-cell epitopes. BMC Genomics. 2005;6:79. DOI: 10.1186/1471-2164-6-79

[55] Huang J, Honda W. CED: a conformational epitope database. BMC Immunol. 2006;7:7. DOI: 10.1186/1471-2172-7-7

[56] Sharma OP, Das AA, Krishna R, Suresh Kumar M, Mathur PP. Structural epitope database (SEDB): a web-based database for the epitope, and its intermolecular interaction along with the tertiary structure information. J Proteomics Bioinform. 2012;5:84–89. DOI: 10.4172/jpb.1000217

[57] The Immune Epitope Database (IEDB). 2010. Available from: http://www.iedb.org [Accessed: 2016-03-07].

[58] Vita R, Zarebski L, Greenbaum JA, Emami H, Hoof I, Salimi N, Damle R, Sette A, Peters B. The immune epitope database 2.0. Nucl Acids Res. 2010;38:D854–D862. DOI: 10.1093/nar/gkp1004

[59] Vaughan K, Peters B, Larche M, Pomes A, Broide D, Sette A. Strategies to query and display allergy-derived epitope data from the immune epitope database. Int Arch Allergy Immunol. 2013;160:334–345. DOI: 10.1159/000343880

[60] Davies V, Vaughan K, Damle R, Peters B, Sette A. Classification of the universe of immune epitope literature: representation and knowledge gaps. PLoS One. 2009;4:e6948. DOI: 10.1371/journal.pone.0006948

[61] AllerBase. Available from: http://bioinfo.net.in/AllerBase/Home.html [Accessed: 2016-03-07]

[62] Brusic V, Petrovsky N, Gendel SM, Millot M, Gigonzac O, Stelman SJ. Computational tools for the study of allergens. Allergy. 2003;58:1083–1092. DOI: 10.1034/j.1398-9995.2003.00224.x

[63] Metcalfe DD, Astwood JD, Townsend R, Sampson HA, Taylor SL, Fuchs RL. Assessment of the allergenic potential of foods derived from genetically engineered crop plants. Crit Rev Food Sci Nutr. 1996;36:165–186. DOI: 10.1080/10408399609527763

[64] FAO/WHO. Allergenicity of Genetically Modified Foods. Rome, Italy. 2001. Available from: http://www.who.int/foodsafety/publications/biotech/en/ec_jan2001.pdf [Accessed: 2016-03-07].

[65] FAO/WHO. Codex Principles and Guidelines on Foods Derived from Biotechnology. Rome, Italy: Joint FAO/WHO Food Standards Programme. 2003. Available from: ftp://ftp.fao.org/codex/Publications/Booklets/Biotech/Biotech_2009e.pdf [Accessed: 2016-03-07].

[66] Fiers MW, Kleter GA, Nijland H, Peijnenburg AA, Nap JP, Van Ham RC. Allermatch™, a webtool for the prediction of potential allergenicity according to current FAO/WHO Codex alimentarius guidelines. BMC Bioinformatics. 2004;5:133. DOI: 10.1186/1471-2105-5-133

[67] Zhang ZH, Koh JL, Zhang GL, Choo KH, Tammi MT, Tong JC. AllerTool: a web server for predicting allergenicity and allergic cross-reactivity in proteins. Bioinformatics. 2007;23:504–506. DOI: 10.1093/bioinformatics/btl621

[68] Verma A, Misra A, Subash S, Das M, Dwivedi P. Computational allergenicity prediction of transgenic proteins expressed in genetically modified crops. Immunopharmacol Immunotoxicol. 2011;33:410–422. DOI: 10.3109/08923973.2010.523704

[69] Hileman RE, Silvanovich A, Goodman RE, Rice EA, Holleschak G, Astwood JD, Hefle SL. Bioinformatic methods for allergenicity assessment using a comprehensive allergen database. Int Arch Allergy Immunol. 2002;128:280–291. DOI: 10.1159/000063861

[70] Kleter GA, Peijnenburg AA. Screening of transgenic proteins expressed in transgenic food crops for the presence of short amino acid sequences identical to potential, IgE–binding linear epitopes of allergens. BMC Struct Biol. 2002;2:8. DOI: 10.1186/1472-6807-2-8

[71] Stadler MB, Stadler BM. Allergenicity prediction by protein sequence. FASEB J. 2003;17:1141–1143. DOI: 10.1096/fj.02-1052fje

[72] The web tool WebAllergen (http://weballergen.bii.a-star.edu.sg/) could not be accessed on 2016-06-01

[73] Riaz T, Hor HL, Krishnan A, Tang F, Li KB. WebAllergen: a web server for predicting allergenic proteins. Bioinformatics. 2005;21:2570–2571. DOI: 10.1093/bioinformatics/bti356

[74] Kong W, Tan TS, Tham L, Choo KW. Improved prediction of allergenicity by combination of multiple sequence motifs. In Silico Biol. 2007;7:77–86.

[75] Saha S, Raghava GP. AlgPred: prediction of allergenic proteins and mapping of IgE epitopes. Nucl Acids Res. 2006;34:W202–W209. DOI: 10.1093/nar/gkl343

[76] Tyagi N, Farnell EJ, Fitzsimmons CM, Ryan S, Tukahebwa E, Maizels RM, Dunne DW, Thornton JM, Furnham N. Comparisons of allergenic and metazoan parasite proteins: allergy the price of immunity. PLoS Comput Biol. 2015;11:e1004546. DOI: 10.1371/journal.pcbi.1004546

[77] Kadam K, Sawant S, Kulkarni-Kale U, Jayaraman VK. Prediction of protein function based on machine learning methods: an overview. In: iConcept Press, editor. Genomics III - Methods, Techniques and Applications. Hong Kong: iConcept Press; 2014 . pp. 125–162. ISBN: 978-1-922227-096

[78] Zorzet A, Gustafsson M, Hammerling U. Prediction of food protein allergenicity: a bioinformatic learning systems approach. In Silico Biol. 2002;2:525–534.

[79] Soeria-Atmadja D, Zorzet A, Gustafsson MG, Hammerling U. Statistical evaluation of local alignment features predicting allergenicity using supervised classification algorithms. Int Arch Allergy Immunol. 2004;133:101–112. DOI: 10.1159/000076382

[80] Li KB, Issac P, Krishnan A. Predicting allergenic proteins using wavelet transform. Bioinformatics. 2004;20:2572–2578. DOI: 10.1093/bioinformatics/bth286

[81] Dimitrov I, Naneva L, Bangov I, Doytchinova I. Allergenicity prediction by artificial neural networks. J Chemom. 2014;28:282–286. DOI: 10.1002/cem.2597

[82] AllerTOP. Available from: http://www.ddg-pharmfac.net/AllerTOP [Accessed: 2016-03-07].

[83] Dimitrov I, Bangov I, Flower DR, Doytchinova I. AllerTOP v. 2—a server for in silico prediction of allergens. J Mol Model. 2014;20:1–6. DOI: 10.1007/s00894-014-2278-5

[84] AlgPred. Available from: http://www.imtech.res.in/raghava/algpred/ [Accessed: 2016-03-07].

[85] The web tool EVALLER (http://bioinformatics.bmc.uu.se/evaller.html) could not be accessed on 2016-06-01

[86] Barrio AM, Soeria-Atmadja D, Nister A, Gustafsson MG, Hammerling U, Bongcam-Rudloff E. EVALLER: a web server for in silico assessment of potential protein allergenicity. Nucl Acids Res. 2007;35:W694–W700. DOI: 10.1093/nar/gkm370

[87] The web tool AllerTool (http://research.i2r.a-star.edu.sg/AllerTool/) could not be accessed on 2016-06-01

[88] AllerHunter. Available from: http://tiger.dbs.nus.edu.sg/AllerHunter/ [Accessed: 2016-03-07].

[89] Muh HC, Tong JC, Tammi MT. AllerHunter: a SVM-pairwise system for assessment of allergenicity and allergic cross-reactivity in proteins. PLoS One. 2009;4:e5861. DOI: 10.1371/journal.pone.0005861

[90] The web tool APPEL (http://jing.cz3.nus.edu.sg/cgi-bin/APPEL) could not be accessed on 2016-06-01

[91] Cui J, Han LY, Li H, Ung CY, Tang ZQ, Zheng CJ, Cao ZW, Chen YZ. Computer prediction of allergen proteins from sequence-derived protein structural and physicochemical properties. Mol Immunol. 2007;44:514–520. DOI: 10.1016/j.molimm.2006.02.010

[92] SORTALLER. Available from: http://sortaller.gzhmu.edu.cn/ [Accessed: 2016-03-07].

[93] Zhang L, Huang Y, Zou Z, He Y, Chen X, Tao A. SORTALLER: predicting allergens using substantially optimized algorithm on allergen family featured peptides. Bioinformatics. 2012;28:2178–2179. DOI: 10.1093/bioinformatics/bts326

[94] Mohabatkar H, Mohammad Beigi M, Abdolahi K, Mohsenzadeh S. Prediction of allergenic proteins by means of the concept of Chou's pseudo amino acid composition and a machine learning approach. Med Chem. 2013;9:133–137. DOI: 10.2174/1573406411309010133

[95] PREAL. Available from: http://gmobl.sjtu.edu.cn/PREAL/index.php [Accessed: 2016-03-07].

[96] Wang J, Zhang D, Li J. PREAL: prediction of allergenic protein by maximum Relevance Minimum Redundancy (mRMR) feature selection. BMC Syst Biol. 2013;7:S9. DOI: 10.1186/1752-0509-7-S5-S9

[97] Chrysostomou C, Seker H. Prediction of protein allergenicity based on signal-processing bioinformatics approach. In: Proceedings of the Engineering in Medicine and Biology Society (EMBC), 36th Annual International Conference of the IEEE 2014. Chicago: IEEE; 2014. pp. 808–811. DOI: 10.1109/EMBC.2014.6943714

[98] Allerdictor. Available from: http://allerdictor.vbi.vt.edu/ [Accessed: 2016-03-07].

[99] Dang HX, Lawrence CB. Allerdictor: fast allergen prediction using text classification techniques. Bioinformatics. 2014;30:1120–1128. DOI: 10.1093/bioinformatics/btu004

[100] Wang J, Yu Y, Zhao Y, Zhang D, Li J. Evaluation and integration of existing methods for computational prediction of allergens. BMC Bioinformatics. 2013;14:S1. DOI: 10.1186/1471-2105-14-S4-S1

[101] proAP. Available from: http://gmobl.sjtu.edu.cn/proAP/main.html [Accessed: 2016-03-07].

[102] AllergenFP. Available from: http://ddg-pharmfac.net/AllergenFP/ [Accessed: 2016-03-07].

[103] Dimitrov I, Naneva L, Doytchinova I, Bangov I. AllergenFP: allergenicity prediction by descriptor fingerprints. Bioinformatics. 2014;30:846–851. DOI: 10.1093/bioinformatics/btt619

[104] Bragin AO, Demenkov PS, Kolchanov NA, Ivanisenko VA. Accuracy of protein allergenicity prediction can be improved by taking into account data on allergenic protein discontinuous peptides. J Biomol Struct Dyn. 2013;31:59–64. DOI: 10.1080/07391102.2012.691362

[105] Vijayakumar S, Lakshmi PTV. A fuzzy inference system for predicting allergenicity and allergic cross-reactivity in proteins. In: Proceedings of the Bioinformatics and Biomedicine (BIBM), IEEE International Conference; 18–21 December 2013. Shanghai: IEEE; 2013. pp. 49–52. DOI: 10.1109/BIBM.2013.6732458

[106] FuzzyApp. Available from: http://fuzzyapp.bicpu.edu.in/fuzzyapp.php [Accessed: 2016-03-07].

[107] Saravanan V, Lakshmi PT. Fuzzy logic for personalized healthcare and diagnostics: FuzzyApp—a fuzzy logic based allergen-protein predictor. Omics. 2014;18:570–581. DOI: 10.1089/omi.2014.0021

[108] Jiang B, Qu H, Hu Y, Ni T, Lin Z. Computational analysis of the relationship between allergenicity and digestibility of allergenic proteins in simulated gastric fluid. BMC Bioinformatics. 2007;8:375. DOI: 10.1186/1471-2105-8-375

[109] Yang X, Yu X. An introduction to epitope prediction methods and software. Rev Med Virol. 2009;19:77–96

[110] Wang HW, Pai TW. Machine learning-based methods for prediction of linear B-cell epitopes. Methods Mol Biol. 2014;1184:217–236. DOI: 10.1007/978-1-4939-1115-8_12

[111] Saha S, Raghava GP. Prediction of continuous B-cell epitopes in an antigen using recurrent neural network. Proteins. 2006;65:40–48. DOI: 10.1002/prot.21078

[112] Shen W, Cao Y, Cha L, Zhang X, Ying X, Zhang W, Ge K, Li W, Zhong L. Predicting

linear B-cell epitopes using amino acid anchoring pair composition. BioData Min. 2015;8:14. DOI: 10.1186/s13040-015-0047-3

[113] EL-Manzalawy Y, Dobbs D, Honavar V. Predicting linear B-cell epitopes using string kernels. J Mol Recognit. 2008;21:243–255. DOI: 10.1002/jmr.893

[114] Saha S, Raghava GP. BcePred: prediction of continuous B-cell epitopes in antigenic sequences using physico-chemical properties. In: Nicosia G, Cutello V, Bentley P, Timmis J, eds. Artificial immune systems. Berlin: Springer; 2004 p. 197–204. DOI: 10.1007/978-3-540-30220-9_16

[115] Larsen JE, Lund O, Nielsen M. Improved method for predicting linear B-cell epitopes. Immunome Res. 2006;2:1. DOI: 10.1186/1745-7580-2-2

[116] Gao J, Faraggi E, Zhou Y, Ruan J, Kurgan L. BEST: improved prediction of B-cell epitopes from antigen sequences. PLoS One. 2012;7:e40104. DOI: 10.1371/journal.pone.0040104

[117] Wee JL, Simarmata D, Kam YW, Ng FL, Tong JC. SVM-based prediction of linear B-cell epitopes using Bayes Feature Extraction. BMC Genomics. 2010;11:1. DOI: 10.1186/1471-2164-11-S4-S21

[118] Sweredoski MJ, Baldi P. COBEpro: a novel system for predicting continuous B-cell epitopes. Protein Eng Des Sel. 2009;22:113–120. DOI: 10.1093/protein/gzn075

[119] Lian Y, Ge M, Pan XM. EPMLR: sequence-based linear B-cell epitope prediction method using multiple linear regression. BMC Bioinformatics. 2014;15:414. DOI: 10.1186/s12859-014-0414-y

[120] Kim Y, Ponomarenko J, Zhu Z, Tamang D, Wang P, Greenbaum J, Lundegaard C, Sette A, Lund O, Bourne PE, Nielsen M. Immune epitope database analysis resource. Nucl Acids Res. 2012;40:W525–W530. DOI: 10.1093/nar/gks438

[121] Singh H, Ansari HR, Raghava GP. Improved method for linear B-cell epitope prediction using antigen's primary sequence. PLoS One. 2013;8:e62216. DOI: 10.1371/journal.pone.0062216

[122] Yao B, Zhang L, Liang S, Zhang C. SVMTriP: a method to predict antigenic epitopes using support vector machine to integrate tri-peptide similarity and propensity. PLoS One. 2012;7:e45152. DOI: 10.1371/journal.pone.0045152

[123] Kolaskar AS, Kulkarni-Kale U. Prediction of three-dimensional structure and mapping of conformational epitopes of envelope glycoprotein of Japanese encephalitis virus. Virology. 1999;261:31–42. DOI: 10.1006/viro.1999.9859

[124] Kulkarni-Kale U, Bhosle S, Kolaskar AS. CEP: a conformational epitope prediction server. Nucl Acids Res. 2005;33:W168–W171. DOI: 10.1093/nar/gki460

[125] Kringelum JV, Lundegaard C, Lund O, Nielsen M. Reliable B cell epitope predictions: impacts of method development and improved benchmarking. PLoS Comput Biol. 2012;8:e1002829. DOI: 10.1371/journal.pcbi.1002829

[126] Qi T, Qiu T, Zhang Q, Tang K, Fan Y, Qiu J, Wu D, Zhang W, Chen Y, Gao J, Zhu R. SEPPA 2.0—more refined server to predict spatial epitope considering species of immune host and subcellular localization of protein antigen. Nucl Acids Res. 2014;42:W59–W63. DOI: 10.1093/nar/gku395

[127] Sweredoski MJ, Baldi P. PEPITO: improved discontinuous B-cell epitope prediction using multiple distance thresholds and half sphere exposure. Bioinformatics. 2008;24:1459–1460. DOI: 10.1093/bioinformatics/btn199

[128] Giacò L, Amicosante M, Fraziano M, Gherardini PF, Ausiello G, Helmer-Citterich M, Colizzi V, Cabibbo A. B-Pred, a structure based B-cell epitopes prediction server. Adv Appl Bioinform Chem. 2012;5:11–21. DOI: 10.2147/AABC.S30620

[129] Ansari HR, Raghava GP. Identification of conformational B-cell epitopes in an antigen from its primary sequence. Immunome Res. 2010;6:6. DOI: 10.1186/1745-7580-6-6

[130] Ponomarenko J, Bui HH, Li W, Fusseder N, Bourne PE, Sette A, Peters B. ElliPro: a new structure-based tool for the prediction of antibody epitopes. BMC Bioinformatics. 2008;9:1. DOI: 10.1186/1471-2105-9-514

[131] Rubinstein ND, Mayrose I, Martz E, Pupko T. Epitopia: a web-server for predicting B-cell epitopes. BMC Bioinformatics. 2009;10:287. DOI: 10.1186/1471-2105-10-287

[132] Liang S, Zheng D, Standley DM, Yao B, Zacharias M, Zhang C. EPSVR and EPMeta: prediction of antigenic epitopes using support vector regression and multiple server results. BMC Bioinformatics. 2010;11:381. DOI: 10.1186/1471-2105-11-381

[133] Liang S, Zheng D, Zhang C, Zacharias M. Prediction of antigenic epitopes on protein surfaces by consensus scoring. BMC Bioinformatics. 2009;10:302. DOI: 10.1186/1471-2105-10-302

[134] Zheng W, Ruan J, Hu G, Wang K, Hanlon M, Gao J. Analysis of conformational B-cell epitopes in the antibody–antigen complex using the depth function and the convex Hull. PLoS One. 2015;10:e0134835. DOI: 10.1371/journal.pone.0134835

[135] AgAbDb. Available from: http://bioinfo.net.in/AgAbDb.htm [Accessed: 2016-03-07].

[136] Weber CA, Mehta PJ, Ardito M, Moise L, Martin B, De Groot AS. T cell epitope: friend or foe? Immunogenicity of biologics in context. Adv Drug Deliv Rev. 2009;61:965–976. DOI: 10.1016/j.addr.2009.07.001

[137] Tong JC, Tan TW, Ranganathan S. Methods and protocols for prediction of immunogenic epitopes. Brief Bioinform. 2007;8:96–108. DOI: 10.1093/bib/bbl038

[138] Ponomarenko JV, Van Regenmortel MH. B cell epitope prediction. In: Gu J, Bourne PE, editors. Structural bioinformatics. New York: John Wiley & Sons; 2009. pp. 849–879.

[139] Desai DV, Kulkarni-Kale U. T-cell epitope prediction methods: an overview. In: De RK, Tomar N, editors. Immunoinformatics. New York: Springer; 2014. pp. 333–364. DOI: 10.1007/978-1-4939-1115-8_19

[140] Bhasin M, Raghava GP. Prediction of CTL epitopes using QM, SVM and ANN techniques. Vaccine. 2004;22:3195–3204. DOI: 10.1016/j.vaccine.2004.02.005

[141] Doytchinova IA, Guan P, Flower DR. EpiJen: a server for multistep T cell epitope prediction. BMC Bioinformatics. 2006;7:131. DOI: 10.1186/1471-2105-7-131

[142] Dimitrov I, Garnev P, Flower DR, Doytchinova I. EpiTOP—a proteochemometric tool for MHC class II binding prediction. Bioinformatics. 2010;26:2066–2068. DOI: 10.1093/bioinformatics/btq324

[143] Stranzl T, Larsen MV, Lundegaard C, Nielsen M. NetCTLpan: pan-specific MHC class I pathway epitope predictions. Immunogenetics. 2010;62:357–368. DOI: 10.1007/s00251-010-0441-4

[144] Karosiene E, Rasmussen M, Blicher T, Lund O, Buus S, Nielsen M. NetMHCIIpan-3.0, a common pan-specific MHC class II prediction method including all three human MHC class II isotypes, HLA-DR, HLA-DP and HLA-DQ. Immunogenetics. 2013;65:711–724. DOI: 10.1007/s00251-013-0720-y

[145] Oyarzún P, Ellis JJ, Bodén M, Kobe B. PREDIVAC: CD4+ T-cell epitope prediction for vaccine design that covers 95% of HLA class II DR protein diversity. BMC Bioinformatics. 2013;14:52. DOI: 10.1186/1471-2105-14-52

[146] Schuler MM, Nastke MD, Stevanović S. SYFPEITHI: database for searching and T-cell epitope prediction. Methods Mol Biol. 2007;409:75–93. DOI: 10.1007/978-1-60327-118-9_5

[147] Zhang L, Chen Y, Wong HS, Zhou S, Mamitsuka H, Zhu S. TEPITOPEpan: extending TEPITOPE for peptide binding prediction covering over 700 HLA-DR molecules. PLoS One. 2012;7:e30483. DOI: 10.1371/journal.pone.0030483

[148] Donnes P, Kohlbacher O. Integrated modeling of the major events in the MHC class I antigen processing pathway. Protein Sci. 2005;14:2132–2140. DOI: 10.1110/ps.051352405

[149] Farrell D, Gordon SV. Epitopemap: a web application for integrated whole proteome epitope prediction. BMC Bioinformatics. 2015;16:221. DOI: 10.1186/s12859-015-0659-0

[150] Sela-Culang I, Ofran Y, Peters B. Antibody specific epitope prediction—emergence of a new paradigm. Curr Opin Virol. 2015;11:98–102. DOI: 10.1016/j.coviro.2015.03.012

[151] Tong JC, Tammi MT. Prediction of protein allergenicity using local description of amino acid sequence. Front Biosci. 2007;13:6072–6078. DOI: 10.2741/3138

[152] Dall'Antonia F, Gieras A, Devanaboyina SC, Valenta R, Keller W. Prediction of IgE-binding epitopes by means of allergen surface comparison and correlation to cross-reactivity. J Allergy Clin Immunol. 2011;128:872–879. DOI: 10.1016/j.jaci.2011.07.007

[153] Lollier V, Denery-Papini S, Brossard C, Tessier D. Meta-analysis of IgE-binding allergen epitopes. Clin Immunol. 2014;153:31–39. DOI: 10.1016/j.clim.2014.03.010

[154] LocAllEpi. Available from: http://wwwappli.nantes.inra.fr:8180/LocAllEpi/ [Accessed: 2016-03-07].

[155] Aalberse RC, Stadler BM. In silico predictability of allergenicity: from amino acid sequence via 3-D structure to allergenicity. Mol Nutr Food Res. 2006;50:625–627. DOI: 10.1002/mnfr.200500270

[156] Canonica GW, Bachert C, Hellings P, Ryan D, Valovirta E, Wickman M, De Beaumont O, Bousquet J. Allergen immunotherapy (AIT): a prototype of precision medicine. World Allergy Organ J. 2015;8:31. DOI: 10.1186/s40413-015-0079-7

7

Bioinformatics Approaches for Predicting Kinase–Substrate Relationships

Daniel A. Bórquez and Christian González-Billault

Abstract

Protein phosphorylation, catalyzed by protein kinases, is the main posttranslational modification in eukaryotes, regulating essential aspects of cellular function. Using mass spectrometry techniques, a profound knowledge has been achieved in the localization of phosphorylated residues at proteomic scale. Although it is still largely unknown, the protein kinases are responsible for such modifications. To fill this gap, many computational algorithms have been developed, which are capable to predict kinase–substrate relationships. The greatest difficulty for these approaches is to model the complex nature that determines kinase–substrate specificity. The vast majority of predictors is based on the linear primary sequence pattern that surrounds phosphorylation sites. However, in the intracellular environment the protein kinase specificity is influenced by contextual factors, such as protein–protein interactions, substrates co-expression patterns, and subcellular localization. Only recently, the development of phosphorylation predictors has begun to incorporate these variables, significantly improving specificity of these methods. An accurate modeling of kinase–substrate relationships could be the greatest contribution of bioinformatics to understand physiological cell signaling and its pathological impairment.

Keywords: protein kinases, phosphorylation, machine learning methods, docking sites

1. Introduction

Protein kinases is the second largest family of enzymes, composed by 518 members in the human genome [1]. These enzymes catalyze the transference of γ-phosphate moiety from adenosine triphosphate (ATP) to the hydroxyl group of serine, threonine, or tyrosine residues present in substrate proteins. The transient nature of this modification (reversed by dephosphorylation reactions, catalyzed by protein phosphatases) generates the main molecular switch, regulating each aspect of protein function, including interactions, conformations, subcellular localization, enzymatic activity, and turnover. Protein phosphorylation is also the most widespread post-translational modification, affecting at least three-quarters of the proteome [2].

The identification of phosphorylated sites (or phosphosites) has experienced an explosion with the utilization of mass spectrometry techniques. PhosphoSitePlus database [3] collects large part of the information obtained in these studies, including the localization of 144,899 serines, 61,654 threonines, and 41,273 phosphorylated tyrosines, but only 12,180 (5%) of them have annotated the protein kinase responsible for such modifications [3]. This is largely due to the expensive and time-consuming methodologies that need to be used in the identification of kinase–substrate relationships (KSRs).

This complex scenario has opened an important field for the development of computational strategies for phosphorylation site labeling with the specific protein kinase(s) responsible for its modifications in a whole proteome scale, in an effort to reconstruct the underlying regulatory networks. These approaches must overcome several challenges including the complexity of the regulatory networks itself, and the scarce information available about the molecular mechanisms that ensure recognition between protein kinases and substrates.

The list of currently developed tools for KSRs prediction is shown in **Table 1**. Of note, most of these tools are mostly based on classifiers designed to assign a phosphorylation site to a particular protein kinase considering only the sequence pattern surrounding the phosphorylation site, which provides an imperfect description of the kinase–substrate specificity. In this chapter, we will discuss the underlying biological rational of these tools and its potential for improvement.

Method name	Approach	Contextual information	Training data	Number of kinase families	References	Website
HeteSim	Heterogeneous information networks	Yes	Phospho. ELM	210	[6]	No web implementation available
SlapRLS	Supervised Laplacian regularized least squares	No	Phospho. ELM	23	[7]	No web implementation available
PUEL	Positive-unlabeled ensemble	Yes	PhosphoSite Plus + Literature	2	[8]	https://github.com/PengyiYang/KSP-PUEL

Method name	Approach	Contextual information	Training data	Number of kinase families	References	Website
	learning (SVM and PSSM-based)					
Scansite 3	PSSM	No	Experimental data (*in vitro*)	62	[9]. Previous developments: [10] (Scansite) [11] (Scansite 2.0)	http://scansite3.mit.edu/
Phospho PICK	Bayesian network	Yes	Phospho. ELM + HPRD	45	[12]	http://bioinf.scmb.uq.edu.au/phosphopick/
No name	SVM	No	dbPTM	16	[13]	No web implementation available
Kinome Xplorer	ANN and PSSM	Yes	Phospho. ELM	222	[14]. Previous developments: [15] (Net Phorest) [16] (NetworKIN)	http://kinomexplorer.info/
RegPhos 2.0	SVM	Yes	dbPTM	122	[17]. Previous developments: [18] (RegPhos)	http://csb.cse.yzu.edu.tw/RegPhos2/
PSEA	Set enrichment analysis (BLOSUM62 similarity)	No	PhosphoSite Plus	27	[19]	http://bioinfo.ncu.edu.cn/PKPred_Home.aspx
PhosNet Construct	Similarity matrix	No	PhosphoSite Plus	61	[20]	http://202.54.249.142/~nikhil/network_reconstruction/index.html
No name	SVM	Yes	Phospho. ELM	21	[21]	No web implementation available
phos_pred	Random forest	Yes	Phospho. ELM	54	[22]	http://bioinformatics.ustc.edu.cn/phos_pred/
PKIS	SVM	No	Phospho. ELM	56	[23]	http://bioinformatics.ustc.edu.cn/pkis/
iGPS	Optimized	Yes	Phospho.ELM	69	[24]. Previous	http://igps.biocuckoo.org/

Method name	Approach	Contextual information	Training data	Number of kinase families	References	Website
	BLOSUM62 similarity		+ literature		developments: [25] (GPS 2.1) [26] (GPS 2.0) [27, 28] (GPS)	
ConDens	Conservation of local motif density	No	Unnecessary	All kinases with known motifs	[29]	http://www.moseslab.csb.utoronto.ca/andyl/
No name	PSSM	No	Unnecessary	492	[30, 31]	No web implementation available
PrediKin	PSSM	No	Unnecessary	All	[32]. Previous developments: [33–35]	http://predikin.biosci.uq.edu.au/
Musite	SVM	No	Phospho.ELM + Swiss-Prot + PhosphoPep	13	[36, 37]	http://musite.sourceforge.net/
Phos3D	SVM	No	Phospho. ELM	6	[38]	http://phos3d.mpimp-golm.mpg.de/
NetphosK	ANN	No		17	[39, 40]	http://www.cbs.dtu.dk/services/NetPhosK/
CRPhos	Conditional random fields	No	Phospho. ELM	18	[41]	http://www.ptools.ua.ac.be/CRPhos/
MetaPred PS	Meta predictor	No	Phospho.ELM + PhosphoSite Plus + Swiss-Prot	15	[42]	Web implementation no longer available
PhoScan	Log odds ratios	No	Phospho. ELM	48	[43]	http://bioinfo.au.tsinghua.edu.cn/phoscan/
Kinase Phos 2.0	SVM	No	Phospho. ELM + Swiss-Prot	71	[44]. Previous development: [45] (Kinase Phos 1.0)	http://kinasephos2.mbc.nctu.edu.tw/
PPSP	Bayesian	No	Phospho.	68	[46]	Web implementation

Method name	Approach	Contextual information	Training data	Number of kinase families	References	Website
	decision theory		ELM			no longer available
Pred Phospho	SVM	No	Phospho Base	4	[47]	Web implementation no longer available

Table 1. Computational methods for kinase–substrate relationships prediction.

2. Comparing prediction tools: data, metrics, and methods

One of the most challenging problems in the field of prediction tools is to establish benchmarks between them, allowing a real assessment of the method itself. Each prediction tool requires for its testing (and often for training) a set of positive (actually phosphorylated sites) and negative data (sites actually not phosphorylated). The sources of phosphorylated sites for most of the predictors are limited to a few databases as Phospho.ELM [4] and PhosphositePlus [3]. These databases include information from different experimental approaches (*in vivo* and/or *in vivo*), which is processed homogeneously for training prediction algorithms. This can lead to a significant bias in the quality of predictions: protein kinases exhibit low specificity at *in vitro* experiments (which constitute the largest proportion of the information in databases), generating simpler motifs than those that may present in cells [5]. Moreover, using information derived only from *in vivo* experiments does not ensure that the observed phosphorylation site was directly phosphorylated by the protein kinase studied. Careful selection of the positive data set for training, including only the sites phosphorylated both *in vivo* (physiological) and *in vitro* (direct) by a protein kinase, can significantly improve the prediction [5].

Another problem is the construction of a negative data set used in machine-learning methods. Although experiments can verify that a protein kinase phosphorylates a given residue, it is very difficult to demonstrate that a particular residue in a protein is not phosphorylated at any situation. A good approximation is made by Neuberger et al. [48], which consider any residue present in a protein which is phosphorylated by a particular protein kinase and which has not been reported as phosphorylated in databases, as part of the negative data set.

The sensitivity (S_n) and specificity (S_p) are commonly used to assess the performance of prediction algorithms. For a set of data predicted as positive, real positive (previously experimentally determined as phosphorylated) are called true positives (TP), while the remaining are called false positives (FP). Concomitantly, for data predicted as no phosphorylated sites, the real ones are called true negatives (TN) whereas phosphorylated sites are considered false negatives (FN). The ratio of positive sites can be correctly classified is named sensitivity (S_n). On the other hand, proportion of negatives sites correctly identified, is called specificity (S_p). Both parameters are calculated as follows (Eq. (1)):

$$S_n = \frac{TP}{TP + FN}$$

$$S_p = \frac{TN}{TN + FP}$$

Other common parameter used to evaluate the predictor performance is the accuracy that denotes the percentage of correct prediction in both negative and positive data sets. Also, Matthews correlation coefficient (MCC) is widely used as a general estimator for the performance of a predictor. It considers the four numbers described in Eq. (1), giving a balanced assessment of the performance of the predictor even if these parameters are very different. Both parameters are calculated as follows:

$$Ac = \frac{TP + TN}{TP + FP + TN + FN}$$

$$MCC = \frac{(TP \times TN) - (FN \times FP)}{\sqrt{(TP + FN) \times (TN + FP) \times (TP + FP) \times (TN + FN)}}$$

The ROC (receiver operating characteristic) curve is commonly used for evaluation and comparison of classifiers. The true positive rate (sensitivity) is plotted against the false positive rate (1-specificity). For a perfect classifier the area under curve (AUC) is 1, while a poor classifier achieved values near 0.5 (which defines a random guess).

It may be considered that the parameters previously described can only be compared when the data sets of TP and TN are similar, which is relatively common for the former, but very uncommon for the later. It should be necessary to define standard data sets that are occupied by the research community or define benchmarks that are independent of the training data sets. One approach is to compare the predictors based on their ability to assign to known phosphorylation sites the lowest ranking in a proteomic search for phosphosites of a given protein kinase [5].

Apart from the difficulties to make quantitative comparisons between predictors, there is a perception that machine-learning algorithms, as artificial neural networks (ANN) or support vector machines (SVM), provide a better predictability of protein kinase substrates that simpler methods such as the position-specific scoring matrices (PSSM). Such an idea is substantiated in the assumption that machine-learning algorithms are capable of classifying highly complex sequences, in which correlations amongst positions are important. Such an assumption was recently questioned by studying the sequence interpositional dependence for ataxia telangiectasia mutated (ATM/ATR) kinase, casein kinase 2 (CK2), and cyclin-dependent kinase 2 (CDK2) substrates. Through statistical analysis, Joughin and colleagues [49] found few pairs of positions in the sequences of the phosphorylated sites that are significantly deviated from the positionwise independence. Accordingly, the predictors that incorporate second-order information were less accurate than those who consider only first-order information, over fitting the training data [49]. This strongly suggests that a good strategy to develop most accurate predictive tools is to integrate simple sequence models with contextual information, such as protein–protein interactions, subcellular localization, and distal recognition sites.

3. Beyond the sequence: improving substrate prediction with contextual information

There are two factors, which are important to determine the specific phosphorylation of substrates by protein kinases: recruitment and phosphorylation site recognition. Recruitment

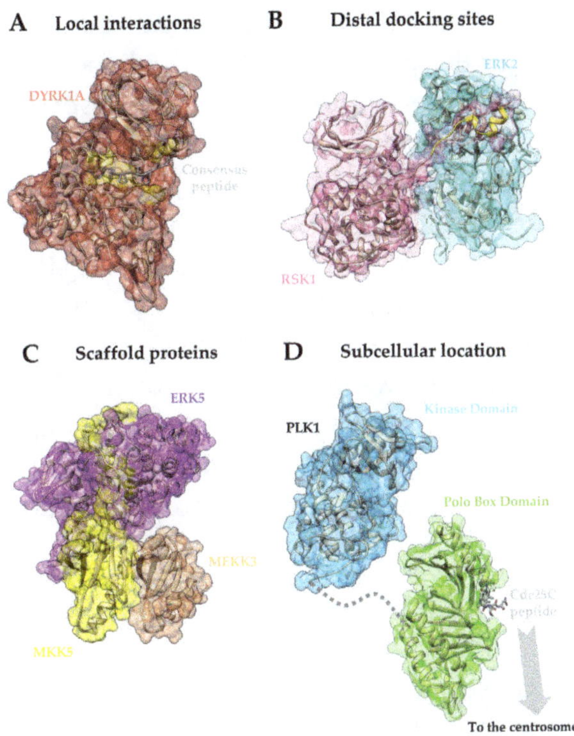

Figure 1. A structural view of protein kinases specificity determinants. (A) Local interactions. Extensive contacts are established between the protein kinase active site and the surrounding region of phophosite, which partly defines the specificity of protein kinases. For example, a complex between Dual-specificity tyrosine-phosphorylation-regulated kinase (DYRK)-1A and a consensus substrate peptide (ARPGT*PAL) is shown. Active site residues (F170, F196, Y246, D287, K289, E291, Y321, S324, Y327, E353; colored in yellow) establish contacts with the peptide substrate. (B) Distal docking sites. Often the interaction of the substrate with the active site of the protein kinase is not enough to ensure specificity. For example, Extracellular signal regulated kinase (ERK)-2 and its substrate Ribosomal S6 kinase (RSK)-1, establish additional contacts distal to the protein kinase active site, through a linear binding motif (colored in yellow). (C) Scaffolds proteins. Many kinases utilize scaffold proteins to be placed close to their substrates. For example, mitogen-activated protein kinase (MAPK) kinase (MKK)-5 organizes MAPK kinase kinase (MEKK)-3, MKK5, and ERK5 in a signaling complex. (D) Subcellular location. Through protein–protein interactions, protein kinases are located in specific subcellular structures, wherein phosphorylate specific substrates. For example, Polo-like kinase (PLK)-1 interacts through its Polo Box Domain with Cdc25C, allowing its centrosomal localization. Molecular graphics were performed with the UCSF Chimera package [56] based on the following structures: DYRK1A-substrate complex (PDB: 2WO6), ERK2-RSK1 complex (PDB: 2Y4I), ERK5-MKK5-MEKK3 ternary complex (made by superimposition of PDB: 4IC7 and PDB: 2O2V) and PLK1 kinase domain (PDB: 2OU7)-Polo box domain-Cdc25c complex (PDB: 2OJX).

is related with a number of determining factors that promote productive interaction between protein kinases and substrates. Phosphorylation site recognitions are related with the preference of individual residues surrounding the modified residue (**Figure 1**).

The relative importance of these factors has been rarely studied experimentally in the functional specificity of protein kinases. For example, in yeast, the high specificity of the mitogen-activated protein kinase kinase (MAPKK) is ensured mainly by the use of docking motifs and scaffolding interactions [50].

3.1. Distal docking sites

A transient physical interaction between protein kinases and their substrates could place them in close proximity and in the correct orientation, creating the opportunity for post-translational modification. These interactions are based on short linear motifs, termed docking sites, that reside in disordered regions of the proteins, and that only adopt a defined structure upon binding. The utilization of docking sites seems to be a widespread strategy to improve the specificity of protein kinases to phosphorylate defined phosphosites, as evidenced by studies of SR protein-specific kinase-1 (SRPK1) [51], Cbk1 [52], Polo-like kinases [53], and Cdks [54, 55]. Due to the lack of structural models of interaction between complete substrates and protein kinases, it is not yet possible to measure the importance of distal interactions to the site of phosphorylation in specificity determination. However, by combining protein–protein docking and adaptive biasing force molecular dynamics simulations, Mottin et al. obtained a structural model of the interaction between an active protein kinase (the complex Cdk5/p25) and a complete substrate: Peroxisome proliferator-activated receptor ^{3} (PPAR^{3}). This model suggests that the protein kinase establishes two distal docking sites with the substrate, pinpointing the importance of those contact sites for proper positioning of the phosphosite in the kinase active site [57].

Mitogen-activated protein kinases (MAPKs) are the prototypical example of using docking sites to enhance the specificity of their promiscuous active sites, which phosphorylate their substrates at most in a weak consensus site (Ser/Thr-Pro). Two types of docking sites have been characterized for MAPKs, called D-sites and F-sites. D-sites interacts with a D-recruitment site (DRS), consisting of a negatively charged region and a shallow hydrophobic pocket, located on the opposite side from the kinase active site. On the other hand, F-sites bind a hydrophobic docking groove (F-recruitment site or FRS). Although all MAPKs have a DRS (hence, it is referred, too, as Common Docking or CD), the FRS seems to be a characteristic only of ERK1/2 and p38α.

Structural studies of the interaction between extracellular-regulated kinase 2 (ERK2) and a peptide containing the F-site from Elk-1 suggests that the main effect is to increase the local phosphosite concentration, enhancing productive encounters that enhance phosphorylation [58].

Although systematic *in silico* exploration of docking sites could be helpful to find new putative substrates for MAPKs, it has been hampered by low stringency of the sequence, generating a high rate of false positives that can stochastically occur.

Whisenant et al. [59] developed D-finder, a computational tool that uses a hybrid pattern matching algorithm/hidden Markov model, to find sequences of docking sites at protein kinase substrates. Trained with 20 experimentally identified D-sites for a c-Jun N-terminal kinase (JNK), they identified 394 proteins with putative D-sites, and experimentally validated some of them [59]. In order to prioritize the predictions made by D-finder, this was combined with a predictor of JNK phosphorylation sites based on position weight matrices, achieving the identification and experimental validation of one additional substrate of JNK [60].

Significant efforts have been done to study the determinants of specificity of the DRS [61, 62], results that undoubtedly will improve the design of predictors. More recently, Zeke and colleagues [63] addressed the problem by separating the D-sites in 4 class of motifs based on experimental studies and structural and evolutionary analysis, generating PSSMs that allowed them to generate specific interactomes for each class of motif [63].

Utilization of docking motifs to enhance the accuracy of the predictions is still incipient for other kinases. Bibi and collaborators [64] conducted a search for new substrates of Plk1, using both the consensus sequence of phosphorylation site as the recognition motif of the Polo-box domain, responsible for the recruitment of the substrates and activation of the protein kinase [64].

3.2. Scaffold proteins

Scaffold proteins are important to ensure the encounter of the protein kinases with its substrates, as suggested by a large number of these signaling proteins associated with phosphorylation [65], especially in the MAPK pathway, where a recent interactome analysis identified 10 scaffold proteins associated [66]. A canonical example of utilization of scaffold proteins is found in the signaling of cAMP-activating protein kinase (PKA), where more than 50 A-kinase anchoring proteins (AKAPs) are responsible for associating the protein kinase with its substrates [67].

The first attempt to integrate protein interactions with PhosphoSite sequence patterns to improve KSR prediction was the development of NetworKIN [16], a two-stage algorithm. In the first term, artificial neural networks (from NetPhosK) and PSSMs (from Scansite) are used to label a given phosphosite sequence with a kinase or kinase family. In a second stage, the contextual information is included, calculating the proximity of the substrate to all kinases (the most likely route connecting them) in a network of functional relationships extracted from the Search Tool for the Retrieval of Interacting Genes/Proteins (STRING) database [68].

For a group of well-studies kinase families (CDK, PKC, PIKK and INSR) NetwortKIN doubled prediction accuracy (to 64%) over only sequence-based methods. However, NetworKIN includes a circular logic system, which can overestimate the values of accuracy: the performance assessment was performed with known phosphosites derived from the literature and the STRING database also includes information from text-mining (as co-occurrence in abstracts), therefore associates many kinases with its substrates. NetworKIN has recently been upgraded into KinomeXplorer platform [14]. This introduces improvements in the accuracy of prediction through a new scoring scheme based on naive Bayes method, to overcome the bias in the network structure caused by highly studied proteins. Another algorithm based on sequence and contextual information is PhosphoPICK. It integrates information from previously known KSRs, protein–protein interactions and protein abundance profiles through the cell cycle in a Bayesian network model. The average AUC for the 59 protein kinases evaluated was 0.86. The main determinant of this good performance was the inclusion of data from protein–protein interactions, while the protein expression throughout the cell cycle had a modest contribution [12].

To our knowledge, no predictor used only physical protein–protein interaction information, which better translates the concept that kinases phosphorylate proteins that are in close proximity. Recently, studies carried out by affinity purification coupled with mass spectrometry revealed the structure of many protein complexes in human cells [69, 70], including some specific for protein kinases [71, 72] or signaling pathways associated with them [73, 74], which can be exploited to improve sequence-only based predictors.

Contextual information can also be given by the location of the protein kinase in a specific subcellular structure, which would have privileged access to certain substrates. Alexander and colleagues elegantly demonstrated that substrate specificity between mitotic kinases, Aurora A/B, Cdk1, Plk1 and Nek2 are based on mutually exclusive motifs in the case of protein kinases presenting overlapping location and for similar motifs in protein kinases showing an exclusive distribution [75].

4. Structural aspects of phosphosites

It has been traditionally assumed that the phosphorylation site can be described as a linear sequence and this assumption underlies almost all predictors developed to date. However, it has been recently identified that the Thr253 of the α-tubulin is located in a nonlinear motif, comprising residues distant in primary sequence but which are folded in a consensus site phosphorylated by PKC [76]. Durek and collaborators [38] had previously addressed this problem by characterizing the structural motifs (3D) present in the phosphorylation sites. Using only the radial distance from the phosphorylated residue and regardless of the angular information, they achieved a modest increase in performance over only linear sequence-predictions [38]. This can be explained if only a limited number of residues are recognized by protein kinases based on a structural epitope or by the low complexity of the model used. Currently, they have developed more sophisticated tools to find patterns in protein 3D structures that could be useful to identify KSRs based on nonlinear motifs. For example, Amino acid pattern Search for Substructures And Motifs (ASSAM) [77]) uses 3D coordinates of a motif comprising up to 12 amino acids, for matching in a structure database as the Protein Data Bank (PDB). ASSAM represents protein structures as a graph in which nodes consist of a vector between two pseudoatoms, representing the side chains of the individual amino acids, while the edges are the distances between the corresponding vectors, providing a more exhaustive representation that used by Durek and collaborators [38].

To date only one predictor, denominated MODPROPEP [78], is based exclusively on structural features of the interaction between the protein kinase and the phosphorylation site. MODPROPEP modulates putative substrate peptides into the active site of the protein kinases, using 49 kinase-peptide complexes with structure available in the Protein Data Bank (PDB) as templates. The scoring function is based on the binding energy of the peptides with the kinase calculated using a statistical-based residue pair potential [79]. Owing to the low accuracy in some families of kinases, a new scoring scheme based on molecular mechanics Poisson-Boltzmann Surface Area (MM-PBSA) binding energy values was generated. The performance with this improved scoring system was similar to predictors based only on sequence as GPS, PPSP, Scansite and NetPhosK, but with the advantage that MODPROPEP not require a training set of known sites, which is a remarkable advantage for prediction of previously uncharacterized protein kinases.

5. The kinase side: determinants of specificity

The combinations of residues in the kinase domain that allow substrate specificity are known determinants of specificity (DoS). For example, discrimination between Ser and Thr by eukaryotic protein kinases appears to depend on the nature of a single residue immediately adjacent to the DFG motif (DFG+1). Whereas most of the Thr-specific kinases have a β-branched aliphatic residue at this position (as Ile, Val or Thr), Ser-specific kinases have large hydrophobic residues (predominantly Leu, Phe, and Met). All non-selective kinases have Leu or Ser as the DFG+1 residue. The mutation of this residue is enough to modify the amino acid preference of multiple protein kinases [80].

The idea of building predictors based on the characteristics of the kinase domain can overcome the problem of building a positive data set of phosphorylated sequences, an aspect especially complicated for poorly characterized protein kinases, and ideally would serve to perform explorations at whole-kinome level. The first DoS-based predictor was Predikin, which allows automatic prediction of peptide substrates using only the amino acid sequence of the protein kinase [33]. This algorithm has been continuously improved [34], achieving the most accurate prediction of experimentally obtained position weight matrices in the category of Domain Recognition Peptide/Kinase protein of Dialogue for Reverse Engineering Assessments and Methods (DREAM4) challenge [32].

On the other hand, Safaei and colleagues [30] using the primary sequence of the kinase catalytic domains, generate PSSMs describing substrate specificity for 492 protein kinases. For this purpose, residues act as DoS are identified using multiple alignments of the kinase domain and the correlation between these and those residues present in the consensus sequences derived from known KSRs [30]. Although the performance of this strategy was similar to NetPhorest, the advantage it does not require information about substrates.

A computational methodology called KINspect (based on learning classifier systems) was recently developed, in order to predict DoS through an iterative process based on randomly generated specificity masks, which are progressively improved in its predictive ability through mutation and crossover [81]. This sophisticated approach, transferred to a predictor of KSRs, possibly will overcome previous advances.

6. Conclusions

In silico prediction of KSRs may become the most important contribution of bioinformatics toward the understanding of cell signaling. To date, there are no high-throughput experimental techniques allowing pairing thousands of phosphorylation sites (identified by mass spectrometry) with protein kinases that catalyzing such reactions. Recent studies provide a deepest characterization of the determinants of specificity of protein kinases for their substrates, enabling a more realistically modeling of the KSRs. These models reduce the rate of false positives and support the construction of feasible regulatory networks.

Author details

Daniel A. Bórquez[1] and Christian González-Billault[2,3*]

*Address all correspondence to: chrgonza@uchile.cl

1 Biomedical Research Center, School of Medicine, Diego Portales University, Santiago, Chile

2 Department of Biology, Faculty of Sciences, University of Chile, Santiago, Chile

3 FONDAP Geroscience Center, Brain Health and Metabolism, Santiago, Chile

References

[1] Manning G, Whyte DB, Martinez R, Hunter T, Sudarsanam S. The protein kinase complement of the human genome. Science. 2002;298:1912–34. DOI:10.1126/science.1075762.

[2] Sharma K, D'Souza RC, Tyanova S, Schaab C, Wisniewski JR, Cox J et al. Ultradeep human phosphoproteome reveals a distinct regulatory nature of Tyr and Ser/Thr-based signaling. Cell Reports. 2014;8:1583–94. DOI:10.1016/j.celrep.2014.07.036.

[3] Hornbeck PV, Zhang B, Murray B, Kornhauser JM, Latham V, Skrzypek E. PhosphoSitePlus, 2014: mutations, PTMs and recalibrations. Nucleic Acids Research. 2015;43:D512–20. DOI:10.1093/nar/gku1267.

[4] Dinkel H, Chica C, Via A, Gould CM, Jensen LJ, Gibson TJ et al. Phospho.ELM: a database of phosphorylation sites—update 2011. Nucleic Acids Research. 2011;39:D261–7. DOI:10.1093/nar/gkq1104.

[5] Borquez DA, Olmos C, Alvarez S, Di Genova A, Maass A, Gonzalez-Billault C. Bioinformatic survey for new physiological substrates of Cyclin-dependent kinase 5. Genomics. 2013;101:221–8. DOI:10.1016/j.ygeno.2013.01.003.

[6] Li H, Wang M, Xu X. Prediction of kinase–substrate relations based on heterogeneous networks. Journal of Bioinformatics and Computational Biology. 2015;13:1542003. DOI: 10.1142/S0219720015420032.

[7] Li A, Xu X, Zhang H, Wang M. Kinase identification with supervised laplacian regularized least squares. PLoS One. 2015;10:e0139676. DOI:10.1371/journal.pone. 0139676.

[8] Yang P, Humphrey SJ, James DE, Yang YH, Jothi R. Positive-unlabeled ensemble learning for kinase substrate prediction from dynamic phosphoproteomics data. Bioinformatics. 2016;32:252–9. DOI:10.1093/bioinformatics/btv550.

[9] Ehrenberger T, Cantley LC, Yaffe MB. Computational prediction of protein–protein interactions. Methods in Molecular Biology. 2015;1278:57–75. DOI: 10.1007/978-1-4939-2425-7_4.

[10] Yaffe MB, Leparc GG, Lai J, Obata T, Volinia S, Cantley LC. A motif-based profile scanning approach for genome-wide prediction of signaling pathways. Nature Biotechnology. 2001;19:348–53. DOI:10.1038/86737.

[11] Obenauer JC, Cantley LC, Yaffe MB. Scansite 2.0: Proteome-wide prediction of cell signaling interactions using short sequence motifs. Nucleic Acids Research. 2003;31:3635–41. DOI:10.1093/nar/gkg584

[12] Patrick R, Le Cao KA, Kobe B, Boden M. PhosphoPICK: modelling cellular context to map kinase–substrate phosphorylation events. Bioinformatics. 2015;31:382–9.DOI: 10.1093/bioinformatics/btu663.

[13] Su M-G, Huang K-Y, Tung C-H, Lee T-Y. A new scheme to predict kinase-specific phosphorylation sites on protein three-dimensional structures. International Journal of Bioscience, Biochemistry and Bioinformatics. 2013;3: 473. DOI:10.7763/IJBBB. 2013.V3.258

[14] Horn H, Schoof EM, Kim J, Robin X, Miller ML, Diella F et al. KinomeXplorer: an integrated platform for kinome biology studies. Nature Methods. 2014;11:603–4. DOI: 10.1038/nmeth.2968.

[15] Miller ML, Jensen LJ, Diella F, Jorgensen C, Tinti M, Li L et al. Linear motif atlas for phosphorylation-dependent signaling. Science Signaling. 2008;1:ra2. DOI:10.1126/ scisignal.1159433.

[16] Linding R, Jensen LJ, Ostheimer GJ, van Vugt MA, Jorgensen C, Miron IM et al. Systematic discovery of in vivo phosphorylation networks. Cell. 2007;129:1415–26. DOI:10.1016/j.cell.2007.05.052.

[17] Huang KY, Wu HY, Chen YJ, Lu CT, Su MG, Hsieh YC et al. RegPhos 2.0: an updated resource to explore protein kinase–substrate phosphorylation networks in mammals. Database: the Journal of Biological Databases and Curation. 2014;2014:bau034. DOI: 10.1093/database/bau034.

[18] Lee TY, Bo-Kai Hsu J, Chang WC, Huang HD. RegPhos: a system to explore the protein kinase–substrate phosphorylation network in humans. Nucleic Acids Research. 2011;39:D777–87. DOI:10.1093/nar/gkq970.

[19] Suo SB, Qiu JD, Shi SP, Chen X, Liang RP. PSEA: Kinase-specific prediction and analysis of human phosphorylation substrates. Scientific Reports. 2014;4:4524. DOI:10.1038/srep04524.

[20] Damle NP, Mohanty D. Deciphering kinase–substrate relationships by analysis of domain-specific phosphorylation network. Bioinformatics. 2014;30:1730–8. DOI: 10.1093/bioinformatics/btu112.

[21] Xu X, Li A, Zou L, Shen Y, Fan W, Wang M. Improving the performance of protein kinase identification via high dimensional protein–protein interactions and substrate structure data. Molecular Biosystems. 2014;10:694–702. DOI:10.1039/c3mb70462a.

[22] Fan W, Xu X, Shen Y, Feng H, Li A, Wang M. Prediction of protein kinase-specific phosphorylation sites in hierarchical structure using functional information and random forest. Amino Acids. 2014;46:1069–78. DOI:10.1007/s00726-014-1669-3.

[23] Zou L, Wang M, Shen Y, Liao J, Li A, Wang M. PKIS: computational identification of protein kinases for experimentally discovered protein phosphorylation sites. BMC Bioinformatics. 2013;14:247. DOI:10.1186/1471-2105-14-247.

[24] Song C, Ye M, Liu Z, Cheng H, Jiang X, Han G et al. Systematic analysis of protein phosphorylation networks from phosphoproteomic data. Molecular & Cellular Proteomics: MCP. 2012;11:1070–83. DOI:10.1074/mcp.M111.012625.

[25] Xue Y, Liu Z, Cao J, Ma Q, Gao X, Wang Q et al. GPS 2.1: enhanced prediction of kinase-specific phosphorylation sites with an algorithm of motif length selection. Protein Engineering, Design & Selection: PEDS. 2011;24:255–60. DOI:10.1093/protein/gzq094.

[26] Xue Y, Ren J, Gao X, Jin C, Wen L, Yao X. GPS 2.0, a tool to predict kinase-specific phosphorylation sites in hierarchy. Molecular & Cellular Proteomics: MCP. 2008;7:1598–608. DOI:10.1074/mcp.M700574-MCP200.

[27] Xue Y, Zhou F, Zhu M, Ahmed K, Chen G, Yao X. GPS: a comprehensive www server for phosphorylation sites prediction. Nucleic Acids Research. 2005;33:W184–7. DOI: 10.1093/nar/gki393.

[28] Zhou FF, Xue Y, Chen GL, Yao X. GPS: a novel group-based phosphorylation predicting and scoring method. Biochemical and Biophysical Research Communications. 2004;325:1443–8. DOI:10.1016/j.bbrc.2004.11.001.

[29] Lai AC, Nguyen Ba AN, Moses AM. Predicting kinase substrates using conservation of local motif density. Bioinformatics. 2012;28:962–9. DOI:10.1093/bioinformatics/bts060.

[30] Safaei J, Manuch J, Gupta A, Stacho L, Pelech S. Prediction of 492 human protein kinase substrate specificities. Proteome Science. 2011;9 Suppl 1:S6. DOI:10.1186/1477-5956-9-S1-S6.

[31] Safaei J, Manuch J, Gupta A, Stacho L, Pelech S, editors. Prediction of human protein kinase substrate specificities. Prediction of human protein kinase substrate specificities; 2010: IEEE.

[32] Ellis JJ, Kobe B. Predicting protein kinase specificity: Predikin update and performance in the DREAM4 challenge. PloS One. 2011;6:e21169. DOI:10.1371/journal.pone.0021169.

[33] Brinkworth RI, Breinl RA, Kobe B. Structural basis and prediction of substrate specificity in protein serine/threonine kinases. Proceedings of the National Academy of Sciences of the United States of America. 2003;100:74–9. DOI:10.1073/pnas.0134224100.

[34] Saunders NF, Brinkworth RI, Huber T, Kemp BE, Kobe B. Predikin and PredikinDB: a computational framework for the prediction of protein kinase peptide specificity and an associated database of phosphorylation sites. BMC Bioinformatics. 2008;9:245. DOI: 10.1186/1471-2105-9-245.

[35] Saunders NF, Kobe B. The Predikin webserver: improved prediction of protein kinase peptide specificity using structural information. Nucleic Acids Research. 2008;36:W286–90. DOI:10.1093/nar/gkn279.

[36] Gao J, Xu D. The Musite open-source framework for phosphorylation-site prediction. BMC Bioinformatics. 2010;11(Suppl 12):S9. DOI:10.1186/1471-2105-11-S12-S9.

[37] Gao J, Thelen JJ, Dunker AK, Xu D. Musite, a tool for global prediction of general and kinase-specific phosphorylation sites. Molecular & Cellular Proteomics: MCP. 2010;9:2586–600. DOI:10.1074/mcp.M110.001388.

[38] Durek P, Schudoma C, Weckwerth W, Selbig J, Walther D. Detection and characterization of 3D-signature phosphorylation site motifs and their contribution towards improved phosphorylation site prediction in proteins. BMC Bioinformatics. 2009;10:117. DOI:10.1186/1471-2105-10-117.

[39] Miller ML, Blom N. Kinase-specific prediction of protein phosphorylation sites. Methods in Molecular Biology. 2009;527:299–310. DOI:10.1007/978-1-60327-834-8_22.

[40] Blom N, Sicheritz-Ponten T, Gupta R, Gammeltoft S, Brunak S. Prediction of post-translational glycosylation and phosphorylation of proteins from the amino acid sequence. Proteomics. 2004;4:1633–49. DOI:10.1002/pmic.200300771.

[41] Dang TH, Van Leemput K, Verschoren A, Laukens K. Prediction of kinase-specific phosphorylation sites using conditional random fields. Bioinformatics. 2008;24:2857–64. DOI:10.1093/bioinformatics/btn546.

[42] Wan J, Kang S, Tang C, Yan J, Ren Y, Liu J et al. Meta-prediction of phosphorylation sites with weighted voting and restricted grid search parameter selection. Nucleic Acids Research. 2008;36:e22. DOI:10.1093/nar/gkm848.

[43] Li T, Li F, Zhang X. Prediction of kinase-specific phosphorylation sites with sequence features by a log-odds ratio approach. Proteins. 2008;70:404–14. DOI:10.1002/prot.21563.

[44] Wong YH, Lee TY, Liang HK, Huang CM, Wang TY, Yang YH et al. KinasePhos 2.0: a web server for identifying protein kinase-specific phosphorylation sites based on sequences and coupling patterns. Nucleic Acids Research. 2007;35:W588–94. DOI:10.1093/nar/gkm322.

[45] Huang HD, Lee TY, Tzeng SW, Horng JT. KinasePhos: a web tool for identifying protein kinase-specific phosphorylation sites. Nucleic Acids Research. 2005;33:W226–9. DOI:10.1093/nar/gki471.

[46] Xue Y, Li A, Wang L, Feng H, Yao X. PPSP: prediction of PK-specific phosphorylation site with Bayesian decision theory. BMC Bioinformatics. 2006;7:163. DOI:10.1186/1471-2105-7-163.

[47] Kim JH, Lee J, Oh B, Kimm K, Koh I. Prediction of phosphorylation sites using SVMs. Bioinformatics. 2004;20:3179–84. DOI:10.1093/bioinformatics/bth382.

[48] Neuberger G, Schneider G, Eisenhaber F. pkaPS: prediction of protein kinase A phosphorylation sites with the simplified kinase–substrate binding model. Biology Direct. 2007;2:1. DOI:10.1186/1745-6150-2-1.

[49] Joughin BA, Liu C, Lauffenburger DA, Hogue CW, Yaffe MB. Protein kinases display minimal interpositional dependence on substrate sequence: potential implications for the evolution of signalling networks. Philosophical transactions of the Royal Society of London Series B, Biological sciences. 2012;367:2574–83. DOI:10.1098/rstb.2012.0010.

[50] Won AP, Garbarino JE, Lim WA. Recruitment interactions can override catalytic interactions in determining the functional identity of a protein kinase. Proceedings of the National Academy of Sciences of the United States of America. 2011;108:9809–14. DOI:10.1073/pnas.1016337108.

[51] Ngo JC, Chakrabarti S, Ding JH, Velazquez-Dones A, Nolen B, Aubol BE et al. Interplay between SRPK and Clk/Sty kinases in phosphorylation of the splicing factor ASF/SF2 is regulated by a docking motif in ASF/SF2. Molecular Cell. 2005;20:77–89. DOI:10.1016/j.molcel.2005.08.025.

[52] Gogl G, Schneider KD, Yeh BJ, Alam N, Nguyen Ba AN, Moses AM et al. The structure of an NDR/LATS Kinase-Mob complex reveals a novel kinase-coactivator system and

substrate docking mechanism. PLoS Biology. 2015;13:e1002146. DOI:10.1371/journal.pbio.1002146.

[53] van de Weerdt BC, Littler DR, Klompmaker R, Huseinovic A, Fish A, Perrakis A et al. Polo-box domains confer target specificity to the Polo-like kinase family. Biochimica et Biophysica Acta. 2008;1783:1015–22. DOI:10.1016/j.bbamcr.2008.02.019.

[54] Koivomagi M, Valk E, Venta R, Iofik A, Lepiku M, Morgan DO et al. Dynamics of Cdk1 substrate specificity during the cell cycle. Molecular Cell. 2011;42:610–23. DOI:10.1016/j.molcel.2011.05.016.

[55] Cheng KY, Noble ME, Skamnaki V, Brown NR, Lowe ED, Kontogiannis L et al. The role of the phospho-CDK2/cyclin A recruitment site in substrate recognition. The Journal of Biological Chemistry. 2006;281:23167–79. DOI:10.1074/jbc.M600480200.

[56] Pettersen EF, Goddard TD, Huang CC, Couch GS, Greenblatt DM, Meng EC et al. UCSF Chimera – a visualization system for exploratory research and analysis. Journal of Computational Chemistry. 2004;25:1605–12. DOI:10.1002/jcc.20084.

[57] Mottin M, Souza PC, Skaf MS. Molecular Recognition of PPARgamma by Kinase Cdk5/p25: Insights from a combination of protein–protein docking and adaptive biasing force simulations. The Journal of Physical Chemistry B. 2015;119:8330–9. DOI:10.1021/acs.jpcb.5b04269.

[58] Piserchio A, Ramakrishan V, Wang H, Kaoud TS, Arshava B, Dutta K et al. Structural and dynamic features of F-recruitment site driven substrate phosphorylation by ERK2. Scientific Reports. 2015;5:11127. DOI:10.1038/srep11127.

[59] Whisenant TC, Ho DT, Benz RW, Rogers JS, Kaake RM, Gordon EA et al. Computational prediction and experimental verification of new MAP kinase docking sites and substrates including Gli transcription factors. PLoS Computational Biology. 2010;6. DOI:10.1371/journal.pcbi.1000908.

[60] Gordon EA, Whisenant TC, Zeller M, Kaake RM, Gordon WM, Krotee P et al. Combining docking site and phosphosite predictions to find new substrates: identification of smoothelin-like-2 (SMTNL2) as a c-Jun N-terminal kinase (JNK) substrate. Cellular Signalling. 2013;25:2518–29. DOI:10.1016/j.cellsig.2013.08.004.

[61] Bardwell AJ, Bardwell L. Two hydrophobic residues can determine the specificity of mitogen-activated protein kinase docking interactions. The Journal of Biological Chemistry. 2015;290:26661–74. DOI:10.1074/jbc.M115.691436.

[62] Garai A, Zeke A, Gogl G, Toro I, Fordos F, Blankenburg H et al. Specificity of linear motifs that bind to a common mitogen-activated protein kinase docking groove. Science Signaling. 2012;5:ra74. DOI:10.1126/scisignal.2003004.

[63] Zeke A, Bastys T, Alexa A, Garai A, Meszaros B, Kirsch K et al. Systematic discovery of linear binding motifs targeting an ancient protein interaction surface on MAP kinases. Molecular Systems Biology. 2015;11:837. DOI:10.15252/msb.20156269.

[64] Bibi N, Parveen Z, Rashid S. Identification of potential Plk1 targets in a cell-cycle specific proteome through structural dynamics of kinase and Polo box-mediated interactions. PloS One. 2013;8:e70843. DOI:10.1371/journal.pone.0070843.

[65] Hu J, Neiswinger J, Zhang J, Zhu H, Qian J. Systematic prediction of scaffold proteins reveals new design principles in scaffold-mediated signal transduction. PLoS Computational Biology. 2015;11:e1004508. DOI:10.1371/journal.pcbi.1004508.

[66] Bandyopadhyay S, Chiang CY, Srivastava J, Gersten M, White S, Bell R et al. A human MAP kinase interactome. Nature Methods. 2010;7:801–5. DOI:10.1038/nmeth.1506

[67] Calejo AI, Tasken K. Targeting protein–protein interactions in complexes organized by A kinase anchoring proteins. Frontiers in Pharmacology. 2015;6:192. DOI:10.3389/fphar.2015.00192.

[68] Szklarczyk D, Franceschini A, Wyder S, Forslund K, Heller D, Huerta-Cepas J et al. STRING v10: protein–protein interaction networks, integrated over the tree of life. Nucleic Acids Research. 2015;43:D447–52. DOI:10.1093/nar/gku1003.

[69] Hein MY, Hubner NC, Poser I, Cox J, Nagaraj N, Toyoda Y et al. A human interactome in three quantitative dimensions organized by stoichiometries and abundances. Cell. 2015;163:712–23. DOI:10.1016/j.cell.2015.09.053.

[70] Huttlin EL, Ting L, Bruckner RJ, Gebreab F, Gygi MP, Szpyt J et al. The BioPlex Network: a systematic exploration of the human interactome. Cell. 2015;162:425–40. DOI:10.1016/j.cell.2015.06.043.

[71] Varjosalo M, Keskitalo S, Van Drogen A, Nurkkala H, Vichalkovski A, Aebersold R et al. The protein interaction landscape of the human CMGC kinase group. Cell Reports. 2013;3:1306–20. DOI:10.1016/j.celrep.2013.03.027.

[72] Varjosalo M, Sacco R, Stukalov A, van Drogen A, Planyavsky M, Hauri S et al. Interlaboratory reproducibility of large-scale human protein-complex analysis by standardized AP-MS. Nature Methods. 2013;10:307–14. DOI:10.1038/nmeth.2400.

[73] Couzens AL, Knight JD, Kean MJ, Teo G, Weiss A, Dunham WH et al. Protein interaction network of the mammalian Hippo pathway reveals mechanisms of kinase-phosphatase interactions. Science Signaling. 2013;6:rs15. DOI:10.1126/scisignal.2004712.

[74] Hauri S, Wepf A, van Drogen A, Varjosalo M, Tapon N, Aebersold R et al. Interaction proteome of human Hippo signaling: modular control of the co-activator YAP1. Molecular Systems Biology. 2013;9:713. DOI:10.1002/msb.201304750.

[75] Alexander J, Lim D, Joughin BA, Hegemann B, Hutchins JR, Ehrenberger T et al. Spatial exclusivity combined with positive and negative selection of phosphorylation motifs is the basis for context-dependent mitotic signaling. Science Signaling. 2011;4:ra42. DOI:10.1126/scisignal.2001796.

[76] Duarte ML, Pena DA, Nunes Ferraz FA, Berti DA, Paschoal Sobreira TJ, Costa-Junior HM et al. Protein folding creates structure-based, noncontiguous consensus phosphorylation motifs recognized by kinases. Science Signaling. 2014;7:ra105. DOI:10.1126/scisignal.2005412.

[77] Nadzirin N, Gardiner EJ, Willett P, Artymiuk PJ, Firdaus-Raih M. SPRITE and ASSAM: web servers for side chain 3D-motif searching in protein structures. Nucleic Acids Research. 2012;40:W380–6. DOI:10.1093/nar/gks401.

[78] Kumar N, Mohanty D. MODPROPEP: a program for knowledge-based modeling of protein–peptide complexes. Nucleic Acids Research. 2007;35:W549–55. DOI:10.1093/nar/gkm266.

[79] Kumar N, Mohanty D. Identification of substrates for Ser/Thr kinases using residue-based statistical pair potentials. Bioinformatics. 2010;26:189–97. DOI:10.1093/bioinformatics/btp633.

[80] Chen C, Ha BH, Thevenin AF, Lou HJ, Zhang R, Yip KY et al. Identification of a major determinant for serine-threonine kinase phosphoacceptor specificity. Molecular Cell. 2014;53:140–7. DOI:10.1016/j.molcel.2013.11.013.

[81] Creixell P, Palmeri A, Miller CJ, Lou HJ, Santini CC, Nielsen M et al. Unmasking determinants of specificity in the human kinome. Cell. 2015;163:187–201. DOI:10.1016/j.cell.2015.08.05

8

Bioinformatics: A Way Forward to Explore "Plant Omics"

Mehboob-ur- Rahman, Tayyaba Shaheen, Mahmood-ur- Rahman, Muhammad Atif Iqbal and Yusuf Zafar

Abstract

Bioinformatics, a computer-assisted science aiming at managing a huge volume of genomic data, is an emerging discipline that combines the power of computers, mathematical algorithms, and statistical concepts to solve multiple genetic/biological puzzles. This science has progressed parallel to the evolution of genome-sequencing tools, for example, the next-generation sequencing technologies, that resulted in arranging and analyzing the genome-sequencing information of large genomes. Synergism of "plant omics" and bioinformatics set a firm foundation for deducing ancestral karyotype of multiple plant families, predicting genes, etc. Second, the huge genomic data can be assembled to acquire maximum information from a voluminous "omics" data. The science of bioinformatics is handicapped due to lack of appropriate computational procedures in assembling sequencing reads of the homologs occurring in complex genomes like cotton (2n = 4x = 52), wheat (2n = 6x = 42), etc., and shortage of multidisciplinary-oriented trained manpower. In addition, the rapid expansion of sequencing data restricts the potential of acquisitioning, storing, distributing, and analyzing the genomic information. In future, inventions of high-tech computational tools and skills together with improved biological expertise would provide better insight into the genomes, and this information would be helpful in sustaining crop productivities on this planet.

Keywords: databases, data mining, comparative genomics, plant genomes, sequence analysis, structure prediction

1. Introduction

Sustainability in agriculture systems is largely challenged by a number of factors including human population increase, environmental changes, and tremendous demands for growing crops to produce biofuels worldwide [1, 2]. In this regard, exploring the plant genomes for determining the function of important genes involved in conferring tolerance to biotic and abiotic stresses, followed by exploiting these genes in the development of resilient cultivars, is one of the durable strategies for bringing sustainability in crop yields [2, 3].

After the genome sequencing of *Arabidopsis thaliana* genome, a project was launched by the National Science Foundation (NSF) for determining the function of 25,000 Arabidopsis genes [4]. Rice was the first genome-sequenced crop (International Rice Genome Sequencing Project 2005) followed by sequencing of a number of genomes of major crops. All these sequencing projects released a large amount of data. For arranging and analyzing these data, a number of bioinformatic tools have been developed, which helped a lot in drawing important biological conclusions, predicting gene functions, etc. Furthermore, development of unconventional mapping populations and online resources of molecular markers [4] facilitate researchers to identify quantitative trait loci (QTLs). A number of databases have been developed to tackle the newly generated genomic data. These databases provided a foundation to build hypothesis, to design experiments, and to infer knowledge about a particular organism. Moreover, the datasets and "omics" resources of numerous species facilitated the assessment of "omics" properties among species, which further allows studying of conserved genes and evolutionary relationships. Bioinformatics is a crucial tool to access datasets of "omics" and to gather a substantial biological knowledge [5].

From the sequence analysis to the identification of genes, clustering of associated sequences and study of evolutionary relationships using phylogenetics are major tasks of bioinformatics. It also includes the identification and functional annotation of all genes, proteins, and active sites of protein structure in the cell [6]. At present, with the advancement in NGS tools, a voluminous sequencing data is emerging. For deducing meaningful information from these data, it is important for the science of bioinformatics to coevolve with the genomic tools. In this regard, the main three components including mathematics, computer science, and biology upon which the whole citadel of bioinformatics is based, should evolve in parallel to the sequencing tools. It would pave the way for deducing useful information (phylogenies, syntenic relationship, predicting genes, and their function) from the data in a shortest possible time [6, 7].

2. Databases

Databases are collection of organized data that can be retrieved from a website easily for addressing different queries. For managing and handling a database, different hardware and software programs in a computer are needed. The data are organized in structured records that can cater the easy retrieval of information. Broadly, biological databases are classified into sequence databases, relevant to protein and nucleic acid sequences, and structure databases, only relevant to proteins. The first database was developed after a short period of sequencing the insulin protein in 1956. The "Protein Data Bank" was the first ever biological database developed in 1971. Biological databases have flourished enormously due to availability of huge amount of data being generated every day [8]. The individual laboratories maintained the preliminary databases of protein sequences; later, the creation of a combined formal database called SWISS-PROT protein sequence database was introduced in 1986. Now a plethora of data resources are available for study and research purposes and CDROMs (on request from), which are constantly being updated with the availability of new data [9].

Biological databases generally offer software tools to analyze the data available on it and to compare new data with already available data. With the help of these computational methods, the laborious and costly "wet lab" work can be avoided. In future, prospects are dealing with some hindrances such as limited awareness of data, complications in data retrieval, availability of limited data analysis tools, and inadequate literature reference accessibility [10]. A number of biological databases are available that can be divided into three categories on the basis of their contents: (1) primary databases—contain raw nucleotide sequences (GenBank, EMBL, and DDBJ), (2) secondary databases—contain highly annotated data (SWISS-PROT and Protein Information Resource), and (3) specialized databases—deal with particular organism and unique data (FlyBase, WormBase, and TAIR). A major problem in interlinking these databases is the lack of format compatibility. This problem is overcome by using a specified language known as Common Object Request Broker Architecture (CORBA) [11].

At National Center for Biotechnology Information (NCBI), text-based search and retrieval of information can be undertaken by deploying Entrez. It deals with all databases, for example, PubMed, Nucleotide and Protein Sequences, Complete Genomes, etc. In sequence retrieval system (SRS), the Boolean operators are used for undertaking complex searching. It is also used for sequence retrieval, abstract searching, references, etc.

2.1. Dedicated databases for plant genomics

A number of databases deal with datasets focused on particular genes and transcription factors (TFs) related to plant issues and cellular processes. First, a genome-wide finding of repertories of TFs encoded by genes in Arabidopsis genome was described [12]. Accessibility of complete genome sequences in the last few years has enabled us to assemble catalogs of TFs based on their function and association of regulatory systems in different plant species. Numerous databases deliver datasets about genes putatively involved in encoding TFs. These databases are based on predictions made by computational methods (sequence similarity search and hidden Markov model (HMM) conserved DNA-binding domains search). In recent years, GRASSIUS was established to compile resources and tools for undertaking comparative genomics of regulatory sequences in grass species [13]. The Grass Transcription Factor Database (GrassTFDB, another database) of GRASSIUS contains combined sequence information on RiceTFDB, MaizeTFDB, CaneTFDB, and SorghumTFDB. These can be searched through a website. Information of the predicted genes coding TF (carried out by doing annotations across the three genome sequences of legumes) is available on the LegumeTFDB [14]—an extended database of the SoybeanTFDB.

The enhancement of the PGSB PlantsDB database framework has been accomplished with new tools, and sufficient new data have been added into the system particularly for the large complex genomes of wheat, barley, and rye. New resources such as GenomeZipper and CrowsNest for the comparative analysis of data RNASeq Expression Browser have been established. The transPLANT project makes available a platform to compile heterogeneous data about plant genome, for example, integrated searches over multiple databases (**Table 1**).

Database	URL	Species
RARTF	http://rarge.gsc.riken.jp/rartf/	Arabidopsis
AGRIS, AtTFDB	http://arabidopsis.med.ohio-state.edu/AtTFDB/	Arabidopsis
DATF	http://datf.cbi.pku.edu.cn/	Arabidopsis
DRTF	http://drtf.cbi.pku.edu.cn/	Rice
DPTF	http://dptf.cbi.pku.edu.cn/	Poplar
TOBFAC	http://compsysbio.achs.virginia.edu/tobfac/	Tobacco
SoybeanTFDB	http://soybeantfdb.psc.riken.jp/	Soybean
PlantTFDB	http://planttfdb.cbi.pku.edu.cn/ 22	Plant species
PlnTFDB	http://plntfdb.bio.uni-potsdam.de/v3.0/ 20	Plant species
GRASSIUS, GrassTFDB	http://grassius.org/grasstfdb.html	Maize, rice, sorghum, and sugarcane
LegumeTFDB	http://legumetfdb.psc.riken.jp/	Soybean, *Lotus japonicas*, and *Medicago truncatula*
DBD	http://dbd.mrc-lmb.cam.ac.uk/DBD/index.cgi?Home	700 species
PlnTFDB	http://plntfdb.bio.uni-potsdam.de/v3.0/	Plant species
STIFDB	http://caps.ncbs.res.in/stifdb2/	Arabidopsis and rice
PlantTFDB	http://planttfdb.cbi.pku.edu.cn/	83 species
PGSB	http://pgsb.helmholtz-muenchen.de/plant/index.jsp	Barley, wheat, and rye

Table 1. Databases can be exploited for undertaking transcription factor studies in plants.

3. Analysis of the "omic" data

3.1. Sequence retrieval

First step is the identification and retrieval of sequences from different databases (NCBI, TAIR, Gramene, Rap-db, TIGR, Phytozome, PlantGDB, UniProt and SwissProt) developed for handling protein, DNA, RNA, and Expressed Sequence Tag(EST) sequences (**Table 2**). Sequence retrieval is not only carried out through query words but it can also be done using BLAST and or their specific accession numbers. To find out similar sequences from databases, BLAST variations according to sequence retrieval could be performed.

S. No.	Database	URL	Description
1.	TRANSFAC	http://transfac.gbf.de/TRANSFAC/	Transcription factor database
2.	TFD	http://www.tfdg.com/Pages/tfddata.html	Transcription factor database
3.	TRRD	http://www.mgs.bionet.nsc.ru/mgs/dbases/trrd4/	Transcription regulatory region database
4.	PlantCARE	http://sphinx.rug.ac.be:8080/PlantCARE/	Plant cis-acting regulatory elements database
5.	PLACE	http://www.dna.affrc.go.jp/htdocs/PLACE/	Plant cis-acting regulatory elements database
6.	RegulonDB	http://www.cifn.unam.mx/ Computational_Genomics/regulondb/	Database on transcriptional regulation in *Escherichia coli*
7.	SCPD	http://cgsigma.cshl.org/jian	Promoter database of yeast
8.	EPD	http://www.epd.isb-sib.ch/	Eukaryotic promoter database
9.	PRATT	http://web.expasy.org/pratt	It is an online server tool used to identify pattern of amino acids
10.	Phobius	http://phobius.sbc.su.se/	Identification of signal peptides
11.	SignalP 4.0	http://www.cbs.dtu.dk/services/SignalP/	Identification of signal peptides
12.	TargetP	http://www.cbs.dtu.dk/services/TargetP/	Subcellular localization of sequences
13.	LOCTREE3	https://www.rostlab.org/services/loctree3/	Subcellular localization of sequences
14.	Wolf PSORT	http://www.omictools.com/**wolf-psort**-tool	Subcellular localization of sequences
15.	Plant-mPLoc	www.csbio.sjtu.edu.cn/bioinf/**plant**-multi/	Subcellular localization of sequences
16.	Cello v2.5	https://bioinformatictools.wordpress.com/tag/**cello**/	Subcellular localization of sequences
17.	PSI-Pred	http://bioinf.cs.ucl.ac.uk/psipred/	Prediction of transmembrane regions of the gene
18.	CIMMiner	http://discover.nci.nih.gov/cimminer/	To explore the expression of a gene or protein on heat map
19.	DNASTAR	http://www.dnastar.com/	Making of sequence assembly
20.	PromPredict	http://nucleix.mbu.iisc.ernet.in/prompredict/ prompredict.html	Promoter analysis
21.	FoldIndex	http://bioportal.weizmann.ac.il/	It is used to predict folding state of proteins

Table 2. Databases which are helpful in studying regulatory elements and promoter sequences of a gene.

3.2. Multiple sequence alignment

Multiple sequence alignment (MSA) deals with aligning three or more biological sequences, which may be DNA, RNA, and/or protein. Primarily, its purpose is to study similarity among sequences that can help to assess the evolutionary linkage and their common ancestry. It can be undertaken by many sequence analysis softwares including but not limited to ClustalW online software [15], ProbCons, and MAFFT [16]. Some other MSA tools are DNAMAN, T-Coffee, M-Coffee, R-Coffee, Expresso, PSI-Coffee, PSAlign, PRRN, MUSCLE, POA, MEME, etc.

A number of algorithms are available to generate MSA of proteins and DNA sequences. The basic approach in producing multiple alignments is to optimize the sum of pairs (SP) score. This approach is practical, and reproduces high-quality MSA dataset [17]. Mathematical approach (also called as probabilistic and stochastic methods) exploits the probability in developing MSA. Hidden Markov model is a masterpiece example of this approach. In this approach, MSA data are modeled as probabilistic models. All possible combination of gaps, mismatches, and matches are assigned with probabilities, and the algorithm finds the most likely MSA [18]. Other approaches are genetic algorithms and simulated annealing, which break a series of possible MSA into segments followed by their rearrangement. It can use an existing MSA and refines it by a series of rearrangements [19].

3.3. Domain and motif study

Domain always refers to a conserved part of protein sequence and structure, which can evolve, function, and exist independently. Whereas motif is a well-maintained sequence of protein or DNA that remains the same to execute certain function [20]. For characterization of a gene, it is always advisable to study its functional domains and motifs. The novel sequences identified can be subjected for analyzing their domains and motifs to predict their functions. For motif analysis, MEME tool can be used, while for domain analysis PFAM, InterProScan, and SMART tools can be used.

Large protein molecules comprise of structural and functional domains. Structural domains regions are either compact, globular modules, or separated clearly from the flanking regions including membrane regions or long coiled-coil helices that are separating the other domains [21]. These domains can be seen in proteins as semi-independent three-dimensional (3D), and have the ability to fold independently [22]. These domains constitute the "units of evolution" [23] and have typical functions [24]. Structural Classification of Proteins (SCOP) database has been used extensively for assigning domains in proteins [25]. Most databases and methods (e.g., Class Architecture Topology Homology database) are not fully automated, which combine several other methods for assigning domains to the proteins [26]. Protein Informatics System for Modeling (PrISM) is the only completely automated method that can be used to assign sequence-continuous domains to proteins of known 3D structures [27]. If the structure (3D) of the protein is not known, then a number of alternative methods and databases are available. For example, one of the most prominent databases is putative protein domains (ProDom) [28].

3.4. Structural analysis

Modeller is used to generate 3D structure [29]. LOMETS server is used to find the best template for comparative modeling. DOPE (discrete optimized protein energy) helps to find best model by calculating each structure's value that is evaluated through PROSAII [30] and PROCHECK [31]. To calculate electrostatic surface and solvation properties of complex compounds, APBS [32] is used. For structure alignment, PDBsum tool [33] is deployed. Structure of gene can also be displayed on GSDS2.0 (Gene Structure Display Server) [34]. YASARA software is used to draw 3D structure, c-terminal, n-terminal, and domains of proteins [35]. Chromosomal position of genes can be located by NCBI map viewer tool, Mapchar 2.1, and cucumber genome database map viewer tool.

3.5. Analysis of regulatory elements

The regulatory elements encode a protein that binds to promoter or operator region of a gene for up- and/or downregulating its expression. For instance, catabolite activator protein (CAP) is a regulatory element present in prokaryotes, which regulates the lac operon [36].

Regulation of gene expression takes place at transcription level by specific sequences known as transcription factors—inhibit or initiate the transcription. These factors can be repressors, activators, or both. It is worth mentioning that repressors inhibit the binding of RNA polymerase with the transcription complex (promoters)—thus blocking the transcription. However, activators are activated by the enabling binding of RNA polymerase with the transcription complex.

These elements can be found *in silico* by deploying PlantCARE [37], and PLACE program. PLACE is repository of motifs occur cis-acting regulatory DNA elements of plants. This database also gives information about the variations in motifs found in different genes or plant species. Relevant literature and comprehensive description of different motifs can be retrieved from this database. Several research groups have identified a number of genes including WRKY genes, Ascorbate Peroxidase, PSY, etc. using different bioinformatics tools [38–40].

3.6. Mutation identification

Mutation alters the nucleotide sequences of a gene that may change the gene expression. These mutations can be identified using conventional as well as NGS tools [41, 42]. Sequencing of cytosine methylome (methylC-seq), transcriptome (RNA-seq), and small RNA transcriptome (small RNA-seq) in Arabidopsis was undertaken by deploying NGS tools. Genome-scale methylation patterns and a direct relationship between the location of sRNAs and DNA methylation were identified [43]. Protein-protein interactions occur in majority cellular processes. The interactome, representing complete set of all protein-protein connections, is vital for studying the molecular networks [44]. Correlated mutation analysis can be harnessed to predict interface residues. Protein-protein interaction can be studied by detecting correlated mutations at interface [45].

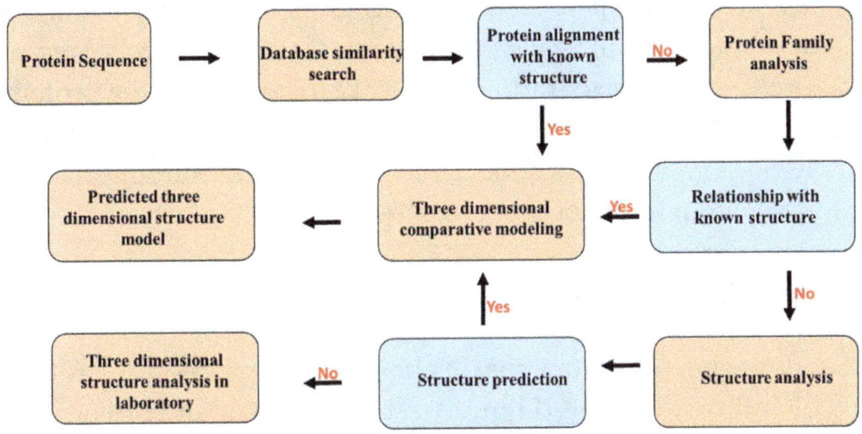

Figure 1. Flow chart diagram for protein structure prediction **(Source: Ref. [117]).**

S. No.	Software/server	Link	Description
1.	MODELLER	http://salilab.org/modeller/	Comparative modeling of protein 3D structures
2.	3DJigsaw	http://bmm.cancerresearchuk.org/~3djigsaw/	Predict structure and function of protein
3.	ESyPred3D	http://www.unamur.be/sciences/biologie/urbm/bioinfo/easypred/	Homology modeling with increased alignment performance
4.	SWISS-MODEL	http://swissmodel.expasy.org/	Automated protein homology modeling server
5.	YASARA	http://www.yasara.org/	Molecular modeling tool
6.	RaptorX	http://raptorx.uchicago.edu/	Protein structure prediction
7.	HHPred	http://toolkit.tuebingen.mpg.de/hhpred	Homology detection and structure prediction server
8.	Phyre[2]	http://www.sbg.bio.ic.ac.uk/~phyre2/html/page.cgi?id=index	3D structure prediction
9.	ROSETTA	http://boinc.bakerlab.org/resetta/	3D structure prediction
10.	I-TASSER	http://zhanglab.ccmb.med.umich.edu/I-TASSER/	Predict structure and function of protein
11.	Bhageerah	http://www.scfbio-iitd.res.in/bhageerath/index.jsp	Energy-based protein structure prediction server

Table 3. Bioinformatics tools which are helpful in predicting protein structure.

3.7. Protein structure prediction

It is the prediction of protein from amino acids. Protein structure can be predicted by undertaking similarity searches, MSAs, secondary structure prediction, identification of domains, solvent accessibility predict, itself protein fold recognition, making 3D models, and model validation [46]. For example, small heat shock proteins (smHSPs, largely present in plants) are ubiquitous in nature, and their size is ranged from 17 to 30 kDa. These proteins are encoded by six nuclear gene families. Every gene family encodes a protein that is present in different part of the cell including cytosol, mitochondria, chloroplast, and endoplasmic reticulum. These proteins protect plants from high temperature stress [47].

3.7.1. Protein structure prediction steps

Following is the flow sheet diagram that elaborates the process of protein 3D structure prediction using bioinformatics tools (**Figure 1**). Various online and offline resources that can be used for the prediction of protein structure are described in **Table 3**.

3.8. Phylogenetic analysis

Phylogenetic analysis is the study of evolutionary relationships among different organisms. Phylogenetic analysis corresponds to the evolutionary interactions that can be presented in branching form. Phylogenetics refers as cladistics is a set of respective descendants such that it evolves from a respective single ancestor (**Figure 2**). Cladistics is a specific methodology of theorizing almost every evolutionary interactions [48]. In order to construct a phylogenetic tree, different methods are used that are based on the nature of the data and algorithms used. Each method is based on certain assumptions. Thus, the method used to draw evolutionary relationship on one kind of dataset may not be equally good for the other kind of dataset. It is therefore suggested that a number of distance-based methods [unweighted pair group method arithmetic mean (UPGMA) and neighbor joining (NJ)] and character-based (CB) methods [maximum parsimony (MP), maximum likelihood (ML)] should be run.

3.8.1. Distance-based method

The distance-based method also called as phonetic method depends upon the extent of dissimilarity (the distance) to derive a tree from the two aligned sequences. This method can rebuild the accurate tree if whole genetic divergence proceedings are precisely verified in the sequence. Tree construction is based on the resultant genetic distances from sequenced data, distances from immunological studies, and Euclidean distance applied in various ways [49].

3.8.1.1. Unweighted pair group method arithmetic mean (UPGMA)

It is the simplest procedure for studying the phylogenetic relationship among different organisms which uses the clustering approach and uncorrected data to make a tree. It joins tree branches based on the criterion of greatest similarity among pairs and averages of joined pairs. UPGMA generates a correct topology with true branch lengths only when the natural mutation is proportional to time (a molecular clock) or approximately equal to raw sequence dissimilarity [50]. However, these conditions are rarely met in practice. Distance matrix is recalculated, and this procedure is continued until the operational taxonomic units [OTU (=

neighbors)] are grouped in one cluster. However, this method does not reflect the evolutionary descents.

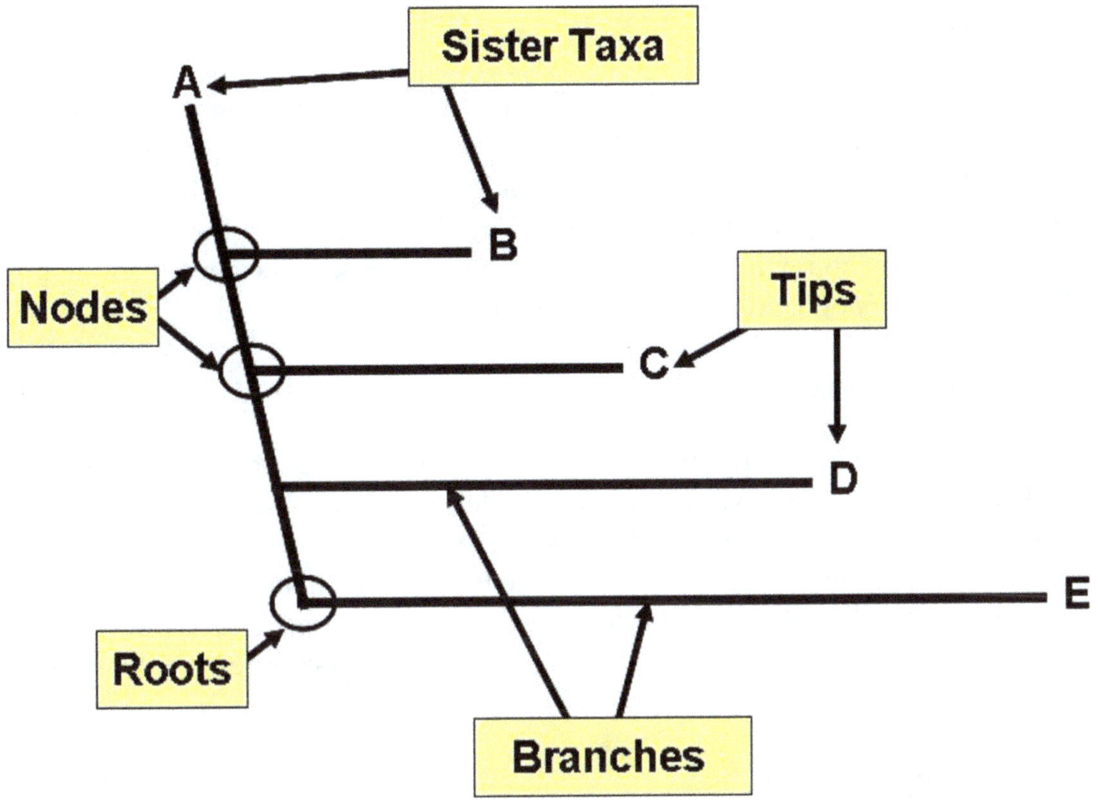

Figure 2. A descriptive diagram of phylogenetic analysis based on biological data.

3.8.1.2. Neighbor joining method (NJ)

This method is usually pragmatic with distance tree making, irrespective of optimization measure. This method works on the principle to discover pairs of OUTs(Operational Taxonomic Units) that curtails the total branch length at respective stage of clustering of OTUs beginning with a star-like tree. Branch length and distance matrix are recalculated until one terminal is found. This method can be used to obtain the branch lengths in addition to the topology of a parsimonious tree speedily [51]. This method is relatively efficient than that of the UPGMA. This method can analyze a large dataset. Construction of one possible tree and also the biased tree are the major drawbacks of this method.

3.8.2. Character-based methods (CB)

These methods are also called cladistic methods that use directly the aligned characters, for instance, DNA or protein sequences, through tree inference. The algorithm based on character takes an aligned set of characters, for example, DNA sequences, and builds a tree relating the changes in discrete characters, desirable to create the observed set of characters. These methods assume that a set of sequences descended from a common ancestor that may change by mutation and/or selection process without involving any kind of hybridization or horizontal gene transfers. Character-based algorithms are comprised of two groups: maximum likelihood and maximum parsimony [50].

3.8.2.1. Maximum likelihood (ML)

Different statistical tools are exploited to assess hypothesis of evolutionary history. It constructs all possible trees of evolutionary history from a given data. Multiple alignment is done in this method. Probability of all possible topologies for each data partition is estimated to identify a tree with the highest probability at all partitions based on the maximum similar phylogeny. In this method, whole sequence information is used to evaluate all the possible trees. This method cannot handle a large amount of data.

3.8.2.2. Maximum parsimony (MP)

This method uses the philosophy of "the simpler hypothesis is better than the complicated ones" [52]. By this criterion, the MP tree is one with few character-state transformations for all the sequences from a common ancestor. It works by selecting trees that minimize the total tree length. For each site in the alignment, all possible trees are evaluated that is not the characteristic of other methods. This method is less dependent on suppositions about the evolution of sequences than the other strategies to construct a tree. This procedure is handicapped when the data are heterogeneous.

3.8.3. Evaluation of trees

Phylogenetic trees can be statistically evaluated for reliability of branches/clades created using (1) skewness test, (2) bootstrapping analysis, and (3) likelihood ratio tests where all have currently computerized algorithms. Skewness test never has approximation with dependability of specific topology; it is subtle to very small amounts of respective signal contemporary in otherwise random information set. Bootstrapping analysis is a resampling or rechecking tree evaluation methodology that works with distance, likelihood, and parsimony method. The outcome of bootstrap examination is a number related with specific branch in phylogenetic tree giving up the amount of bootstrap duplicates that ropes the monophyly of particular clade. Likelihood ratio tests support the likelihood ratio (tests) that is easily applicable to ML (maximum likelihood) examination. Value of likelihood is calculated for implication against normal circulation of fault in optimal models [50].

3.8.4. Software mostly used for phylogenetic analysis

Phylogeny inference package (PHYLIP) [53] contains 30 programs that cover the main flows of phylogenetic analysis. It is a freely available software and is accessible to almost all kinds of computer platforms (Mac, UNIX, DOC, etc.). In addition, phylogenetic analysis using parsimony (PAUP) software is widely used to infer and interpret the evolutionary tree. Now the old version has been upgraded (PAUP*) after the inclusion of maximum likelihood and distance methods. Other than those described above, some phylogenetic programs have unique proficiencies but mostly inadequate in their respective actions, and movability. These include molecular phylogenetics (MOLPHY) [54], TREE-PUZZLE [55], FastDNAml [56], and MACCLADE [57].

3.9. Molecular dynamics simulations for plant molecules

Molecular dynamics simulations are the principal methods for elaborating the physical foundation of structure, function, and interaction of biological macromolecules (e.g., proteins and nucleic acids). Earlier, proteins were considered as comparatively rigid structures that now have been changed by a dynamic model in which the internal movements and conformational changes are key players in determining their functions. Computer simulations are carried out in comprehending the characteristics and arrangements of different molecules related to physical structure and interactions, otherwise not possible to observe by other means. There are two major classes of simulation techniques, i.e., molecular dynamics and Monte Carlo. These simulations have been used extensively in characterizing plant compounds (natural distillates) followed by finding optical counter parts with identical efficiency [58].

3.10. Proteomics and transcriptomics

Study of proteins along with mRNA transcripts is referred as proteomics and transcriptomics, respectively [59]. Due to intrinsic complexity, experimental workflows and variety of data types, storage, and open depository of proteomics data based on mass spectrometry (MS) are still insufficiently established. Many public sources with particular purposes for MS proteomics research have been established to fulfill this need. These databases are Global Proteome Machine Database (GPMDB), PRIDE, PeptideAtlas, ProteomicsDB, Mass Spectrometry Interactive Virtual Environment (MassIVE), PeptideAtlas SRM Experiment Library (PASSEL), etc. Moreover, for the purpose of enhanced integration and harmonized sharing of public warehouses, the ProteomeXchange consortium has been developed recently to capitalize on its advantage for the scientific community [60].

For transcriptomics studies, there are numerous databases comprising microarray data: NASCArrays, ArrayExpress, Genevestigator, Stanford Microarray Database, and the Gene Expression Omnibus, which are freely available [61]. An example of the transcriptome database is Chickpea Transcriptome Database (CTDB), which has information about the tools used for transcriptome sequence, conserved domain(s), molecular markers, transcription factor families, and complete gene expression information [62].

3.11. Protein-protein interactions

The protein-protein interactions (PPIs) control the expensive scope of biological procedures that include interactions between cells, metabolic as well as developmental pathways. This noncovalent bonding brings a range of interactions and associations between proteins. PPIs can be classified in several ways depending upon their contrasting structural and functional characteristics [63]. There are several *in vivo* and *in vitro* methods for finding PPIs but our focus is on computational approaches. Computer modeling assisted with mathematical methods facilitates the study of different processes [64]. *In silico* methods combining the computational modeling are being used to study protein interactions. The *in silico* analysis integrates multiple data types including gene coexpression, colocalization, functional category, and the occurrence of orthologs or interologs to derive a global network in a species [65]. A list of webservers can be used to predict protein-protein interaction (**Table 4**).

S. No.	Web server	Description	URL
1.	Coev2Net	Coev2Net is a general framework to predict, assess, and boost confidence in individual interactions inferred from a high-throughput experiment	http://groups.csail.mit.edu/cb/coev2net/
2.	InterPreTS	InterPreTS uses tertiary structure to predict interactions	http://gabrmn.uab.es/interpret/
3.	PrePPI	PrePPI predicts protein interactions using both structural and nonstructural information	http://technology.sbkb.org/portal/page/350/
4.	iWARP	iWARP is a threading-based method to predict protein interaction from protein sequences	http://groups.csail.mit.edu/cb/iwrap/
5.	PoiNet	PoiNet provides PPI filtering and network topology from different databases	http://poinet.bioinformatics.tw/
6.	PreSPI	PreSPI predicts protein interactions using a combination of domains	http://code.google.com/p/prespi/
7.	PIPE2	PIPE2 queries the protein interactions between two proteins based on specificity and sensitivity	http://cgmlab.carleton.ca/PIPE2
8.	HomoMINT	HomoMINT predicts interaction in human based on ortholog information in model organisms	http://mint.bio.uniroma2.it/HomoMINT
9.	SPPS	SPPS searches protein partners of a source protein in other species	http://mdl.shsmu.edu.cn/SPPS/
10.	InPrePPI	InPrePPI predicts protein interactions in prokaryotes based on genomic context	http://inpreppi.biosino.org/InPrePPI/index.jsp
11.	STRING	STRING database includes protein interactions containing both physical and functional associations	http://string.embl.de
12.	MirrorTree	The MirrorTree allows graphical and interactive study of the coevolution of two protein families and assess their interactions in a taxonomic context	http://csbg.cnb.csic.es/mtserver/
13.	TSEMA	TSEMA predicts the interaction between two families of proteins based on Monte Carlo approach	http://tsema.bioinfo.cnio.es/
14.	COG	COG shows phylogenetic classification of proteins encoded in genomes	http://www.ncbi.nlm.nih.gov/COG/

Table 4. A number of important computational tools use to study protein-protein interactions.

3.11.1. *Arabidopsis protein interaction analysis*

More than 10 freely accessible protein interaction databases are available for *A. thaliana*. An intelligent bioinformatics web device, ANAP (Arabidopsis Network Analysis Pipeline) has been created for incorporating Arabidopsis protein collaboration databases. A total of 11 Arabidopsis protein collaboration databases having 201,699 protein association sets, 15,208 identifiers, 89 connection discovery routines, 73 species that interface with Arabidopsis, and 6161 references were incorporated in ANAP [66].

3.11.2. *Computational identification of protein-protein interactions in rice*

Complexity of plant molecules always hinders progress toward exploring the protein-protein interaction networks on large scale. A total of 5049 proteins with 76,585 interactions were predicted in rice using Predicted Rice Interactome Network (PRIN). The prolonged molecular network in PRIN has greatly improved the ability to analyze the function and organization of genes and gene networks [67].

3.11.3. *iPlants: the world's plant online*

This database has been designed to develop a comprehensive working list of scientific names of all plant species. Through this database, authenticated names of plant species (agreed by the scientific community) with their alternative synonyms can be found. This type of list will empower untrained botanists to get useful information about different plant species. *i*Plants will also resolve the existing confusions found in the published taxonomies. A total of 422,000 known plant species and 1,500,000–1,700,000 scientific names are used to refer these plant species are present in this database.

This database will help in exploiting plant biodiversity information in different breeding as well as gene cloning programs [68].

3.11.4. *Reactome*

Reactome database provides access without any restriction about the peer-reviewed pathways [69]. This database is equipped with bioinformatics tools, which can be used to examine, visualize, interpret, and analyze knowledge about pathway. The information in this database is generated by the experts (curators and software developers) and cross-referenced to other databases, for example, NCBI, Ensembl, UniProt, UCSC Genome Browser, HapMap, KEGG, ChEBI, PubMed, and GO. In this database, orthologous reaction for over 20 nonhuman species including rice, Arabidopsis, and *Escherichia coli* can be found. This database can be accessed in the form of online text book [70]. Biological pathways and reaction can be viewed in a number of formats, comprising of PDF, SBML, and BioPax [71]. Recent version "v55" of Reactome was released in December 2015.

3.12. Metabolomics

Study of all or utmost metabolites in an organism are denoted as metabolomics. It is a complex research field that involves interdisciplinary interaction of different sciences. One of the numerous methods is soft independent modeling of class analogy (SIMCA). Besides this, an effective protocol for data mining in metabolomics has also been developed [72]. In recent years, numerous databases containing data about compound names and structures, mass spectra, metabolic pathways, metabolite profile, and statistical/mathematical models are established. These databases are extremely useful for metabolomics research [73].

The MeRy-B (http://bit.ly/meryb) is dedicated to plants, and it provides information related to metabolites detected using NMR(Nuclear magnetic resonance), together with related analytical and experimental metadata. MeRy-B is equipped with a list of many plant metabolites along with the data of their experimental conditions, the features studied, and concentration of metabolites of 19 different species including the model plant species such as Arabidopsis [74].

4. Implications of bioinformatics in plant omics

Bioinformatics is an essential part of omics providing techniques to analyze large biological data sets and interpreting them into applications of "omic". Tools dealing with "omics" generate massive data that assist system biology to combine multivariate information into systems and models. The omics tools including high-throughput genome-scale genotyping platforms such as whole-genome resequencing, proteomics, and metabolomics offer better prospects for gene identification and exploration of molecular mechanisms. This information can be used to develop ideal genotypes suitable for varying climatic conditions [75].

4.1. Plant genome sequencing

With the advancements in high-throughput techniques, whole-genome characterization of a wide range of organisms has been possible. Nevertheless, the storage and management of this massive genomic data is a major challenge. Revolution in sequencing technologies has made it possible to sequence large and complex genomes at extremely low cost and in much less time period. Presently, the most popular methods of genome sequencing are shotgun sequencing and NGS. The NGS is very popular tool for the identification of housekeeping genes in crop plants. Many tools such as Genome Analyzer, the Applied Biosystems SOLiD System, Roche/454 FLX, and the Illumina/Solexa are commercially available for NGS [76]. NGS can be utilized for whole-genome sequencing, isolation of transcription factor binding sites, and expression of noncoding RNA and targeted resequencing [77]. Various software packages are available to assemble sequences, for example, Phred/Phrap/Consed [78], GAP4 [79], and chromaseq [80]. Another software called AMOS was developed by TIGR, which is useful for comparative genome assemblage [81].

4.2. Plant whole-genome resequencing

The most effective approach in functional genomics is the whole-genome resequencing. For reducing cost, target region can be sequenced. Microarray is also a common way of target region sequencing, which is based on hybridization to arrays comprising of synthetic oligonucleotides that match the target DNA sequence [82]. Recent NGS technology has made it possible to discover differences between individuals and populations especially of the crop species whose genomes have already been sequenced and assembled. Similar projects in Arabidopsis [83] and rice [84] generated a huge data of natural variations occurring within different accessions.

4.3. Plant comparative genomics and databases

Using comparative genomic approaches, functions to different genes (especially representing the less studied species) have been assigned. The developments in RNA interference and other technologies like mutagenesis have allowed phenotypic screens for genes—known as phenomics [85]. The field of phenomics is heavily dependent upon the interaction of plant genome with the prevailing environments. This science is largely dependent upon intensive collaboration between three disciplines including plant science, computer science, and engineering. Currently, there are yearly plant-focused image-processing tasks [86] that have positively stimulated the community and invigorated computer scientists to focus on developing joint plant datasets. Though there is limited accessibility to high-throughput phenotyping platforms. A current list of accessible image datasets can be accessed at the website [85].

4.4. Important information source of plant species

The most prevalent and unified information collection source is TAIR that maintains data of molecular biology, genetic and genomic of Arabidopsis [87]. Similarly, Salk Institute Genomic Analysis Laboratory (SIGnAL) deals with the omics research of Arabidopsis.

Gramene is an integrated source of information for grasses. It exploits the rice-genome-sequencing information as a foundation source for comparing the information of other members of grass family [88]. At this website, information about DNA and mRNA sequences, genome assembly and annotations, genes, genetic maps and physical maps, QTLs, and many more are available. These interesting features make this website more attractive for researchers, and it is being updated regularly with new attributes like genetic diversity data, comparison of genomes of *Oryza sativa* with its wild relatives or with the other taxa for undertaking evolutionary studies, etc. [89].

The portal site SoyBase [90] provides information about whole-genome sequence data. The portal site for Solanaceae genome is the Sol genomics network. It also provides information about the tomato-genome-sequencing project [91]. The MaizeGDB is a public database for *Zea mays* [92]. GreenPhylDB is a broad platform intended for facilitation of comparative functional genomics in *O. sativa* and *A. thaliana* genomes [93]. PLAZA 3.0 has been established to develop comparative genomics data of plants accessible via user-friendly web interface. Structural and functional annotation, phylogenetic trees protein domains, gene families, and detailed data about genome organization can simply be inquired and envisioned [94]. A comparative genomics database named PIECE was established to accommodate information pertaining to gene structure comparisons and evolution. This database covers all the annotated genes mined from 25 plant species [95].

4.5. Use of bioinformatics for comparative genomics in plants

Availability of whole-genome sequences and bioinformatics have accelerated the process for identifying specific gene families in different plant species. These tools were also used to study the duplications as well as deletions in different plant genomes [96]. These results are helpful in phylogenetic studies [97], study of synteny and collinearity relationship, and inference of shared ancestry of genes [98]. The plant genome duplication database (PGDD) provides important data for studying the syntenic relationships of intragenome or cross-genome identified in the genome-sequenced species [99]. Analysis of orthologous clusters at genome level is a significant element in elucidating comparative genomics. Recognizing overlap between orthologous clusters can permit us to clarify the utility and evolution of proteins among multiple species. OrthoVenn is a web platform that is freely accessible and can be used for making comparisons and annotations of orthologous gene clusters. It can be accessed at [100]. Information regarding orthologs of plants and green algae can be searched at PlantOrDB [101].

4.6. Gene prediction and genome annotation

The characterization of introns and exons in a sequenced genome is referred as gene prediction. These predictions can be undertaken computationally or combination of manual as well as computational annotations. Numerous computer programs to find protein-coding genes are accessible through OMIC TOOLs website [102], which has been extensively used for genome annotations and genes prediction.

For structural annotations of a genome, a number of software packages were described [102, 103]. Additionally, tools (SynBrowse and VISTA) of genome comparison can be used to improve precision of gene identification. Repeat-Masker [104] was designed to find interspersed repeats and low complexity sequences in whole sequenced genome. Through this program, the repetitive sequences can also be masked. Similarly, a number of software programs (Repeat Finder, RECON, etc.) are available that can be used to find repeats in a sequenced genome.

4.7. Genome mapping and bioinformatics

Selecting suitable mapping tool and sequences search may claim adjustments in specificity and sensitivity of the search statistics. The process of finding candidate genes conferring traits can be accelerated for those crops where genetic and physical maps and annotated genome assemblies are available. A wide range of tools have been developed recently for illustrating maps and imagining genomes primarily to facilitate genome assembly.

NCBI is a source to assess all types of information regarding genomes. Access to various biological databases is possible using "Entrez." For aligned genetic, physical, and sequence information of eukaryotes including plants, a genome browser "Map viewer" has been developed. To display aligned map from various species entered in Map Viewer, a special plant query page can be accessed. Customized plant basic local alignment search tool (Plant BLAST) facilitates the process of exploring sequence similarity from the collection of mapped plants sequence data, and the resulting alignment can be visualized in genomic text using "Map viewer" [105], R/qtl [106], JoinMap [107], OneMap [108], MSTMap [109], Lep-MAP [110], and HighMap [111], which can be used to develop genetic linkage maps [112].

Numerous databases offer data for exploring markers in multiple crop species. DNA markers including Single nucleotide polymorphism (SNP), Simple sequence repeat (SSR), and conserved ortholog set (COS) markers can be predicted using PlantMarkers [113]. A famous site for Triticeae genome is GrainGenes that contains information about linkage maps and DNA markers of wheat, rye, barley, and oat [114]. Gramene, a database for comparative genomics, contains genetic maps of multiple plant species [89]. The Triticeae Mapped EST database (TriMEDB) gives information of mapped cDNA markers related to barley and wheat [115]. The CottonGen web-based database provides information and open access to genetic, genomic, and breeding data of cotton. CottonGen has improved tools for sharing, mining, retrieval, and visualization of data as compared with the CottonDB and Cotton Marker Database [116].

5. Conclusions

In this chapter, we described comprehensively the available resources and tools of bioinformatics pertaining to gene expression, databases, protein, and metabolite analyses and genome sequencing. Bioinformatics has been evolved rapidly over the last 15 years—emerged as a new key discipline of biology. A huge amount of genetic and genomic data have been generated using next-generation sequencing technologies that provide opportunities for generating huge genetic and genomic data. However, drawing useful genetic information is handicapped due to unavailability of skilled bioinformaticians. Still, there is room for some unsolved problems in bioinformatics like computerized data mining, vigorous inference of phenotype from genotype, trainings of students and recognized researchers in bioinformatics, etc. Bioinformatics is generating job opportunities for brilliant and skilled researchers in biology, statistics, and computer science. The remarkable evolution of bioinformatics has been confronted by a number of troublesome revolutions in science and technology. Even though, bioinformatics has developed possibly itself to a level above recognition. Today's bioinformatics is a luxury to biological scientists, generating huge data in all fields of biological sciences. In near future, bioinformatics will be an indispensable part of plant research, and novel tools and methods will be incorporated by every plant scientist. The next half century is the era of "data integration." Both basic and applied research will replenish the society for renewable energy, dropping world hunger and poverty, and protecting the environment.

Author details

Mehboob-ur- Rahman[1*], Tayyaba Shaheen[2], Mahmood-ur- Rahman[2],
Muhammad Atif Iqbal[1] and Yusuf Zafar[3]

*Address all correspondence to: mehboob_pbd@yahoo.com and mehboob@nibge.org

1 National Institute for Biotechnology and Genetic Engineering, Faisalabad, Pakistan

2 Department of Bioinformatics and Biotechnology, Government College University, Faisalabad, Pakistan

3 Department of Technical Co-operation, IAEA, Vienna International Centre, Vienna, Austria

References

[1] Turner WR, Oppenheimer M, Wilcove DS. A force to fight global warming. Nature. 2009;462:278–279. DOI: 10.1038/462278a

[2] Azadi H, Ghanian M, Ghoochani OM, Taning CNT, Hajivand RY, Dogot T. Genetically modified crops: towards agricultural growth, agricultural development, or agricultural sustainability?. Food Reviews International. 2015;31:195–221. DOI: 10.1080/87559129.2014.994816

[3] Takeda S, Matsuoka M. Genetic approaches to crop improvement: responding to environmental and population changes. Nature Reviews Genetics. 2008;9:444–457. DOI: 10.1038/nrg2342

[4] Ruperao P, Edwards D. Identification of markers from next-generation sequence data plant genotyping. Methods in Molecular Biology. Bioinformatics. 2014;1245:29–47. DOI: 10.1007/978-1-4939-1966-6_3

[5] Mochida K, Shinozaki K. Genomics and bioinformatics resources for crop improvement. Plant Cell Physiology. 2010;51:497–523. DOI: 10.1093/pcp/pcq027

[6] Rao VS, Das SK, Rao VJ, Srinubabu G. Recent developments in life sciences research: role of Bioinformatics. African Journal of Biotechnology. 2008;7:495–503

[7] Alemu K. The role and application of bioinformatics in plant disease management. Advances in Life Science and Technology. 2015;28:28–33

[8] Cooray MPNS. Molecular biological databases: evolutionary history, data modeling, implementation and ethical background. Sri Lanka Journal of Bio-Medical Informatics. 2012;3:2–11. DOI: 10.4038/sljbmi.v3i1.2489

[9] Babu MM. Biological databases and protein sequence analysis [Internet]. 1997. Available from http://www.mrc-lmb.cam.ac.uk/genomes/madanm/pdfs/biodbseq.pdf [Accessed: 2016-04-26]

[10] Bry FK, Kröger P. A computational biology database digest: data, data analysis, and data management. Distributed Parallel Databases 2003;13:7–42. DOI: 10.1023/A: 1021540705916

[11] Brown JR, editor. Handbook of Comparative Genomics: Basic and Applied Research. 1st ed. Boca Raton, FL: CRC Press; 2007. 400 p.

[12] Riechmann JL, Heard J, Martin G, Reuber L, Jiang C, Keddie J, et al. Arabidopsis transcription factors: genome-wide comparative analysis among eukaryotes. Science. 2000;290:2105–2110. DOI: 10.1126/science.290.5499.2105

[13] Yilmaz A, Nishiyama MY, Jr., Fuentes BG, Souza GM, Janies D, Gray J, et al. GRASSIUS: a platform for comparative regulatory genomics across the grasses. Plant Physiology. 2009;149:171–180. DOI: 10.1104/pp.108.128579

[14] Mochida K, Yoshida T, Sakurai T, Yamaguchi-Shinozaki K, Shinozaki K, Lam-Son Phan Tran. LegumeTFDB: an integrative database of *Glycine max, Lotus japonicus* and *Medicago truncatula* transcription factors [Internet]. 2009. Available from: http://legumetfdb.psc.riken.jp [Accessed: 2016-04-28]

[15] Chenna R, Sugawara H, Koike T, Lopez R, Gibson TJ, Higgins DG, Thompson JD. Multiple sequence alignment with the Clustal series of programs. Nucleic Acids Research. 2003;31:3497–3500. DOI: 10.1093/nar/gkg500

[16] Nuin PAS, Wang Z, Tillier ERM. The accuracy of several multiple sequence alignment programs for proteins. BMC Bioinformatics. 2006;7:471–488. DOI: 10.1186/1471-2105-7-471

[17] Waterman MS, Smith TF, Beyer WA. Some biological sequence metrics. Advances in Mathematics. 1976;20:367–387. DOI: 10.1016/0001-8708(76)90202-4

[18] Edgar RC. MUSCLE: multiple sequence alignment with high accuracy and high throughput. Nucleic Acids Research. 2004;32:1792–1797. DOI: 10.1093/nar/gkh340

[19] Goldberg DE, editor. Handbook of Genetic Algorithms in Search, Optimization and Machine Learning. 1st ed. Reading, MA: Addison-Wesley Longman; 1989. 372 p.

[20] Roey KV, Davey NE. Motif co-regulation and co-operativity are common mechanisms in transcriptional, post-transcriptional and post-translational regulation. Cell Communication Signal. 2015;13:45–60. DOI: 10.1186/s12964-015-0123-9

[21] Abrahams JP, Leslie AGW, Lutter R, Walker JE. Structure at 2.8 Å resolution of F_1-ATPase from bovine heart mitochondria. Nature. 1994;370:621–628. DOI: 10.1038/370621a0

[22] Jaenicke R. Folding and association of proteins. Progress in Biophysics and Molecular Biology.1987;49:117–237. DOI: 10.1016/0079-6107(87)90011-3

[23] Holm L, Sander C. The FSSP database of structurally aligned protein fold families. Nucleic Acids Research. 1994;22:3600–3609.

[24] Teichmann SA, et al. The evolution and structural anatomy of the small molecule metabolic pathways in *Escherichia coli.* Journal of Molecular Biology. 2001;311:693–708. DOI: 10.1006/jmbi.2001.4912

[25] Lo Conte L, Brenner SE, Hubbard TJ, Chothia C, Murzin AG. SCOP database in 2002: refinements accommodate structural genomics. Nucleic Acids Research. 2002;30:264–267. DOI: 10.1093/nar/30.1.264

[26] Orengo CA, et al. The CATH protein family database: a resource for structural and functional annotation of genomes. Proteomics. 2002;2:11–21. DOI: 10.1002/1615-9861(200201)2

[27] Yang AS, Honig B. An integrated approach to the analysis and modeling of protein sequences and structures. III. A comparative study of sequence conservation in protein structural families using multiple structural alignments. Journal of Molecular Biology. 2000;301:691–711. DOI: 10.1006/jmbi.2000.3975

[28] Corpet F, Gouzy F, Kahn D. The ProDom database of protein domain families. Nucleic Acids Research. 1998;26:323–326. DOI: 10.1093/nar/26.1.323

[29] Sali A. Modeller: program for comparative protein structure modeling by satisfaction of spatial restraints [Internet]. Available from: https://salilab.org/modeller/ [Accessed: 2016-04-21]

[30] Wiederstein M, Sippl MJ. ProSA-web: interactive web service for the recognition of errors in three-dimensional structures of proteins [Internet]. 2007. Available from: https://prosa.services.came.sbg.ac.at/prosa.php [Accessed: 2016-04-21]

[31] Roman AL, MacArthur MW, Smith DK, Jones DT, Hutchinson EG, Morris AL, Moss DS, Thornton JM. PROCHECK-Programs to check the Stereochemical Quality of Protein Structures [Internet]. 1998. Available from: http://www.ebi.ac.uk/thornton-srv/28 software/PROCHECK [Accessed:2016-07-11]

[32] Baker NA, Sept D, Joseph S, Holst MJ, McCammon JA. Electrostatics of nanosystems: application to microtubules and the ribosome. Proceedings of the National Academy of Sciences. 2001:98;10037–10041. DOI: 10.1073/pnas.181342398

[33] The European Bioinformatics Institute [Internet]. Available from: http://www.ebi.ac.uk/Tools/structure/ [Accessed: 2016-04-18]

[34] Hu B, Jin J, Guo AY, Zhang H, Luo J, Gao G. GSDS 2.0: an upgraded gene feature visualization server. Bioinformatics. 2015:31;1296–1297.

[35] Ratan A. ASSEMBLY algorithms for next generation sequence data [Internet]. 2009. Available from: http://www.yasara.org/products.htm#structure

[36] Maston GA, Evans K, Green MR. Transcriptional regulatory elements in the human genome. Annual Review of Genomics and Human Genetics. 2006;7:29–59. DOI: 10.1146/annurev.genom.7.080505.115623

[37] Lescot M, Déhais P, Thijs G, Marchal K, Moreau Y, Van de Peer Y, Rouzé P, Rombauts S. PlantCARE, a database of plant cis-acting regulatory elements and a portal to tools for in silico analysis of promoter sequences [Internet]. 2002. Available from: http://hdl.handle.net/1854/LU-153936 [Accessed: 2004-01-14]

[38] Pandey S, Negi YK, Marla SS, Arora S. Comparative in silico analysis of ascorbate peroxidase protein sequences from different plant species. Journal of Bioengineering & Biomedical Science. 2011;1:103–107. DOI: 10.4172/2155-9538.1000103

[39] Han Y, Zheng QS, Wei YP, Chen J, Liu R, Wan HJ. *In silico* identification and analysis of phytoene synthase genes in plants. Genetics and Molecular Research. 2015;14:9412–9422. DOI: 10.4238/2015

[40] Rao G, Sui J, Zhang J. In silico genome-wide analysis of the WRKY gene family in Salix arbutifolia. Plant Omics Journal. 2015;8:353–360.

[41] He G, Elling AA, Deng XW. The epigenome and plant development. Annual Review of Plant Biology. 2011;62:411–435. DOI: 10.1146/annurev-arplant-042110-103806

[42] Law JA, Jacobsen SE. Establishing, maintaining and modifying DNA methylation patterns in plants and animals. Nature Reviews Genetics. 2010;11:204–220. DOI: 10.1038/nrg2719

[43] Lister R, Gregory BD, Ecker JR. Next is now: new technologies for sequencing of genomes, transcriptomes, and beyond. Current Opinion in Plant Biology. 2009;12:107–118. DOI: 10.1016/j.pbi.2008.11.004

[44] Morsy M, Gouthu S, Orchard S, Thorneycroft D, Harper JF, Mittler R, Cushman JC. Charting plant interactomes: possibilities and challenges. Trends in Plant Sciences. 2008; 13:183–191. DOI: 10.1016/j.tplants.2008.01.006

[45] Guo F, Ding Y, Li Z, Tang J. Identification of protein-protein interactions by detecting correlated mutation at the interface. Journal of Chemical Information and Modeling. 2015;55:2042–2049. DOI: 10.1021/acs.jcim.5b00320

[46] Marks DS, Colwell LJ, Sheridan R, Hopf TA, Pagnani A, Zecchina R, Sander C. Protein 3D structure computed from evolutionary sequence variation. PLoS One. 2011;6:e28766. DOI: 10.1371/journal.pone.0028766

[47] Waters ER, Lee GJ, Vierling E. Evolution, structure and function of the small heat shock proteins in plants. Journal of Experimental Botany. 1996;47:325–338. DOI: 10.1093/jxb/47.3.325

[48] Brinkman FS, Leipe DD. Phylogenetic analysis. Methods of Biochemical Analysis. 2001;43:323–358.

[49] Wiley EO, Lieberman BS, editors. Handbook of Phylogenetics: Theory and Practice of Phylogenetic Systematics. 2nd ed. Chichester: Wiley; 2011. 432 p.

[50] Peng C. Distance based methods in phylogenetic tree construction. Neural Parallel and Scientific Computations. 2007;15:547–560.

[51] Saitou N, Nei M. The neighbor-joining method: a new method for reconstructing phylogenetic trees. Molecular Biology and Evolution. 1987;4:406–425.

[52] Swofford DL, Olsen GJ, Waddel PJ, Hillis DM. Phylogenetic inference. In: Hillis DM, Moritz C, Mableeds BK, editors. Hand Book of Molecular Systematics. 2nd ed. Sunderland: Sinauer Associates;1996. pp. 407–446.

[53] Felsenstein, J. PHYLIP (Phylogeny Inference Package) [Internet]. 1993. Available from: http://evolution.genetics.washington.edu/phylip.html [Accessed: 2016-04-11]

[54] Adachi J, Hasegawa M. MOLPHY version 2.3: programs for molecular phylogenetics based on maximum likelihood [Internet]. 1996. Available from: http://www.ism.ac.jp/ismlib/softother.e.html [Accessed: 2016-04-21]

[55] Schmidt HA, Strimmer K, Vingron M, Haeseler A. TREE-PUZZLE: maximum likelihood phylogenetic analysis using quartets and parallel computing. Bioinformatics. 2002;18:502–504. DOI: 10.1093/bioinformatics/18.3.502

[56] Olsen GJ, Matsuda H, Hagstrom R, Overbeek R. fastDNAml: a tool for construction of phylogenetic trees of DNA sequences using maximum likelihood. Computer Applications in the Biosciences. 1994;10:41–48. DOI: 10.1093/bioinformatics/10.1.41

[57] Maddison DR, Maddison WP. MacClade 4: analysis of phylogeny and character evolution [Internet]. 2005. Available from: http://macclade.org

[58] Kopec W, Telenius J, Khandelia H. Molecular dynamics simulations of the interactions of medicinal plant extracts and drugs with lipid bilayer membranes. FEBS Journal. 2013;280:2785–2805. DOI: 10.1111/febs.12286

[59] Turenne N. Role of a web-based software platform for systems biology. Journal of Computer Science and System Biology. 2011;4:35–41. DOI: 10.4172/jcsb.1000101e

[60] Perez-Riverol Y, Alpi E, Wang R, Hermjakob H, Vizcaíno JA. Making proteomics data accessible and reusable: current state of proteomics databases and repositories. Proteomics. 2015;15:930–949. DOI: 10.1002/pmic.201400302

[61] Bhardwaj T, Somvanshi P. Plant systems biology: insights and advancements. In: Barh D, Khan MS, Davies E, editors. Hand Book of Plant Omics: The Omics of Plant Science. New Delhi:Springer; 2015. pp. 791–819. DOI: 10.1007/978-81-322-2172-2_28

[62] Verma M, Kumar V, Patel RK, Garg R, Jain M. CTDB: an integrated chickpea transcriptome database for functional and applied genomics. PLoS One. 2015;10:0136880. DOI: 10.1371/journal.pone.0136880

[63] Nooren IMA, Thornton JM. Diversity of protein-protein interactions. The EMBO Journal. 2003;22:3486–3492. DOI 10.1093/emboj/cdg359

[64] You L. Toward computational systems biology. Cell Biochemistry and Biophysics. 2004;40:167–184. DOI: 10.1385/CBB:40:2:167

[65] Sharan R, Ulitsky I, Shamir R. Network-based prediction of protein function. Molecular System Biology. 2007;3:88–100. DOI 10.1038/msb4100129

[66] Wang C, Marshall A, Zhang D, Wilson ZA. ANAP: an integrated knowledge base for Arabidopsis protein interaction network analysis. Plant Physiology. 2012;158:1523–1533. DOI: 10.1104/pp.111.192203

[67] Zhu P, Gu H, Jiao Y, Huang D, Chen M. Computational identification of protein-protein interactions in rice based on the predicted rice interactome network. Genomics Proteomics Bioinformatics. 2011;9:128–137. DOI: 10.1016/S1672-0229(11)60016-8

[68] Alkin B. iPlants-the world's plant on line [Internet]. 2004. Available from: iPlants document library: www.iplants.intranets.com

[69] Joshi-Tope G, Gillespie M, Vastrik I, D'Eustachio P, Schmidt E, de Bono B, Jassal B, Gopinath GR, Wu GR, Matthews L, Lewis S, Birney E, Stein L. Reactome: a knowledgebase of biological pathways. Nucleic Acids Research. 2005;33:428–432. DOI: 10.1093/nar/gki072

[70] Haw R, Stein L. Using the reactome database. In: Andreas D, et al editors. Hand book of Current protocols in bioinformatics. Chichester:Wiley; 2012. DOI: 10.1002/0471250953.bi0807s38

[71] Croft D. Building models using Reactome pathways as templates. Methods in Molecular Biology. 2013;1021:273–283. DOI: 10.1007/978-1-62703-450-0_14

[72] Fukusaki E, Kobayashi A. Plant metabolomics: potential for practical operation. Journal of Bioscience and Bioengineering. 2005;100:347–354. DOI: 10.1263/jbb.100.347

[73] Fukushima A, Kusano M. Recent progress in the development of metabolome databases for plant systems biology. Frontiers in Plant Science. 2013;4:73. DOI: 10.3389/fpls.2013.00073

[74] Deborde C, Jacob D, MeRy-B. A metabolomic database and knowledge base for exploring plant primary metabolism. Methods in Molecular Biology. 2014;1083:3–16. DOI: 10.1007/978-1-62703-661-0_1

[75] Iquebal MA, Jaiswal S, Mukhopadhyay CS, Sarkar C, Rai A, Kumar D. Applications of bioinformatics in plant and agriculture. In: Barh D, Khan MS, Davies E, editors. Hand

Book of Plant Omics: The Omics of Plant Science. Springer; 2015. pp. 755–790. DOI: 10.1007/978-81-322-2172-2_1

[76] Agrawal PK, Babu BK, Saini N. Omics of model plants. In: Barh D, Khan MS, Davies E, editors. Hand Book of Plant Omics: The Omics of Plant Science. New Delhi: Springer; 2015. pp 1–32. DOI: 10.1007/978-81-322-2172-2_1

[77] Tiwary BK. Next-Generation Sequencing and Assembly of Plant Genomes. In: Barh D, Khan MS, Davies E, editors. Hand Book of Plant Omics: The Omics of Plant Science. New Delhi:Springer; 2015. pp. 53–64. DOI: 10.1007/978-81-322-2172-2_1

[78] Gordon D, Abajian C, Green P. Consed: a graphical tool for sequence finishing. Genome Research.1998;8:195–202.

[79] The GAP Group. GAP—groups, algorithms, and programming, version 4.8.3 [Internet]. 2016. Available from: http://www.gap-system.Org[Accessed:2016-07-11]

[80] Maddison DR, Maddison WP. Chromaseq: a Mesquite package for analyzing sequence chromatograms [Internet]. 2014. Available from: http://mesquiteproject.org/packages/Chromaseq[Accessed:2016-07-11]

[81] Rhee SY, Dickerson J, Dong Xu. Bioinformatics and its applications in plant biology. Annual Reviews of Plant Biology. 2006;57:335–360. DOI: 10.1146/ annurev.arplant.56.032604.144103

[82] Liu L, Li Y, Li S, Hu N, He Y, Pong R, Lin D, Lu L, Law M . Comparison of next-generation sequencing systems. Journal of Biomedicine and Biotechnology. 2012; DOI: 10.1155/2012/251364

[83] Weigel D, Mott R. The 1001 genomes project for *Arabidopsis thaliana*. Genome Biology. 2009;10:107–111. DOI: 10.1186/gb-2009-10-5-107

[84] Li JY, Wang J, Zeigler RS, The 3,000 rice genomes project: new opportunities and challenges for future rice research. Gigascience. 2014;3:8–10. DOI: 10.1186/2047-217X-3-8

[85] Fahlgren N, Gehan MA, Baxter I. Lights, camera, action: high-throughput plant phenotyping is ready for a close-up. Current Opinion in Plant Biology. 2015;24;93–99. DOI: 10.1016/j.pbi.2015.02.006

[86] Rick HJ, de Zedde V. IPPN: International Plant Phenotyping Network [Internet]. 2015. Available from: http://www.plant-phenotyping.org [Accessed: 2016-04-21]

[87] Swarbreck D, Wilks C, Lamesch P, Berardini TZ, Garcia-Hernandez M, Foerster H, et al. The Arabidopsis Information Resource (TAIR): gene structure and function annotation. Nucleic Acids Research. 2008;36:1009–1014. DOI: 10.1093/nar/gkm965

[88] Ware D, Jaiswal P, Ni, J et al. Gramene: a resource for comparative grass genomics. Nucleic Acids Research. 2002;30:103–105. DOI: 10.1093/nar/30.1.103

[89] Liang C, Jaiswal P, Hebbard C, Avraham S, Buckler ES, Casstevens T, et al. Gramene: a growing plant comparative genomics resource. Nucleic Acids Research. 2008;36:947–953. DOI: 10.1093/nar/gkm968

[90] Grant, D, Nelson, RT, Cannon, SB, Shoemaker, RC. SoyBase, the USDA-ARS soybean genetics and genomics database. Nucleic Acids Research. 2010;38:843–846. DOI: 10.1093/nar/gkp798

[91] Fernandez-Pozo N, Menda N, Edwards JD, Saha S, Tecle IY, Strickler SR, Bombarely A, Fisher-York T, Pujar A, Foerster H, Yan A, Mueller LA. The Sol Genomics Network (SGN)—from genotype to phenotype to breeding. Nucleic Acids Research. 2015;43:1036–1041.

[92] Lawrence CJ, Dong Q, Polacco ML, Seigfried TE, Brendel V. MaizeGDB, the community database for maize genetics and genomics. Nucleic Acids Research. 2004;32:393–397. DOI: 10.1093/nar/gkh011

[93] Conte MG, Gaillard S, Lanau N, Rouard M, Périn C. GreenPhylDB: a database for plant comparative genomics. Nucleic Acids Research. 2008;36:991–998. DOI: 10.1093/nar/gkm934

[94] Proost S, Van Bel MV, Vaneechoutte D, de Peer YV, Inze D, Mueller-Roeber B, Vandepoele K. PLAZA 3.0: an access point for plant comparative genomics Nucleic Acids Research. 2015;43:974–981. DOI: 10.1093/nar/gku986

[95] Wang Y, You FM, Lazo GR, Luo M-C, Thilmony R, Gordon S, Shahryar F, Kianian SF, Gu YQ. PIECE: a database for plant gene structure comparison and evolution. Nucleic Acids Research. 2012;41:1159–1166. DOI: 10.1093/nar/gks1109

[96] Sterck L, Rombauts S, Vandepoele K, Rouze P, Van de Peer Y. How many genes are there in plants (and why are they there)? Current Opinion in Plant Biology. 2007;10:199–203. DOI: 10.1016/j.pbi.2007.01.004

[97] Wall PK, Leebens-Mack J, Muller KF, Field D, Altman NS, dePamphilis CW. PlantTribes: a gene and gene family resource for comparative genomics in plants. Nucleic Acids Research. 2008;36:970–976. DOI: 10.1093/nar/gkm972

[98] Tang H, Bowers JE, Wang X, Ming R, Alam M, Paterson AH. Synteny and collinearity in plant genomes. Science. 2008;320:486–488. DOI: 10.1126/science.1153917

[99] Lee TH, Tang H, Wang X, Paterson AH. PGDD: a database of gene and genome duplication in plants. Nucleic Acids Research. 2012;41:1152–1158. DOI: 10.1093/nar/gks1104

[100] Wang Y, Coleman-Derr D, Chen G, Gu YQ. OrthoVenn: a web server for genome wide comparison and annotation of orthologous clusters across multiple species. Nucleic Acids Research. 2015;43:78–84. DOI: 10.1093/nar/gkv487

[101] Li F, Fan G, Lu C, Xiao G, Zou C, Kohel R, Ma Z, Shang H, Ma X, Wu J, Liang X, Huang G, G Percy R, Liu K, Yang W, et al. Genome sequence of cultivated upland cotton

(*Gossypium hirsutum* TM-1) provides insights into genome evolution. Nature Biotechnology. 2015;33:524–530. DOI: 10.1038/nbt.3208

[102] Hudek AK, Cheung J, Boright AP, Scherer SW. Genescript: DNA sequence annotation pipeline. Bioinformatics. 2003;19:1177–1178. DOI: 10.1093/bioinformatics/btg134

[103] Allen JE, Pertea M, Salzberg SL. Computational gene prediction using multiple sources of evidence. Genome Research. 2004;14:142–148. DOI: 10.1101/gr.1562804

[104] Smit AFA, Hubley R, Green P. Repeat Masker Open-3.0 [Internet]. 1996–2010. Available from: http://www.repeatmasker.org [Accessed: 2016-04-21]

[105] Tatusova T, Smith White B, Ostell J. A collection of plant-specific genomic data and resources at NCBI. In: Edwards E, editor. Hand Book of Plant Bioinformatics, Methods and Protocols. New York:Humana Press; 2007. pp. 61–87.

[106] Broman KW, Wu H, Sen S, Churchill G A. R/qtl: QTL mapping in experimental crosses. Bioinformatics. 2003;19:889–890. DOI: 10.1093/bioinformatics/btg112

[107] Jansen J, de Jong AG, van Ooijen JW. Constructing dense genetic linkage maps. Theoretical and Applied Genetics. 2001;102:1113–1122. DOI: 10.1007/s001220000489

[108] Margarido GRA, Souza AP, Garcia AAF. One map: software for genetic mapping in outcrossing species. Hereditas. 2007;144:78–79. DOI: 10.1111/j.2007.0018-0661.02000.x

[109] Wu Y, Bhat PR, Close TJ, Lonardi S. Efficient and accurate construction of genetic linkage maps from the minimum spanning tree of a graph. PLoS Genetics. 2008;4:e1000212. DOI: 10.1371/journal.pgen.1000212

[110] Rastas P, Paulin L, Hanski I, Lehtonen R, Auvinen P. Lep-MAP: fast and accurate linkage map construction for large SNP datasets. Bioinformatics. 2013;29:3128–3134. DOI: 10.1093/bioinformatics/btt563

[111] Liu D, Ma C, Hong W, Huang L, Liu M, Liu H, et al. Construction and analysis of high-density linkage map using high-throughput sequencing data. PLoS One. 2014;9:e98855. DOI: 10.1371/journal.pone.0098855

[112] Fierst J. Using linkage maps to correct and scaffold de novo genome assemblies: methods, challenges, and computational tools. Frontiers in Genetics. 2015;6:220. DOI: 10.3389/fgene.2015.00220

[113] Heesacker A, Kishore VK, Gao W, Tang S, Kolkman JM, Gingle A, et al. SSRs and INDELs mined from the sunflower EST database: abundance, polymorphisms, and cross-taxa utility. Theoretical and Applied Genetics. 2008;117:1021–1029. DOI: 10.1007/s00122-008-0841-0

[114] Carollo V, Matthews DE, Lazo GR, Blake TK, Hummel DD, Lui N. Grain genes. An improved resource for the small grains community. Plant Physiology. 2005;139:643–651. DOI: 10.1104/pp.105.064485

[115] Mochida K, Saisho D, Yoshida T, Sakurai T, Shinozaki K. TriMEDB: a database to integrate transcribed markers and facilitate genetic studies of the tribe Triticeae. BMC Plant Biology. 2008;8:72. DOI: 10.1186/1471-2229-8-72

[116] Yu J, Jung S, Cheng CH, Ficklin SP, Lee T, Zheng P, Jones D, Percy RG, Main D. CottonGen: a genomics, genetics and breeding database for cotton research. Nucleic Acids Research. 2014;42:1229–1236. DOI: 10.1093/nar/gkt1064

[117] Mount DM. Hand Book of Bioinformatics: Sequence and Genome Analysis. 2nd ed. New York: Cold Spring Harbor Laboratory Press; 2001.

9

Bioinformatics: Basics, Development, and Future

Ibrokhim Y. Abdurakhmonov

Abstract

Bioinformatics is an interdisciplinary scientific field of life sciences. Bioinformatics research and application include the analysis of molecular sequence and genomics data; genome annotation, gene/protein prediction, and expression profiling; molecular folding, modeling, and design; building biological networks; development of databases and data management systems; development of software and analysis tools; bioinformatics services and workflow; mining of biomedical literature and text; and bioinformatics education and training. Astronomical accumulation of genomics, proteomics, and metabolomics data as well as a need for their storage, analysis, annotation, organization, systematization, and integration into biological networks and database systems were the main driving forces for the emergence and development of bioinformatics. Current critical needs for bioinformatics among others highlighted in this chapter, however, are to understand basics and specifics of bioinformatics as well as to prepare new generation scientists and specialists with integrated, interdisciplinary, and multilingual knowledge who can use modern bioinformatics resources powered with sophisticated operating systems, software, and database/networking technologies. In this introductory chapter, I aim to give an overall picture on basics and developments of the bioinformatics field for readers with some future perspectives, highlighting chapters published in this book.

Keywords: bioinformatics, databases, molecular sequence analysis, software and analysis tools, bioinformatics training

1. Introduction

Biological data can be described as molecular sequence information and "wet-bench" experimented content of genome and gene product analyses [1]. Being an interdisciplinary branch of the life sciences, bioinformatics targets to develop methodology and analysis tools to explore large volumes of biological data, helping to store, organize, systematize, annotate, visualize, query, mine, understand, and interpret complex data volumes. It uses conventional, modern computer science and cloud computing, statistics, and mathematics, as well as pattern recognition, reconstruction, machine learning, simulation and iterative approaches, and molecular modeling/folding algorithms [1, 2]. The emergence and advances of the bioinformatics field, however, are tightly associated with the computerized programming and software developments needed for the handling and structural and functional analysis of large volumes of molecular sequences of DNA, RNA, proteins, and metabolites.

Presently, although still core for genomics and genetics field, bioinformatics became an umbrella for wider range of biological studies analyzing variety types of biological data, structuring, systemizing, annotating, querying, mining, and visualizing available biological information and a variety of biomedical text records [1–3]. Although drawing a fine line between bioinformatics and some other related fields is difficult because of increased applications of computers, statistics, and mathematics to scientific problem solving and experiments of life sciences, there should not be a misperception about bioinformatics description and objectives. Bioinformatics should not be mixed with, for example, biometry and biostatistics, development of DNA computers, or computerized generation and filing of data from imaging.

Bioinformatics also should be differentiated from related scientific fields such as biological computation and computational biology [1, 2]. Biological computation aims to develop biological computers using advances of bioengineering, cybernetics, robotics, and molecular cell biology. In contrast, bioinformatics develops and utilizes computational algorithms to understand and interpret biological processes based on genome-derived molecular sequences and their interactions [2]. Therefore, in many aspects, bioinformatics seems similar to computational biology objectives. A computational biology is concentrated on building and/or developing theoretical models for biological analyses [1, 2], whereas bioinformatics focuses on providing practical tools to organize and analyze basic genomic, proteomic and other "omics" data, including sequence analysis and its visualization [1, 2]. Admittedly, computational biology and bioinformatics both target to use genome data, for example, multiple sequence alignments and/or genome assembly tools. This makes distinctive boundaries of these two fields less distinguishable if their theoretical and practical scales are forgotten [2]. Thus, as mentioned above, the common core aims of bioinformatics are to handle, analyze, and interpret the genome-derived molecular sequence data and its organizational principles in broad scales/spectra of comparative, simulative, and evolutionary/phylogenetics perspectives. These tools are applicable and widely used for studies related to genetics, genomics, biochemistry, physiology, biophysics, all agricultural, medical, and environmental sciences as well as evolution, system biology, and artificial intelligence [1–10].

For instance, bioinformatics tools such as the comparative analysis of genomic and genetic data and/or signal processing help to interpret and understand the molecular and evolutionary processes [9] and interactions from large volumes of raw data in the field of wet-bench experimental molecular biology [1, 2]. In the "omics" fields, it helps to sequence and annotate genomes, and identify distinct patterns, mutation profiles, genetic epistasis, gene/protein expression and regulation, and gene ontologies [1, 2, 4, 8–11] as well as be instrumental in mining and querying the biological data and biomedical literature text [3, 4, 7]. When applied for system biology [2, 6], bioinformatics is a key instrument to analyze and catalogue the biochemical/genetic pathways and networks, which helps to integrate pieces of analyzed information to depict and model a full picture of the life processes. Application of reconstruction, pattern recognition, folding, simulation, and molecular modeling with bioinformatic tools can identify structural peculiarities and interactions of molecular sequences important for structural biology and medicinal drug design [12, 13]. All of these large scale, genome-derived, molecular sequence analyses of raw "Big Data" are impossible to be analyzed manually [1, 2]. This prompted the biology science research community to apply interdisciplinary methods and tools for "Big Data" analysis in combination with modern computing knowledge, which resulted in the emergence of novel interdisciplinary bioinformatics science. Let us, first, take a look the historic developments in the bioinformatics field.

1.1. History of emergence and development

Bioinformatics term was coined by Paulien Hogeweg and Ben Hesper in 1970 [2, 14]. Its meaning was very different from current description and referred to the study of information processes in biotic systems like biochemistry and biophysics [14–16]. However, the emergence of bioinformatics tracks back to the 1960s. It was appeared in concordance with the development of protein sequencing methods from a variety of organisms and with the availability of protein sequences after Frederick Sanger determined the sequence of insulin in the early 1950s [17, 18]. New computer methods to analyze and compare a large number of protein sequences of different organisms were needed because handling many amino acid sequences manually was impractical. This led in compiling the first "Protein Information Resources" (PIR) [1, 19, 20] by Margaret Oakley Dayhoff and her collaborators at the National Biomedical Research Foundation [1]. Dayhoff's team successfully organized the protein sequences into distinct groups and sub-groups based on sequence similarity and percent accepted mutation (PAM) matrices [1]. This was published as protein sequences atlas [21, 22] that has been widely used in performing protein sequence alignments and database similarity searches [1, 2, 23]. This was pioneered methods of protein sequence alignment and molecular evolution [22]. In the 1970s, Elvin A. Kabat further contributed to bioinformatics development by his extended protein sequence analysis of comprehensive volumes of antibody sequences, released in collaboration with Tai Te Wu between 1980 and 1991 [2, 24].

With the objective of providing the theoretical background to immunology experiments in 1974, George Bell and colleagues initiated the collection of DNA sequences into GenBank [1]. During 1982–1992, the first version of GenBank was prepared by Walter Goad's group [1] and the efforts resulted in the development of presently known and widely used DNA sequence databases of GenBank [25], "The European Molecular Biology Laboratory (EMBL) [26], and DNA DataBank of Japan (DDBJ) [27] in 1979, 1980, and 1984, respectively [1]. Most important development in DNA sequence databases, however, was incorporation of web-based searching algorithms allowing researchers to find and compare the target DNA sequences. Such first developments and resulting computer software called "GENEINFO" and its derivative version of "Entrez" were developed by David Benson and David Lipman and colleagues [1].

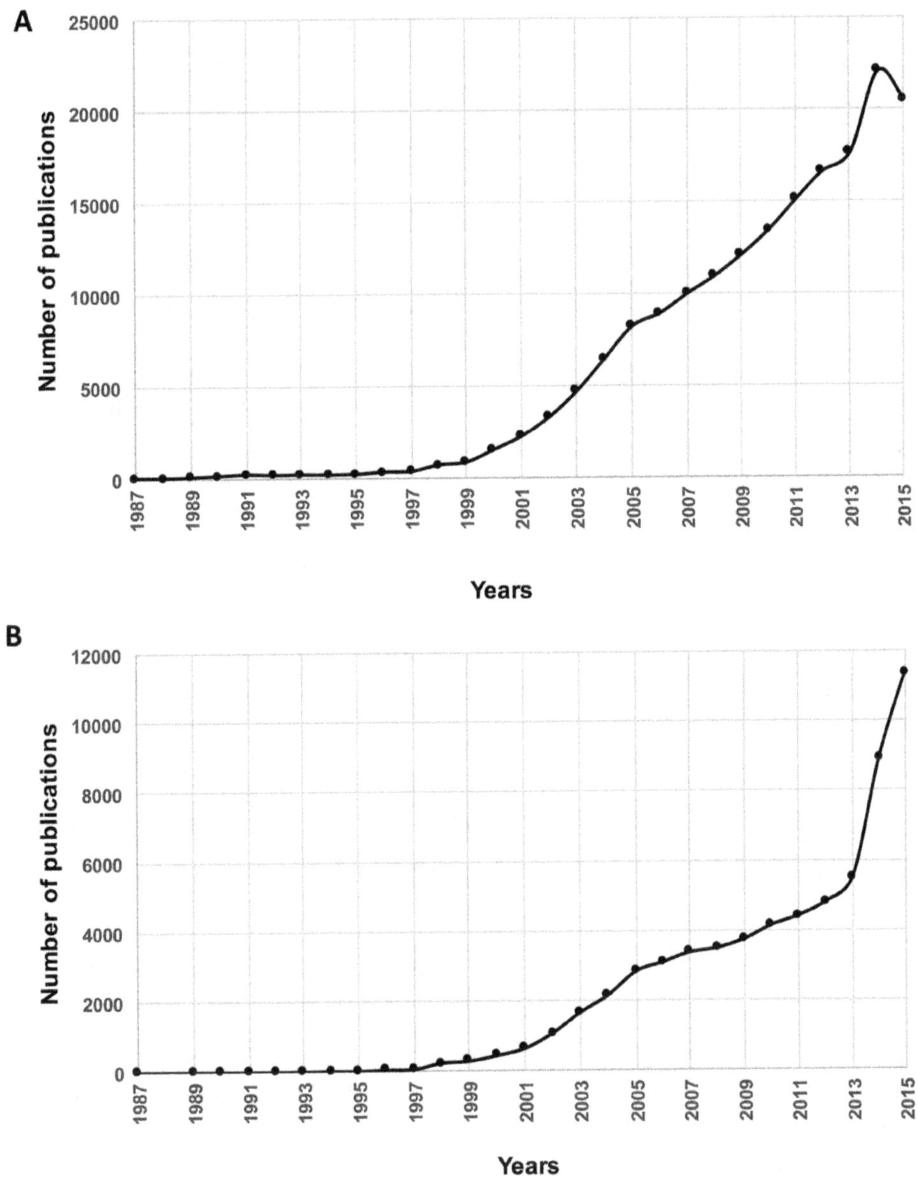

Figure 1. Dynamics of bioinformatics-related publications over the past four decades. (A) Unquoted and (B) quoted keyword retrieved scientific publications from PubMed [74].

This software allowed researchers to rapidly search database-indexed sequences and match them with queried sequence. Software became readily available through web-based interface of the National Center of Biotechnology Information (NCBI) database [28]. Molecular sequence analysis, comparison, and visualization methods have been improved, and many different methodologies have been contributed to bioinformatics advancements in this direction. Such advancements can be exemplified by the development of dot matrix and diagram methods [29], alignment of sequences by dynamic programming [30], finding of local alignments between sequences [31], multiple sequence alignment tools [32–35], predicting the secondary structures of RNAs [36, 37], determination of evolutionary relationships of sequences [38, 39], and assigning the gene function based on sequence similarity of known function from models [40]. Development of FASTA [41, 42], BLAST [43, 44], and their various modifications [45–47] has further powered the bioinformatics field and greatly improved the biological data analysis. Development of tools for predicting the putative protein sequences, structure, and function of proteins/genes based on DNA sequences [48–58], completing full genome sequences, and building web-based genome databases for many prokaryotic and eukaryotic organisms [58] has provided a great advance in the bioinformatics field. In addition, rapid genome-wide gene expression profiling and analysis opportunities [59–62], biological pathway assignment and identification, data storing, and mining and querying for large volume of biological datasets [63–73] have further provided unprecedented popularity of bioinformatics in the scene of world science, which has been briefly reviewed below.

Since its emergence as an interdisciplinary scientific field in 1970, bioinformatics research has continuously increased over the past four-decade period. Unquoted search of keyword of bioinformatics in the PubMed database [74] has found nearly 181,000 scientific publications covering the period of 1958 to March of 2016. Repeating the search with the quoted keyword found 62,402 scientific publications over the four-decade period, demonstrating the starting point of increased publication efforts in the end of 1990s with its first raise in 2000/2001, following significant peaks in 2003/2004 and after 2013 (**Figure 1**). In this introductory chapter, I aim to give a brief highlight of these four-decade developments introducing the chapters presented in this book.

2. Bioinformatics help in handling and analysis of the genomics data, genome annotation, and expression profiling

Rapid and reliable determination of DNA molecules, because of the introduction of the sequencing technique of Sanger and Coulson [75] and Maxam and Gilbert [76], provided large-scale DNA sequence data that needed to be analyzed by computerized programming. This prompted the development of efficient bioinformatics methodologies. For example, a seminal effort of the Phage Φ-X174 [2, 77] and the *Haemophilus influenza* [2, 78] genome sequencing using shotgun sequencing techniques generated the sequences of many thousands of small DNA fragments, ranging from 35 to 900 nucleotides [2] and required the assembly of a complete bacterial genome. The ends of sequenced shotgun clones overlap and can be assembled using computerized similarity search algorithms into the complete genome although the assembly tasks are challenging due to the requirement for powerful computers with sufficient memory and issues of generating multiple gaps in assembled genome. Genome assembly algorithms are a critical area of bioinformatics research as fragmented genome sequencing methods have been the core approach for virtually all genomes sequenced today [1, 2].

Therefore, without bioinformatics tools, it is not possible to think about genome sequencing as present bioinformatics programs such as BLAST/sequence alignments not only provide rapid practical tools to handle, analyze, compare, relate, and visualize DNA sequences but also offer help with the sequencing process itself. The development of cost-effective, next generation sequencing (NGS) platforms [79, 80] has helped to completely decode nearly the entire genome of many different organisms including human and many other model and specialty organisms, or crop genomes with complex polyploidy levels within a short period. For example, according to the listings in the Genomes OnLine Database (GOLD) as of March 8, 2016, there were 79,650 genome sequencing projects of which 8018 were completed projects, 33,489 were permanent drafts, 35,609 were incomplete projects, and 1553 were targeted projects [81]. There are 73,000 organism, including archaea (1201), bacteria (55,303), eukaryotes (11,990), and viruses (4473), listed for sequencing. These numbers should be increased if the sequencing of the 100,000 whole-human genomes [82] is added.

Bioinformatics tools are needed in annotation and prediction of genes from sequenced genomes that requires computerized approaches because genomes are large to be manually annotated as mentioned above. Bioinformatics-based gene finding and annotation including a search for protein-coding genes, RNA transcripts, and other functional sequences within a genome is possible because there are patterns to recognize the start, stop regions, introns, exons, motifs, repeats, and other regulatory and sensory as well as signaling regions with some variations between genes and among organisms. With the availability and need for analysis of *H. influenza* genome, the first genome annotation computer program system was designed in 1995 by Owen White [2, 78], which provided tools to find the genes and identify putative functions of annotated sequences. White's effort was basic for all currently available gene annotation and prediction software, which keep periodically improving [2].

Bioinformatics tools are very important to analyze gene and protein expression profiles. Large-scale sequencing of cDNA libraries has generated large volumes of serial analysis of gene expression (SAGE), expressed sequences tags (ESTs), massively parallel signature sequencing (MPSS), transcriptome profiling, or RNA-Seq, and various applications of multiplexed in-situ hybridization (microarray) profile data [83–95]. All of these gene expression techniques are extremely noise-prone and/or subject to bias in the biological measurement, which requires application of statistical tools to separate signal from noise in high-throughput gene expression studies. In this context, chapter by Zhao et al. in this book reviews and discusses the main tools and algorithms currently available for RNAseq data analyses, discussing rapidly evolving RNAseq technologies such as stranded RNAseq, targeted RNAseq, and single cell RNA-seq. Moreover, Sripathy et al. have comprehensively discussed transcriptome profiling, RNAseq, and micro-RNA expression studies in cotton (*Gossypium* species), whereas Younis et al. present a chapter on skin microbiome, transcriptome, and microarray data analyses. In this book, readers can find an interesting chapter on bioinformatics challenges and tools for Hepatitis B genome analysis written by Bell and Kramvis, which highlight features of this small genome virus for bioinformatics analysis.

Similarly, protein microarrays and high-throughput mass spectrometry require bioinformatics analysis to identify proteins through the complex sequence similarity searches using protein sequence databases [96–103]. Bioinformatics is a great help for analysis of gene regulation through searching and comparing the sequence motifs related to promoters and other regulatory elements. Using bioinformatics tools and sequence motifs/regulatory elements genes can be clustered by function, and the co-expression characteristics can be determined. Examples of such bioinformatics tools include k-means clustering, hierarchical clustering, and consensus clustering methods such as the Bi-CoPaM, and self-organizing maps (SOMs) that can identify functionally active sequences from very complex microarray datasets [104–107]

Not only just these, bioinformatics plays a major role in data collection of the functional elements of sequenced genomes that use the next-generation DNA-sequencing technologies and genomic tiling arrays. This is best exemplified "Encyclopedia of DNA Elements (ENCODE)" [108] project developed by the National Human Genome Research Institute that describes the functional elements of the human genome. Thanks to bioinformatics and applications of its tools, genomes and genes, and protein sequences of different organisms can be rapidly compared, searched, and interpreted. In addition, mutations can be identified that help to judge and diagnose many complex human and plant diseases, crop traits, and interpret complex evolutionary process, such as genome duplications, polyploidization, adaptation, and speciation.

3. Structural bioinformatics: molecular folding, modeling, and design

One of the widely used applications of bioinformatics is identification of three-dimensional protein structures, molecular modeling, and folding to predict the possible function of proteins or other molecular structures, model behavior of molecules, fold the molecule to its native biologically functional three-dimensional structure, and design biomedical drugs for many complex human diseases. It helps *de novo* protein design, enzyme design, protein-ligand/drug docking, protein-peptide interaction, and structure prediction of biological macromolecules and macromolecular complexes [1, 2, 109].

From the coding DNA sequences, the primary structure of proteins can be easily determined that is vital in understanding the function of the protein(s). Further, based on homology patterns in primary structure of proteins and using homology modeling, important structural formations and interaction sites with other proteins can be determined. This helps to predict reliably the structure of a protein based on known structure of a homologous protein(s). Moreover, the identification of secondary, tertiary, and quaternary structures of proteins is very important to understand the function of proteins. The exact three-dimensional structure is essential for correct function, and a failure to fold into native structure generally produces inactive proteins or misfolded proteins that can be toxic [108]. Bioinformatics of protein folding includes (1) energy landscape of protein folding and (2) modeling of protein folding approaches [12, 13, 109].

One of the freely available and leading web server/stand-alone software tools for automated protein structure prediction and structure-based functional annotation can be exemplified by the "Iterative Threading ASSEmbly Refinement" (I-TASSER), which "first generates full-length atomic structural models from multiple threading alignments and iterative structural assembly simulations followed by atomic-level structure refinement" [110]. Using the I-TASSER, all above-mentioned functional and structural characteristics of proteins, including ligand-binding sites, enzyme commission number, and gene ontology terms can be explored in a comparative scale [110, 111].

Molecular modeling through molecular mechanistic and/or the quantum chemistry approaches is the key bioinformatics approaches to study the behavior of molecules. These are routinely used to investigate the structure, dynamics, surface properties, and thermodynamics of inorganic, biological and polymeric systems. It helps to explore conformational changes associated with biomolecular function, and molecular recognition of proteins, and membrane complexes. The protein folding, identification of catalysis sites of enzymes, and protein stability can be studied using molecular modeling. Vast different bioinformatics tools for modeling of biomolecules and designing are available [110–112]. In this book, the chapter by Leong et al. presents bioinformatics modeling and tools for biological membranes using molecular dynamic simulations, all-atom, united-atom, and coarse-grained membrane models of lipids and proteins. In addition, in this book, by Filntisi et al. a computational method for the generation of antibody-drug through site-specific cysteine conjugation using structural prediction methods based on PDB files of a drug, linker, and antibody. Moreover, Bórquez and González-Billault have presented an interesting chapter on computational algorithms of predicting kinase-substrate relationships in protein kinases; this chapter compares prediction tools and methods and discusses improving substrate prediction with contextual information.

4. Biological networks and system biology

Watts and Strogatz in 1998 [113, 114] and Barabási and Albert in 1999 [115–117] fueled the opinion that complex systems can be viewed as networks where components can be represented as nodes and they are linked through their interactions (i.e., edges). The properties of nodes and edges form the network topology. This approach has widely been applied to many scientific fields including bioinformatics that resulted in construction of large-scale biological networks denoted as "omes" like biome, interactome, microbiome [2, 6].

Above highlighted molecular sequence analysis, prediction and annotation, and molecular modeling-related bioinformatics approaches are also the core for building, organizing, and systematizing biological networks of molecules (e.g., metabolic, protein-protein interactions, etc.), and genetic and biochemical pathways of complex cellular processes. These include reception, signal transduction, and gene regulation and gene co-expression. Such molecular networks integrate many different data types including DNA sequences, regulatory RNA, proteins, secondary metabolites, gene expression data, and other small molecules, which may be all connected physically and functionally. The construction and organization of such physically and functionally connected molecular networks of cellular processes can be achieved only by applying the combination of simulative, iterative, and model-oriented bioinformatics approaches. Such biological networks are useful to analyze and visualize the complex connections of these cellular processes, helping understand other biological networks such as neuronal networks, food webs, between/within-species interaction networks, which are the central component of modern system biology [2, 6]. Examples of "omes"-related networks are the Kyoto Encyclopedia of Genes and Genomes (KEGG), BioCyc database collection, BRaunschweig ENzyme DAtabase (BRENDA), Reactome, Comparative Toxicogenomics Database, and many other [118] biological networks. Some biological network databases and their utilization in plant genomics/epigenomics have been discussed by the chapters of Sripathi et al. and Rahman et al. in this book.

5. Databases

An organized collection of data is referred to as database that aims to collect schemes, tables, queries, reports, images, and other objects. An access to information in the databases is provided by an integrated set of computer software, which is referred to as a "database management system" (DBMS) [119]. The DBMS allows users to access all of the data contained in the databases. It has general functions for data definition, entry, storage, update, administration, and retrieval of large quantities of information in an organized way that requires modeling (hierarchical and network models), clustering, query languages and query optimization, and visualization algorithms [1, 2, 119].

Development of databases, therefore, is significantly dependent on bioinformatics tools, advances, research, and applications. There is a large number of different types of databases available, which cover all aspects of biological data storage and organization. Some aforementioned databases such as GenBank, EMBL, DDJB belong to primary nucleotide sequence databases. There are meta-databases that incorporate data compiled from multiple other databases such as Entrez, mGen, Metascape, etc. Some others are specialized databases such as those specific to an organism, for example, TAIR, the p53 Knowledgebase (p53), the plant alternative splicing database (PASD); the plant secretome, and subcellular proteome knowledgebase (PlantSecKB) [119]. All databases vary in their data definition, usage, format, and access types. In this book, the chapter by Kadam et al. specifically describes databases and bioinformatics algorithms related to allergen informatics, discussing the concepts of allergen bioinformatics and the key areas for potential development in the allergology, whereas Bell and Kramvis highlight public sequence database for Hepatitis B virus. In this book, readers can find a comprehensive discussion for bioinformatics resources, including databases for plant "omics," written by Rahman et al.

6. Software, analysis tools, services, and workflow

As mentioned above, astronomical accumulation of genomic and proteomic as well as metabolomic data, and their expression profiles and annotation, storage, organization, systematization, and integration into biological networks as well as database systems and their wide utilization by the science research community *a priori* required computer programming algorithms, analysis tools, services, and workflow systems. Therefore, software and analysis tools, and bioinformatics services and workflow have been the main fields and core targets of bioinformatics since its emergence. Because of the contributions of various bioinformatics companies or public institutions, bioinformatics software, and tools started to exist as simple command-line tools, but later improved to more complex graphical programs standalone packages, and web services. Since development of the first bioinformatics software and analysis tools for molecular sequence evaluations in the early 1980s, many free and open-source software tools have been developed and continue to grow and improve with the advancement made in genomics sciences [2, 120].

The main driving forces for the current and future development of bioinformatics software and tools have been made on the past-decade advances of genome decoding technologies, accumulation of large volume biological data, consequent need for their analyses, as well as advancements of computer technologies, graphics, visualization, and molecular modeling and networking techniques. Moreover, the availability of various open-source codes, shared object models, and community-supported plug-ins has facilitated gathering innovative ideas from the community and performing innovative *in silico* experiments on existing "Big Data." These all-created golden opportunities for all research groups and bioinformatics companies to work, experiment, and develop more new generation of bioinformatics software and tools that are user friendly, capable of performing extended and integrated analysis with better visualization and graphical outputs. The range of open-source software packages includes titles such as UGENE, EMBOSS, GenGIS, GENtle, MOTHUR, BioPerl, PathVisio, BioJava, GenoCAD, Biopython, GeWorkbench, GenomeSpace, Bioclipse, .NET Bio, Apache Taverna, BioJS, Bioconductor, and BioRuby [121, 122].

Development of sharing models and web access tools is also an important bioinformatics objective that allows users to utilize and access bioinformatics tools over the internet and from their computer systems to the main computing resources via servers in other parts of the world. Simple Object Access Protocol (SOAP) [123] and Representational State Transfer (REST) [124–126] are two bioinformatics tools to provide web services. SOAP is a standard-based web service access protocol, originally developed by Microsoft. REST, providing very simple web service access, has been developed to fix the problems with SOAP [127]. Both tools share similarities over the HTTP protocol and have its own issues and challenges, differ in messaging patterns, rules, architecture style, and flexibility. The main advantages derive from the fact that end users do not have to deal with software and database maintenance overheads [127].

There are several basic bioinformatics services, for example, "Sequence Search Services" (SSS), "Multiple Sequence Alignment" (MSA), and "Biological Sequence Analysis" (BSA) [2, 128]. These web service-based bioinformatics analysis resources represent a collection of standalone or web-based interface data analysis tools as well as integrative, distributed, and extensible bioinformatics workflow management systems (BWMS). The BWMSs are designed specifically to compose and execute a series of interactive computational or data manipulation steps (i.e., a workflow) in a bioinformatics analyses. Such systems provide interactive analysis of biological data, build the specific workflows for the analysis, enable the visualization of the analysis outputs in real time, and simplify the process of sharing and reusing workflows between scientists. Some of the platforms giving this service: Galaxy, UGENE, Taverna, etc. [2, 121]. Several chapters of this book cover bioinformatics software, web-based analysis tools, and bioinformatics services for membrane analysis (see Leong et al.), in plant science and crop genomics (see chapters by Rahman et al. and Sripathi et al.), medicine, viral genome analysis and drug design (see chapters by Younis et al., Bell and Kramvis, and Filntisi et al.).

7. Text mining

Part of objectives in bioinformatics research and application is the utilization of computational algorithms and bioinformatics tools to collect, organize, and structure the growing body of biomedical literature allowing scientists to query, mine, read, and synthesize the specific literature and published articles of their research interest [2–4, 7, 129, 130]. Biomedical literature and text mining, therefore, are very important for scientific development, innovations, and integration and application of discoveries to society through extracting information (EI) and assessing the relationships of publications [3, 4]. Analysis of world literature demonstrates that more than 80% of text data remain unstructured that what makes it challenging to read every paper, resulting in disjointed sub-fields of research [3]. Biomedical literature text mining uses a variety of "text mining & data mining" tools, applying techniques such as data clustering, visualization and navigation, information retrieval, and extraction, and text categorization and summarization [3]. The use of IE and "Natural Language Generation and Understanding" (NLG and NLU) that have tokenizing, morphological or lexical, and syntactic analysis components helps to build structured text, and extract, collect, organize structured information [129, 130]. Pattern recognition and matching such as the recognition of biological abbreviations, terms, and interactions are important methods in text mining [2–4].

8. Education

Advances of life sciences and high-throughput biology fields in particular "omics" disciplines, the scale, and complexity of "Big Data," and growing demand for specialists with multilingual and cross-field expertise to understand and solve multidisciplinary scientific concepts and tasks underlie a great need for training and education in the field of bioinformatics. Bioinformatics training and education aim to create, collect, deliver, and share educational and training materials and techniques as well as develop university degree-program curricula on bioinformatics. This is to prepare scientists and specialists, who can utilize modern bioinformatics tools with the sophisticated operating systems, software and algorithms, and database/networking technologies to handle, analyze, interpret, and publish high-throughput complex biological data. This is a great bottleneck and critical need of current life sciences and bioinformatics field, especially in all developing countries, for example, analyzed by some recent reports for African [82] and Central American [131] countries.

To address this, bioinformatics research community has put specific efforts to develop local and global platforms for bioinformatics training and education. Such examples include "Bioinformatics Training Network" (BTN) [132] and "The Global Organization for Bioinformatics Learning, Education, and Training" (GOBLET) [133] that provide a community educational and training resource for bioinformatics trainers and trainees. As an outcome of European 7th Framework grant, BTN targeted to develop and share educational materials, short courses, and training delivery methods as well as discuss the challenge, issues, and needed requirements for bioinformatics training [132]. Furthermore, GOBLET continues similar efforts beyond Europe, aiming to coordinate efforts at the global scales with concentrated strategy and within the frame of single, dedicated foundation although it requires much time, focused strategic efforts, and modern innovative approaches [133].

"The Swiss Institute of Bioinformatics" training portal [134] also provides online courses for software platforms designed to teach bioinformatics concepts and methods including Rosalind [135]. There are open-access website videos and slides from the "Canadian Bioinformatics Workshops" [136]. Similarly, many different, large bioinformatics conferences, and seminars contribute for training and education on bioinformatics such as Intelligent Systems for Molecular Biology (ISMB), European Conference on Computational Biology (ECCB), Research in Computational Molecular Biology (RECOMB), and the annual Bioinformatics Open Source Conference (BOSC) of the non-profit Open Bioinformatics Foundation [2, 128]. As public bioinformatics databases, the MediaWiki engine with the WikiOpener extension, extensively referenced in this chapter, also contributes for training and education of bioinformatics through gathering research materials and descriptions of tools that can be accessed and updated by all experts in the field [128].

With the specific objectives to develop bioinformatics research and application, its integration to genomics research, and training and education as well as to prepare well-qualified new generation scientists to life sciences, we established a dedicated organization—Center of Genomics and Bioinformatics in the developing country Uzbekistan [137]. As in other developing countries, there are many challenges and limitations in funding and in accessing to sophisticated bioinformatics tools and computer operating systems as well as lack of sufficient experience to carry bioinformatics research and resource development. However, our first step goal is to integrate genomics and bioinformatics curricula to the higher education system of Uzbekistan, develop training and educational materials, provide basic training and research practices to the university students and biology field specialists, and establish international collaborations on this direction. The long-term objective is to efficiently and broadly apply genomics and bioinformatics approaches to all areas of life sciences in national and regional levels that would contribute the development of biological sciences in Central Asia. Some efforts are ongoing regarding the establishment of international collaborations [138] and providing training and education in both national and regional levels.

9. Conclusions and future perspectives

Bioinformatics has become an essential interdisciplinary scientific field to the life science helping to "omics" field and technologies and mainly handling and analyzing "omes" data. Accumulation of high-throughput biological data due to the technological advances in "omics" fields required and prioritized the use of bioinformatics resources, and research and application for the analysis of complex and even further enlarging "Big Data" volumes, which would be impractical and useless without bioinformatics. Therefore, as highlighted herein, there is a critical need for the preparation of well-qualified, new generation scientists with integrated knowledge, multilingual ability, and cross-field experience who are capable of using sophisticated operating systems, software and algorithms, and database/networking technologies to handle, analyze, and interpret high-throughput and increasing volume of complex biological data.

Community resources and a globally coordinated foundation of bioinformatics training and education platforms as well as research conferences, workshops, short online training, and web-based educational courses and materials are available to accomplish toward this goal. However, there is an urgent need for the development of bioinformatics education and training, in particular in developing countries, which requires innovative platforms, training techniques, better funding, web and network access, and high-performance computing systems.

In the research side, bioinformatics tools need to be improved for analysis of the growing body of high-throughput pangenomics, metagenomics, proteomics, and metabolomics data. There are needs for "effective tools" to perform better genome assembly and annotation with high accuracy; however, it requires the improvement of quality of sequenced genomes without gaps, and sequencing of more genome representatives, sub-genomes, polyploidy species, genomes of single cells, and specific tissues that would generate information to work, modify, and correct bioinformatics algorithms and programming approaches.

The use of third generation sequencing approaches and platforms as well as efforts on whole genome sequencing of, for example, 1000 or 100,000 human genome representatives [82] or transcriptome/exon sequencing of 1000 distinct plant species (e.g., 1KP) [139] will ultimately improve and advance the bioinformatics analysis tools. These efforts also help to improve orthologous gene identification tools that currently need attention [120]. There is a great need for sampling and handling diverse strains in pangenomic analysis, integration of prokaryotic genome-organization frameworks (GOFs) as well as integration of non-coding RNAs, pseudogenes, and epigenetics elements into the bioinformatics annotation and ontology tools and software [120]. There is a need to make sequenced genome data more functional and integrated through the construction of more organized, user friendly, cell-wide biological networks, and metabolic pathways [140] with better visualization effects, graphics outputs [120], and knowledge base construction (KB) [141]. This, however, requires the development of real-time imaging systems and high throughput phenotyping (referred to as "phenomics") tools that would help for efficiently determining biologically meaningful associations between genomic and phenotypic data, advancing the translational sciences, personal genomics, and personalized medicine [7] and/or agriculture [142].

Acknowledgements

I thank Academy of Sciences of Uzbekistan and Committee for Coordination Science and Technology Development of Uzbekistan, the Office of International Research Programs (OIRP) of the United States Department of Agriculture (USDA)—Agricultural Research Service (ARS) and U.S. Civilian Research & Development Foundation (CRDF) for research Grants FA-F5-T030, FA-A6-T081, FA-A6-T085, I-2015-6-15/2, I5-FQ-0-89-870, P120, P120A, P121, P121B, UZB-

TA-31016, UZB-TA-31017, and UZB-TA-2992, which have been the key factors for development of plant genomics and bioinformatics in Uzbekistan. I greatly acknowledge the Uzbekistan government support and investments/guide from Academy of Sciences of Uzbekistan, Ministry of Agriculture and Water Resources of Uzbekistan, Cotton Industry Joint Stock Company of Uzbekistan, Ministry Foreign Economic Relations, Investments and Trade of Uzbekistan, USDA-ARS, and Texas A&M University for establishment of Center of Genomics and Bioinformatics in Uzbekistan. I also thank Prof. Gilbert S. Omenn, Center for Computational Medicine & Bioinformatics, University of Michigan, USA for critical reading of this introductory chapter, and Mr. Mirzakamol Ayubov and Mr. Muhammad Mirzahmedov, Center of Genomics and Bioinformatics, Uzbekistan, for their technical assistance while preparing this chapter material.

Author details

Ibrokhim Y. Abdurakhmonov

Address all correspondence to: ibrokhim.abdurakhmonov@genomics.uz and genomics@uzsci.net

Center of Genomics and Bioinformatics, Academy of Science of the Republic of Uzbekistan, Tashkent, Uzbekistan

References

[1] Mount WD. Bioinformatics: sequence and genome analysis. 2nd ed. New York: Cold Spring Harbor Laboratory Press; 2004. 692 p. doi:10.1086/431054

[2] Bioinformatics [Internet]. 2016. https://en.wikipedia.org/wiki/Bioinformatics. Accessed: 2016-03-10

[3] Vijaya S, Radha R. Text mining in biosciences-a review. International Journal of Scientific & Engineering Research. 2015;6:769–776.

[4] Raza K. Application of data mining in bioinformatics. Indian Journal of Computer Science and Engineering.2010;1:114–118.

[5] Shah VA, Rathod DN, Basuri T, Modi VS, Parmar IJ. Applications of bioinformatics in pharmaceutical product designing: a review. World Journal of Pharmacy and Pharmaceutical Sciences. 2015;4:477–493.

[6] Ma'ayan A. Introduction to network analysis in systems biology. Science Signaling. 2011;4:tr5. doi:10.1126/scisignal.2001965

[7] Soualmia LF, Lecroq T. Bioinformatics methods and tools to advance clinical care. Yearbook of medical informatics. 2015;10:170–173. doi:10.15265/IY-2015-026

[8] Altman RB, Miller KS. 2010 Translational bioinformatics year in review. Journal of the American Medical Informatics Association. 2011;18:358–366. doi:10.1136/amiajnl-2011-000328

[9] Saxena A., Soni B.P., Gupta V. A chronological review and comparison of four evolutionary based algorithms. European Journal of Advances in Engineering and Technology, 2015;2:35–41.

[10] Ezziane Z. Applications of artificial intelligence in bioinformatics: A review. Expert Systems with Applications. 2006;30:2–10. doi:10.1016/j.eswa.2005.09.042

[11] Upton A, Trelles O, Cornejo-GarcíaJA, Perkins JR. Review: High-performance computing to detect epistasis in genome scale data sets. Briefings in Bioinformatics. 2015;pii: 1–12. DOI: 10.1093/bib/bbv058

[12] Molecular Modelling [Internet]. 2016. https://en.wikipedia.org/wiki/Molecular_modelling. Accessed: 2016-03-10

[13] Protein Folding [Internet]. 2016. https://en.wikipedia.org/wiki/Protein_folding. Accessed: 2016-03-10

[14] Hesper B, Hogeweg P. Bioinformatica: een werkconcept. Kameleon. 1970;1:28–29 (in Dutch).

[15] Hogeweg P. The roots of bioinformatics in theoretical biology. PLoS Computational Biology.2011;7:e1002021. doi:10.1371/journal.pcbi.1002021

[16] Hogeweg P. Simulating the growth of cellular forms. Simulation. 1978;31:90–96. doi: 10.1177/003754977803100305

[17] Sanger F, Tuppy H. The amino-acid sequence in the phenylalanyl chain of insulin. 2. The identification of lower peptides from partial hydrolysates. Biochemical Journal. 1951;49:463–481. doi:10.1042/bj0490463

[18] Sanger F, Tuppy H. The amino-acid sequence in the phenylalanyl chain of insulin. 2. The investigation of peptides from enzymic hydrolysates. Biochemical Journal. 1951;49:481–490. doi:10.1042/bj0490481

[19] Wu CH, Yeh LS, Huang H, Arminski L, Castro-Alvear J, Chen Y, Hu Z, KourtesisP, Ledley RS, Suzek BE, Vinayaka CR, Zhang J, Barker WC. The protein information resource. Nucleic Acids Research. 2003;31:345–347. doi:10.1093/nar/gkg040

[20] Protein Information Resources (PIR) [Internet]. 2016. http://pir.georgetown.edu. Accessed: 2016-03-10

[21] Dayhoff M, Eck R. Atlas of protein sequence and structure 1967–1968. Maryland (Silver Spring): National Biomedical Research Foundation; 1968. 356 p.

[22] Eck RV, Dayhoff MO. Evolution of the structure of ferredoxin based on living relics of primitive amino acid sequences. Science. 1966;152:363–366. doi:10.1126/science.152.3720.363

[23] Moody, G. Digital code of life: how bioinformatics is revolutionizing science, medicine, and business. Chichester: Wiley; 2004. 400 p.

[24] Johnson G, Wu TT. Kabat Database and its applications: 30 years after the first variability plot. Nucleic Acids Research. 2000;28:214–218. doi:10.1093/nar/28.1.214

[25] GenBank [Internet]. 2016. http://www.ncbi.nlm.nih.gov/genbank. Accessed: 2016-03-10

[26] The European Molecular Biology Laboratory [Internet]. 2016. http://www.embl.org. Accessed: 2016-03-10

[27] DataBank of Japan [Internet]. 2016. http://www.ddbj.nig.ac.jp. Accessed: 2016-03-10

[28] The National Center of Biotechnology Information (NCBI) [Internet]. 2016. http://www.ncbi.nlm.nih.gov. Accessed: 2016-03-10

[29] Gibbs AJ, McIntyre GA. The diagram, a method for comparing sequences. Its use with amino acid and nucleotide sequences. European Journal of Biochemistry. 1970;16:1–11. doi:10.1111/j.1432-1033.1970.tb01046.x

[30] Pearson WR, Miller W. Dynamic programming algorithms for biological sequence comparison. Methods in Enzymology. 1992;210:575–601. doi: 10.1016/0076-6879(92)10029-D

[31] Smith TF, Waterman MS. Identification of common molecular subsequences. Journal of Molecular Biology. 1981;147:195–197. doi:10.1016/0022-2836(81)90087-5

[32] Johnson MS, Doolittle RF. A method for the simultaneous alignment of three or more amino acid sequences. Journal of Molecular Evolution. 1986;23:267–278 doi:10.1007/BF02115583

[33] Thompson JD, Higgins DG, Gibson TJ. CLUSTAL W: improving the sensitivity of progressive multiple sequence alignment through sequence weighting, position-specific gap penalties and weight matrix choice. Nucleic Acids Research. 1994;22:4673–4680. doi:10.1093/nar/22.22.4673

[34] Notredame C, Higgins DG, Heringa J. T-Coffee: A novel method for fast and accurate multiple sequence alignment. Journal of Molecular Biology. 2000;302:205–217. doi: 10.1006/jmbi.2000.4042

[35] Rius J, Cores F, Solsona F, van Hemert JI, Koetsier J, Notredame C. A user-friendly web portal for T-Coffee on supercomputers. BMC Bioinformatics. 2011;12:150. doi: 10.1186/1471-2105-12-150

[36] Zuker M, Stiegler P. Optimal computer folding of large RNA sequences using thermodynamics and auxiliary information. Nucleic Acids Research. 1981;9:133–148. doi: 10.1093/nar/9.1.133

[37] Jacobson AB, Good L, Simonetti J, Zuker M. Some simple computational methods to improve the folding of large RNAs. Nucleic Acids Research. 1984;12:45–52. doi:10.1093/nar/12.1Part1.45

[38] Felsenstein J. Phylogenies from molecular sequences: inference and reliability. Annual Review of Genetics. 1988;22:521–565. doi:10.1146/annurev.ge.22.120188.002513

[39] Barns SM, Delwiche CF, Palmer JD, Pace NR. Perspectives on archaeal diversity, thermophily and monophyly from environmental rRNA sequences. Proceedings of the National Academy of Sciences of the USA. 1996;93:9188–9193. doi:10.1073/pnas.93.17.9188

[40] Barker WC, Dayhoff MO. Viralsrc gene products are related to the catalytic chain of mammalian cAMP-dependent protein kinase. Proceedings of the National Academy of Sciences of the USA.1982;79:2836–2839. doi:10.1073/pnas.79.9.2836

[41] Pearson WR, Lipman DJ. Improved tools for biological sequence comparison. Proceedings of the National Academy of Sciences of the USA. 1988;85:2444–2448. doi:10.1073/pnas.85.8.2444

[42] Pearson WR. Flexible sequence similarity searching with the FASTA3 program package. Methods in Molecular Biology. 2000;132:185–219. doi: 10.1385/1-59259-192-2:185

[43] Altschul SF, Gish W, Miller W, Myers EW, Lipman DJ. Basic local alignment search tool. Journal of Molecular Biology. 1990;215:403–410. doi:10.1016/S0022-2836(05)80360-2

[44] Pearson WR. BLAST and FASTA similarity searching for multiple sequence alignment. Methods in Molecular Biology. 2014;1079:75–101. doi:10.1007/978-1-62703-646-7_5

[45] Altschul SF, Madden TL, Schäffer AA, Zhang J, Zhang Z, Miller W, Lipman DJ. Gapped BLAST and PSI-BLAST: a new generation of protein database search programs. Nucleic Acids Research. 1997;25:3389–3402. doi:10.1093/nar/25.17.3389

[46] Ladunga I, Wiese BA, Smith RF. FASTA-SWAP and FASTA-PAT: pattern database searches using combinations of aligned amino acids, and a novel scoring theory. Journal of Molecular Biology. 1996;259:840–854. doi:10.1006/jmbi.1996.0362

[47] Worley KC, Wiese BA, Smith RF. BEAUTY: an enhanced BLAST-based search tool that integrates multiple biological information resources into sequence similarity search results. Genome Research. 1995;5:173–184. doi:10.1101/gr.5.2.173

[48] Tsunoda T, Takagi T. Estimating transcription factor bindability on DNA. Bioinformatics. 1999;15:622–630. doi:10.1093/bioinformatics/15.7.622

[49] Loh SK, Low ST, Mohamad MS, Deris S, Kasim S, Wen CY, Wardani AK. A review of software for predicting gene function. International Journal of Bio-Science and Bio-Technology. 2015;7:57–70. doi:10.14257/ijbsbt.2015.7.2.06

[50] Höglund A, Kohlbacher O. From sequence to structure and back again: approaches for predicting protein-DNA binding. Proteome Sciences. 2004;2:3. doi: 10.1186/1477-5956-2-3

[51] Liu Y, Wei L, Batzoglou S, Brutlag DL, Liu JS, Liu XS. A suite of web-based programs to search for transcriptional regulatory motifs. Nucleic Acids Research. 2004;32:204–207. doi:10.1093/nar/gkh461

[52] Nagarajan V, Elasri MO. Structure and function predictions of the Msa protein in Staphylococcus aureus. BMC Bioinformatics. 2007;8:S5. doi:10.1186/1471-2105-8-S7-S5

[53] Pavlopoulou A, Michalopoulos I. State-of-the-art bioinformatics protein structure prediction tools. International Journal of Molecular Medicine. 2011;28:295–310. doi: 10.3892/ijmm.2011.705

[54] Ma X, Guo J, Liu HD, Xie JM, Sun X. Sequence-based prediction of DNA-binding residues in proteins with conservation and correlation information. IEEE/ACM Transactions on Computational Biology and Bioinformatics. 2012;9:1766–1775. doi: 10.1109/TCBB.2012.106

[55] Lv H, Han J, Liu J, Zheng J, Zhong D, Liu R. ISDTool: a computational model for predicting immunosuppressive domain of HERVs. Computational Biology and Chemistry. 2014;49:45–50. doi:10.1016/j.compbiolchem.2014.02.001

[56] Tuvshinjargal N, Lee W, Park B, Han K. Predicting protein-binding RNA nucleotides with consideration of binding partners. Computer Methods and Programs in Biomedicine. 2015;120:3–15. doi:10.1016/j.cmpb.2015.03.010

[57] Fujimoto MS, Suvorov A, Jensen NO, Clement MJ, Bybee SM. Detecting false positive sequence homology: a machine learning approach. BMC Bioinformatics. 2016;17:101. doi:10.1186/s12859-016-0955-3

[58] Reddy TBK, Thomas A, Stamatis D, Bertsch J, Isbandi M, Jansson J, Mallajosyula J, Pagani I, Lobos E and Kyrpides N. The Genomes OnLine Database (GOLD) v.5: a metadata management system based on a four level (meta) genome project classification. Nucleic Acids Research. 2014;43:D1099–106. doi:10.1093/nar/gku950

[59] Lin E, Tsai SJ. Genome-wide microarray analysis of gene expression profiling in major depression and antidepressant therapy. Progress in Neuro-Psychopharmacology and Biological Psychiatry. 2016;64:334–40. doi:10.1016/j.pnpbp.2015.02.008

[60] Lim do H, Kim WS, Kim SJ, Yoo HY, Ko YH. Microarray Gene-expression profiling analysis comparing PCNSL and non-CNS diffuse large B-cell lymphoma. Anticancer Research. 2015;35:3333–3340.

[61] Kim HS, Lee NK. Gene expression profiling in osteoclast precursors by insulin using microarray analysis. Molecules and Cells. 2014;37:827–832. doi:10.14348/molcells.2014.0223

[62] Xing C, Zhang R, Cui J, Li Y, Li G, Yang Y, Pang L, Ruan X, Li J. Pathway crosstalk analysis of non-small cell lung cancer based on microarray gene expression profiling. Tumori. 2015;101:111–116. doi:10.5301/tj.5000225

[63] Steiger S, Perez-Fons L, Cutting SM, Fraser PD, Sandmann G. Annotation and functional assignment of the genes for the C30 carotenoid pathways from the genomes of two bacteria: *Bacillus indicus* and *Bacillus firmus*. Microbiology. 2015;161:194–202. doi: 10.1099/mic.0.083519-0

[64] Chitale M, Palakodety S, Kihara D. Quantification of protein group coherence and pathway assignment using functional association. BMC Bioinformatics. 2011;12:373. doi:10.1186/1471-2105-12-373

[65] Haw R, Hermjakob H, D'Eustachio P, Stein L. Reactome pathway analysis to enrich biological discovery in proteomics data sets. Proteomics. 2011;11:3598–3613. doi: 10.1002/pmic.201100066

[66] Seoane JA, Day IN, Gaunt TR, Campbell C. A pathway-based data integration framework for prediction of disease progression. Bioinformatics. 2014;30:838–45. doi:10.1093/bioinformatics/btt610

[67] Veldhoven A, de Lange D, Smid M, de Jager V, Kors JA, Jenster G. Storing, linking, and mining microarray databases using SRS. BMC Bioinformatics. 2005;6:192. doi: 10.1186/1471-2105-6-192

[68] Wolstencroft K, Owen S, Krebs O, Nguyen Q, Stanford NJ, Golebiewski M, Weidemann A, Bittkowski M, An L, Shockley D, Snoep JL, Mueller W, Goble C. SEEK: a systems biology data and model management platform. BMC Systems Biology. 2015;9:33. doi: 10.1186/s12918-015-0174-y

[69] Cremaschi P, Carriero R, Astrologo S, Colì C, Lisa A, Parolo S, Bione S. An association rule mining approach to discover lncRNAs expression patterns in cancer datasets. BioMed Research International. 2015;2015:146250. doi:10.1155/2015/146250

[70] Cheng CP, DeBoever C, Frazer KA, Liu YC, Tseng VS. MiningABs: mining associated biomarkers across multi-connected gene expression datasets. BMC Bioinformatics. 2014;15:173. doi:10.1186/1471-2105-15-173

[71] Zhao W, Zou W, Chen JJ. Topic modeling for cluster analysis of large biological and medical datasets. BMC Bioinformatics. 2014;15:S11. doi:10.1186/1471-2105-15-S11-S11

[72] Wang M, Shang X, Li X, Liu W, Li Z. Efficient mining differential co-expression biclusters in microarray datasets. Gene. 2013;518:59–69. doi:10.1016/j.gene.2012.11.085.

[73] Wu WS, Wang CC, Jhou MJ, Wang YC. YAGM: a web tool for mining associated genes in yeast based on diverse biological associations. BMC Systems Biology. 2015;9:S1. doi: 10.1186/1752-0509-9-S6-S1

[74] PubMed database [Internet]. 2015. http://www.ncbi.nlm.nih.gov/pubmed. Accessed from 2016-03-10

[75] Sanger F, Coulson AR. A rapid method for determining sequences in DNA byprimed synthesis with DNA polymerase. Journal of Molecular Biology. 1975;94:441–448. doi: 10.1258/jrsm.99.2.62

[76] Maxam AM, Gilbert W. A new method for sequencing DNA. Proceeding of National Academy Sciences of USA. 1977;74:560–564. doi:10.1073/pnas.74.2.560

[77] Sanger F, Air GM, Barrell BG, Brown NL, Coulson AR, Fiddes CA, Hutchison CA, Slocombe PM, Smith M. Nucleotide sequence of bacteriophage phi X174 DNA. Nature. 1977;265:687–695. doi:10.1038/265687a0

[78] Fleischmann RD, Adams MD, White O, Clayton RA, Kirkness EF, Kerlavage AR, Bult CJ, Tomb JF, Dougherty BA, Merrick JM. Whole-genome random sequencing and assembly of *Haemophilus influenzae* Rd. Science. 1995;269:496–512. doi:10.1126/science.7542800

[79] Oliver GR, Hart SN, Klee EW. Bioinformatics for clinical next generation sequencing. Clinical Chemistry. 2015;61:124–135. doi:10.1373/clinchem.2014.224360

[80] Krampis K, Wultsch C. A review of cloud computing bioinformatics solutions for next-gen sequencing data analysis and research. Methods in Next Generation Sequencing. 2015;2:2084–7173, DOI: 10.1515/mngs-2015-0003

[81] The Genomes OnLine Database (GOLD) [Internet]. 2016. https://gold.jgi.doe.gov/index. Accessed: 2016-03-08

[82] Karikari TK. Bioinformatics in Africa: The Rise of Ghana? PLoS Computational Biology. 2015;11:e1004308. doi:10.1371/journal.pcbi.1004308.

[83] Zhang D, Choi DW, Wanamaker S, Fenton RD, Chin A, Malatrasi M, Turuspekov Y, Walia H, Akhunov ED, Kianian P, Otto C, Simons K, Deal KR, Echenique V, Stamova B, Ross K, Butler GE, Strader L, Verhey SD, Johnson R, Altenbach S, Kothari K, Tanaka C, Shah MM, Laudencia-Chingcuanco D, Han P, Miller RE, Crossman CC, Chao S, Lazo GR, Klueva N, Gustafson JP, Kianian SF, Dubcovsky J, Walker-Simmons MK, Gill KS, Dvorák J, Anderson OD, Sorrells ME, McGuire PE, Qualset CO, Nguyen HT, Close TJ. Construction and evaluation of cDNA libraries for large-scale expressed sequence tag sequencing in wheat (*Triticum aestivum* L.). Genetics. 2004;168:595–608. doi:10.1534/genetics.104.034785

[84] Henry RJ, Edwards M, Waters DL, Gopala Krishnan S, Bundock P, Sexton TR, Masouleh AK, Nock CJ, Pattemore J. Application of large-scale sequencing to marker discovery in plants. Journal of Biosciences. 2012;37:829–41. doi:10.1007/s12038-012-9253-z

[85] Uenishi H, Morozumi T, Toki D, Eguchi-Ogawa T, Rund LA, Schook LB. Large-scale sequencing based on full-length-enriched cDNA libraries in pigs: contribution to annotation of the pig genome draft sequence. BMC Genomics. 2012;13:581. doi: 10.1186/1471-2164-13-581

[86] Rai KM, Singh SK, Bhardwaj A, Kumar V, Lakhwani D, Srivastava A, Jena SN, Yadav HK, Bag SK, Sawant SV. Large-scale resource development in *Gossypium hirsutum* L. by 454 sequencing of genic-enriched libraries from six diverse genotypes. Plant Biotechnology Journal. 2013;11:953–963. doi:10.1111/pbi.12088

[87] Morozova O, Marra MA. Applications of next-generation sequencing technologies in functional genomics. Genomics. 2008;92:255–64. DOI:10.1016/j.ygeno.2008.07.001

[88] Archer SK, Shirokikh NE, Preiss T. Selective and flexible depletion of problematic sequences from RNA-seq libraries at the cDNA stage. BMC Genomics. 2014;15:401. doi: 10.1186/1471-2164-15-401

[89] Salipante SJ, Roach DJ, Kitzman JO, Snyder MW, Stackhouse B, Butler-Wu SM, Lee C, Cookson BT, Shendure J. Large-scale genomic sequencing of extra intestinal pathogenic *Escherichia coli* strains. Genome Research. 2015;25:119–128. doi:10.1101/gr.180190.114

[90] Hou Z, Jiang P, Swanson SA, Elwell AL, Nguyen BK, Bolin JM, Stewart R, ThomsonJA. A cost-effective RNA sequencing protocol for large-scale gene expression studies. Scientific Reports. 2015;5:9570. doi:10.1038/srep09570

[91] Hough CD, Sherman-Baust CA, Pizer ES, Montz FJ, Im DD, Rosenshein NB, Cho KR, Riggins GJ, Morin PJ. Large-scale serial analysis of gene expression reveals genes differentially expressed in ovarian cancer. Cancer Research. 2000;60:6281–6287

[92] El-Meanawy MA, Schelling JR, Pozuelo F, Churpek MM, Ficker EK, Iyengar S, Sedor JR. Use of serial analysis of gene expression to generate kidney expression libraries. American Journal of Physiology Renal Physiology. 2000;279:383–392.

[93] Zhang XL, Gao F, Li W, Tang WZ, Zhang S. Serial analysis of gene expression in adenocarcinoma samples and normal colonic mucosa in a Chinese population. Genetics and Molecular Research. 2015;14:12903–12911. doi:10.4238/2015.October.21.11

[94] Han X, Wei YB, Tian G, Tang Z, Gao JY, Xu XG. Screening of crucial long non-coding RNAs in oral epithelial dysplasia by serial analysis of gene expression. Genetics and Molecular Research. 2015;14:11729–11738. doi:10.4238/2015.October.2.6

[95] Mackintosh CG, Griffin JF, Scott IC, O'Brien R, Stanton JL, MacLean P, Brauning R. SOLiD SAGE sequencing shows differential gene expression in jejunal lymph node samples of resistant and susceptible red deer (*Cervus elaphus*) challenged with Mycobacterium avium subsp. paratuberculosis. Veterinary Immunology and Immunopathology. 2016;169:102–110. doi:10.1016/j.vetimm.2015.10.009

[96] Loo JA, Brown J, Critchley G, Mitchell C, Andrews PC, Ogorzalek Loo RR. High sensitivity mass spectrometric methods for obtaining intact molecular weights from gel-separated proteins. Electrophoresis. 1999;20:743–748.

[97] Figeys D, Pinto D. Proteomics on a chip: promising developments. Electrophoresis. 2001;22:208–216.

[98] Wark AW, Lee HJ, Corn RM. Multiplexed detection methods for profiling microRNA expression in biological samples. Angewandte Chemie International Edition in English. 2008;47:644–652. doi:10.1002/anie.200702450

[99] Tom I, Lewin-Koh N, Ramani SR, Gonzalez LC. Protein microarrays for identification of novel extracellular protein-protein interactions. Current Protocol in Protein Sciences. 2013;Chapter 27:Unit 27.3. doi:10.1002/0471140864.ps2703s72.

[100] McKee CJ, Hines HB, Ulrich RG. Analysis of protein tyrosine phosphatase interactions with micro arrayed phosphopeptide substrates using imaging mass spectrometry. Analytical Biochemistry. 2013;442:62–67. doi:10.1016/j.ab.2013.07.031

[101] Choi HM, Beck VA, Pierce NA. Next-generation in situ hybridization chain reaction: higher gain, lower cost, greater durability. ACS Nano. 2014;8:4284–4294. doi:10.1021/nn405717p

[102] Omenn GS, Lane L, Lundberg EK, Beavis RC, Nesvizhskii AI, Deutsch EW. Metrics for the Human Proteome Project 2015: progress on the human proteome and guidelines for high-confidence protein identification. J. Proteome Res. 2015;14:3452–3460. doi:10.1021/acs.jproteome.5b00499

[103] Strack R. Highly multiplexed transcriptome imaging. Nature Methods. 2015;12:486–487

[104] Abu-Jamous B, Fa R, Roberts DJ, Nandi AK. Paradigm of tunable clustering using binarization of consensus partition matrices (Bi-CoPaM) for gene discovery. PLoS One. 2013;8:e56432. doi:10.1371/journal.pone.0056432

[105] Lord E, Diallo AB, Makarenkov V. Classification of bioinformatics workflows using weighted versions of partitioning and hierarchical clustering algorithms. BMC Bioinformatics. 2015;16:68. doi:10.1186/s12859-015-0508-1

[106] Chen GK, Chi EC, Ranola JM, Lange K. Convex clustering: an attractive alternative to hierarchical clustering. PLoS Computational Biology. 2015;11:e1004228. doi:10.1371/journal.pcbi.1004228

[107] Bouvier G, Desdouits N, Ferber M, Blondel A, Nilges M. An automatic tool to analyze and cluster macromolecular conformations based on self-organizing maps. Bioinformatics. 2015;31:1490–1492. doi:10.1093/bioinformatics/btu849

[108] Sloan CA, Chan ET, Davidson JM, Malladi VS, Strattan JS, Hitz BC, Gabdank I, Narayanan AK, Ho M, Lee BT, Rowe LD, Dreszer TR, Roe G, Podduturi NR, Tanaka F, Hong EL, Cherry JM. ENCODE data at the ENCODE portal. Nucleic Acids Research. 2016;44:726–732. doi:10.1093/nar/gkv1160

[109] Selkoe DJ. Folding proteins in fatal ways. Nature. 2013;426:900–904. doi:10.1038/nature02264

[110] Yang J, Zhang Y. Protein structure and function prediction using I-TASSER. Current Protocols in Bioinformatics. 2015; 52:5.8.1–5.8.15. doi:10.1002/0471250953.bi0508s52

[111] I-TASSER [Internet]. 2016. http://zhanglab.ccmb.med.umich.edu/I-TASSER. Accessed: 2016-04-02

[112] List of software for molecular mechanics modeling [Internet]. 2016. https://en.wikipedia.org/wiki/List_of_software_for_molecular_mechanics_modeling. Accessed: 2016-02-11

[113] de Moura AP. Thin Watts-Strogatz networks. Physical Review. E, Statistical, Nonlinear, and Soft Matter Physics. 2006;73:016110. doi:10.1103/PhysRevE.73.016110

[114] Watts DJ, Strogatz SH. Collective dynamics of 'small-world' networks. Nature. 1998;393:440–442. doi:10.1038/30918

[115] Campbell C, Shea K, Albert R. Network models. Comment on control profiles of complex networks. Science. 2014;346:561. doi:10.1126/science.1256492

[116] Albert R, Barabasi AL. Topology of evolving networks: local events and universality. Physical Review Letters. 2000;85:5234–5237. doi:10.1103/PhysRevLett.85.5234

[117] Barabasi AL, Albert R. Emergence of scaling in random networks. Science. 1999;286:509–12. doi:10.1126/science.286.5439.509

[118] Biological pathway [internet]. 2016. https://en.wikipedia.org/wiki/Biological_pathway. Accessed: 2016-03-11

[119] Database and list of biological databases [Internet]. 2016. https://en.wikipedia.org/wiki/Database; https://en.wikipedia.org/wiki/List_of_biological_databases. Accessed: 2016-03-12

[120] Xiao J, Zhang Z, Wu J, Yu J. A brief review of software tools for pangenomics. Genomics Proteomics Bioinformatics. 2015;13:73–76. doi:10.1016/j.gpb.2015.01.007

[121] List of open-source bioinformatics software [Internet]. 2016. https://en.wikipedia.org/wiki/List_of_open-source_bioinformatics_software. Accessed: 2016–03-11

[122] Stajich JE, Lapp H. Open source tools and toolkits for bioinformatics: significance, and where are we? Briefings in Bioinformatics. 2006;7:287–296. doi:10.1093/bib/bbl026

[123] Sugawara H, Miyazaki S. Biological SOAP servers and web services provided by the public sequence data bank. Nucleic Acids Research. 2003;31:3836–3839. doi: 10.1093/nar/gkg558

[124] Yates A, Beal K, Keenan S, McLaren W, Pignatelli M, Ritchie GR, Ruffier M, Taylor K, Vullo A, Flicek P. The Ensembl REST API: Ensembl Data for Any Language. Bioinformatics. 2015;31:143–145. doi:10.1093/bioinformatics/btu613

[125] Genereaux BW, Dennison DK. REST enabling the report template library. Journal of Digital Imaging. 2014;27:331–336. doi:10.1007/s10278-013-9668-6

[126] Sundvall E, Nyström M, Karlsson D, Eneling M, Chen R, Örman H. Applying representational state transfer (REST) architecture to archetype-based electronic health record systems. BMC Medical Informatics and Decision Making. 2013;13:57. doi: 10.1186/1472-6947-13-57

[127] Understanding SOAP and REST Basics and Differences [Internet]. 2016. http://blog.smartbear.com/apis/understanding-soap-and-rest-basics. Accessed; 2016–03-11

[128] Open Bioinformatics Foundation [Internet]. 2016. https://en.wikipedia.org/wiki/Open Bioinformatics Foundation. Accessed: 2016-03-11

[129] Coulet A, Garten Y, Dumontier M, Altman RB, Musen MA, Shah NH. Integration and publication of heterogeneous text-mined relationships on the Semantic Web. J Biomed Semantics. 2011;2 Suppl 2:S10. doi:10.1186/2041-1480-2-S2-S10

[130] Ahmed A, Xing EP, Cohen WW, Murphy RF. Structured correspondence topic models for mining captioned figures in biological literature. KDD. 2009;2009:39–48. doi: 10.1145/1557019.1557031

[131] Orozco A, Morera J, Jiménez S, Boza R. A review of bioinformatics training applied to research in molecular medicine, agriculture and biodiversity in Costa Rica and Central America. Briefings in Bioinformatics. 2013;14:661–670. doi:10.1093/bib/bbt033

[132] Schneider MV, Walter P, Blatter MC, Watson J, Brazas MD, Rother K, Budd A, Via A, van Gelder CW, Jacob J, Fernandes P, Nyrönen TH, De Las Rivas J, Blicher T, Jimenez RC, Loveland J, McDowall J, Jones P, Vaughan BW, Lopez R, Attwood TK, Brooksbank C. Bioinformatics Training Network (BTN): a community resource for bioinformatics trainers. Briefings in Bioinformatics. 2012;13:383–389. doi:10.1093/bib/bbr064

[133] Attwood TK, Bongcam-Rudloff E, Brazas ME, Corpas M, Gaudet P, Lewitter F, Mulder N, Palagi PM, Schneider MV, van Gelder CW; GOBLET Consortium. GOBLET: the Global Organisation for Bioinformatics Learning, Education and Training. PLoS Computational Biology. 2015;11:e1004143. doi:10.1371/journal.pcbi.1004143

[134] The Swiss Institute of Bioinformatics training portal [internet]. 2016. http://www.isb-sib.ch/training. Accessed: 2016-03-11

[135] Nunes R, Barbosa de Almeida Júnior E, Pessoa Pinto de Menezes I, Malafaia G. Learning nucleic acids solving by bioinformatics problems. Biochemistry and Molecular Biology Education. 2015;43:377–383. doi:10.1002/bmb.20886

[136] Canadian Bioinformatics Workshops [Internet]. 2016. http://bioinformatics.ca. Accessed: 2016-03-11

[137] The Center of Genomics and Bioinformatics, Academy of Sciences of Uzbekistan [Internet]. 2016. http://en.genomics.uz. Accessed: 2016-03-12

[138] Uzbekistan-US Life Sciences Report [Internet]. 2016. http://www.aaas.org/sites/default/files/migrate/uploads/Uzbekistan-US-Life-Sciences-Report_2013.pdf. Accessed: 2016-03-12

[139] Leebens-Mack J, OneKp Capstone Wiki [Internet]. 2015. https://pods.iplantcollaborative.org/wiki/display/iptol/OneKP+Capstone+Wiki. Accessed: 2015-11-10

[140] Arakawa K, Yamada Y, Shinoda K, Nakayama Y, Tomita M. GEM System: automatic prototyping of cell-wide metabolic pathway models from genomes. BMC Bioinformatics. 2006;7:168

[141] Kim JD, Kim JJ, Han X, Rebholz-Schuhmann D. Extending the evaluation of Genia Event task toward knowledge base construction and comparison to Gene Regulation Ontology task. BMC Bioinformatics. 2015;16 Suppl 10:S3. doi:10.1186/1471-2105-16-S10-S3.

[142] Stokes ME, McCourt P. Towards personalized agriculture: what chemical genomics can bring to plant biotechnology. Frontiers in Plant Science. 2014;5:344. doi:10.3389/fpls.2014.00344

10

Bioinformatics for RNA-Seq Data Analysis

Shanrong Zhao, Baohong Zhang, Ying Zhang,
William Gordon, Sarah Du, Theresa Paradis,
Michael Vincent and David von Schack

Abstract

While RNA sequencing (RNA-seq) has become increasingly popular for transcriptome profiling, the analysis of the massive amount of data generated by large-scale RNA-seq still remains a challenge. RNA-seq data analyses typically consist of (1) accurate mapping of millions of short sequencing reads to a reference genome, including the identification of splicing events; (2) quantifying expression levels of genes, transcripts, and exons; (3) differential analysis of gene expression among different biological conditions; and (4) biological interpretation of differentially expressed genes. Despite the fact that multiple algorithms pertinent to basic analyses have been developed, there are still a variety of unresolved questions. In this chapter, we review the main tools and algorithms currently available for RNA-seq data analyses, and our goal is to help RNA-seq data analysts to make an informed choice of tools in practical RNA-seq data analysis. In the meantime, RNA-seq is evolving rapidly, and newer sequencing technologies are briefly introduced, including stranded RNA-seq, targeted RNA-seq, and single-cell RNA-seq.

Keywords: data analysis, gene quantification, pipeline, RNA-seq, workflow

1. Introduction

In recent years, RNA sequencing (RNA-seq) has emerged as a powerful technology for transcriptome profiling [1–4]. Compared with microarrays, it not only avoids some of the technical limitations of this approach including varying probe performance and nonspecific hybridization, and dynamic range issues, but can also detect alternative splicing isoforms and subtle changes of splicing under different conditions. The overview of current RNA-seq approaches using shotgun sequencing technologies such as Illumina and the corresponding data analysis workflow is summarized in **Figure 1**. Polyadenylated (Poly-A) RNA transcripts (for the so-called mRNA-seq) are enriched with oligo (dT) primers and then fragmented. After size selection, millions or even billions of short sequence reads are generated from a randomly fragmented cDNA library. For most RNA-seq studies, the data analyses consist of the following key steps [5, 6]: (1) quality check and preprocessing of raw sequence reads, (2) mapping reads to a reference genome or transcriptome, (3) counting reads mapped to individual genes or transcripts, (4) identification of differential expression (DE) genes between different biological conditions, and (5) biological interpretation of DE genes and functional enrichment analysis. Despite the fact that a large number of algorithms [7] have been developed for RNA-seq data analysis in recent years, there are still many open questions for accurate read mapping, gene quantification, and data normalization.

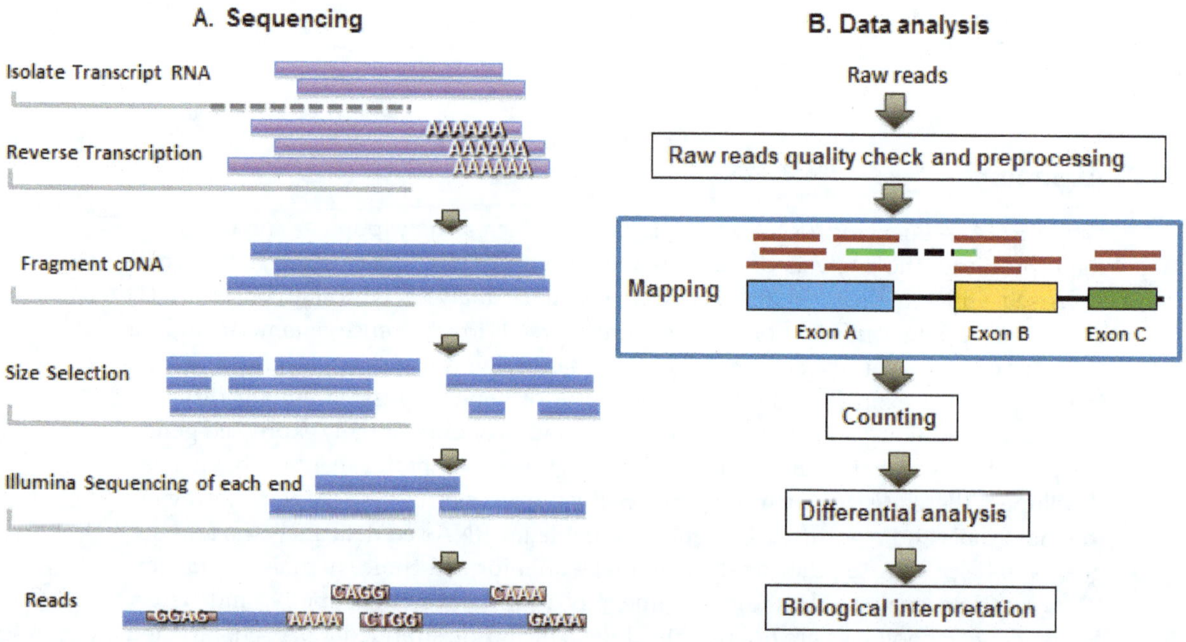

Figure 1. Overview of mRNA-seq laboratory flowchart and data analysis pipeline.

In complex mammalian genomes, both DNA strands encode genes. As a result, if two genes transcribed from opposite strands overlap, nonstranded RNA-seq cannot tell the true origin if a read falls into overlapping regions. Most recently, a stranded RNA-seq protocol has been developed for more accurate gene quantification [8, 9]. RNA-seq is a powerful tool when used to profile the entire transcriptome. However, this method can be inefficient when interested in only a small subset of genes that are involved in particular pathways, or associated with specific diseases. To meet this demand, targeted RNA-seq technology has been developed. Traditionally, gene expression measurements were performed on "bulk" samples containing populations of thousands or millions of cells. Recent advances in genomic technologies have made it possible to measure gene expression in individual cells. Accordingly, cellular properties that were previously masked in "bulk" measurements can now be observed directly [10–

12]. Single-cell RNA-seq (scRNA-seq) introduces more challenges for data analyses due to technical noise, including coverage nonuniformity, data sparsity, amplification biases, and so on. In this chapter, we cover the topics related to RNA-seq technology and data analyses.

2. RNA-seq versus microarray for gene expression profiling analysis

Microarrays and RNA-seq have been the two technologies of choice for large-scale studies of gene expression, and the side-by-side comparisons of RNA-seq and hybridization-based arrays have been performed [12–15]. Malone et al. [16] compared its ability to identify differentially expressed genes with existing array technologies and found that RNA-seq data are highly reproducible with relatively little technical variation. The DE genes identified from RNA-seq had a high overlapping with those identified by microarray. Fu et al. [17] designed a study in which they evaluated the accuracy of both microarrays and RNA-seq for mRNA quantification by using protein expression measurements as ground truths. In that study, they assessed the relative accuracy of the two transcriptome quantification approaches with respect to absolute transcript level measurements and found that RNA-seq provides better estimates of transcript expressions.

Previously, we performed a side-by-side comparison of RNA-seq and microarrays in investigating T-cell activation [18]. A comparison of data sets derived from RNA-seq and Affymetrix platforms using the same set of samples revealed a very high concordance between expression profiles generated by the two platforms. In the meantime, it was also demonstrated that RNA-seq is superior in differentiating biologically critical isoforms, detecting low abundance transcripts, and allowing the identification of genetic variants. Analysis of the two data sets also showed the benefit derived from avoidance of technical issues inherent to microarray probe performance such as cross-hybridization, nonspecific hybridization, and limited detection range of individual probes. In addition, RNA-seq has a much broader dynamic range than microarray technologies, which allows for the detection of more differentially expressed genes with higher fold-change. Thus, RNA-seq delivers both less biased and previously unknown information about the transcriptome. Because RNA-seq does not rely on a predesigned complementary sequence detection probe, it is not limited to the interrogation of selected probes on an array and can also be applied in species, for which the whole reference genome is not yet assembled.

RNA-seq allows for the detection of novel transcript species in well-studied organisms, such as unique transcripts in certain tissues or in rare cell types, and has been instrumental to catalog the diversity of novel transcript species including long noncoding RNA, miRNA, siRNA, and other small RNA classes [19]. Additionally, RNA-seq technology proves to be an invaluable tool for deciphering the extensive alternative splicing of the transcriptome [20, 21]. Alternative splicing creates two to potentially hundreds of variants in more than 90% of human genes. Furthermore, RNA-seq can identify allele-specific expression and gene fusion events [22, 23].

3. RNA-seq library preparation and sequencing platforms

3.1. General considerations for RNA-seq

The quality and quantity of the starting RNA material are likely the most important aspects to consider when deciding on the methods to generate RNA-seq libraries. For high-quality samples with mostly intact RNA, a wide variety of sequencing library preparation methods is available. For lower quality samples with partially or highly degraded RNA, there are considerably fewer methods to choose from. The amount of RNA may limit the choice of library preparation as well, since the majority of standard RNA-seq kits require a minimum of 10–100 ng of the total RNA. If the RNA amount is below this threshold it will require the use of a more specialized kit and/or the sample may require some form of amplification.

To obtain reliable and reproducible RNA-seq data, RNA quality is of paramount importance. Because of the inherent instability of the RNA molecule, the quality of RNA from samples collected in clinical settings and field studies is often impacted by tissue necrosis, which quickly degrades RNA if the sample is not frozen or chemically preserved within minutes after surgery. For assessing RNA quality, most labs utilize an electrophoretic-based system, including the Agilent BioAnalyzer, the Agilent TapeStation, or the Advanced Analytics Fragment Analyzer. All of these instruments produce an RNA integrity score. On the BioAnalyzer, the score ranges from 1 to 10 (10 being perfect) and most labs would consider a RIN > 6 to be acceptable for standard RNA-seq methods. For degraded samples where the RIN is below 6, it is beneficial to calculate the DV200 score (the percent of the sample that is larger than 200 bp in size). Illumina has shown this to be an important metric for successful library preparation with kits designed for degraded RNA.

After quality and quantity of the RNA samples have been addressed, one must then choose whether to profile the total RNA space or the mRNA space only. This choice determines the main split between most RNA-seq library kit types, those which target the removal of ribosomal RNA (rRNA) versus those which utilize Poly-T beads to isolate the Poly-A tail of intact mRNAs. This choice will affect the downstream sequencing, influencing the depth to be targeted per sample, with rRNA depletion libraries requiring significantly deeper sequencing than Poly-A-based libraries in order to generate sufficient reads to capture all of the different RNA species in these samples. Additionally, if you are dealing with degraded RNA samples, you will not be able to utilize the Poly-A method, but the rRNA depletion methodologies will work for these samples. However, if you still want to focus only on the mRNAs, another option would be the more recent exon capture kits such as Illuminas RNA Access, which depend on hybridization with probes designed against known exons.

When considering the sequencing depth, a few considerations should be taken into account. mRNA libraries can be sequenced shallower than the total RNA libraries as they have less diversity and are focused only on the mature transcripts. In general a total RNA prep will only have 15–30% of the sample as mRNA, thus to get the same level of read depth for the mRNAs you will need to sequence those libraries between 3 and 10× as deep as a mRNA only library. Additionally, whether the goal of the experiment is targeted toward DE, isoform discovery, or novel RNA discovery will impact the choice of read depth to target, with the latter requiring more reads. In general, targeting 30–40 million paired reads for an mRNA-seq and 70–100 million for a total RNA-seq would be the starting recommendation.

3.2. From RNA sample to raw sequence reads

The most common workflow for RNA-seq is stranded mRNA-seq. To generate a stranded mRNA-seq library, RNA from different sources (blood, tissue, and cell lines) should be purified utilizing a consistent methodology (kits with columns or manually using Trizol are both acceptable). The RNA should be treated either in solution or on-column (depending on your extraction choice) with DNASE to remove traces of genomic DNA and to prevent contamination of RNA-seq libraries by DNA. DNA-free RNA should then be assessed for both quality and quantity. RNA of sufficient quality can then be passed over Oligo-dT beads to capture mRNAs removing the non-Poly-A RNA species. The captured mRNA is then fragmented (enzymatically or mechanically) through ultrasonic shearing (with devices such as a Covaris ultrasonicator), followed by a two-step conversion to cDNA using first-strand synthesis and then second-strand synthesis (during which the cDNA is marked retaining the strand information) protocols. After cDNA conversion, the ends are repaired making them amenable to adapter ligation. Indexing of the libraries can be utilized at this point allowing for sample pooling before sequencing and a final enrichment of the indexed cDNA fragments by PCR (Illumina recommends 10–15 cycles of PCR depending on the RNA input), which is performed to generate the final RNA-seq library.

After the libraries are complete, their quality is assessed by an electrophoretic assay, if a peak at the size of dimerized adapters (80–100 bp) is observed, the libraries should be repurified before sequencing (adapter dimers will cluster very efficiently and many reads can be lost if they are not fully removed). The sequencing-ready libraries are quantified to allow for equimolar pooling typically through quantitative PCR. Method such as KAPA Library Quantification kits are recommended for best results. After quantification, libraries with different indices are pooled. The number of samples in the pool will vary depending on the read depth desired and sequencer used. Sequencing is then performed on a final diluted sample (following the manufacturer's recommendations). After the run is complete, the raw reads can be converted to fastq files (if Illumina sequencing was performed, Illumina offers an algorithm for this step called bcltofastq) and passed on for QC, alignment, and analysis.

4. Algorithms for RNA-seq data analysis

Millions, or even billions, of short reads are the starting point of RNA-seq computational data analyses [5, 6]. First, reads are QC checked and then mapped to a reference genome or transcriptome. The mapped reads for each sample are subsequently counted on gene, transcript, or exon level to assess the abundance of each category depending on the experimental purpose. The summarized data are then assessed by statistical models to identify differentially expressed genes. Finally, pathway or network level analyses are performed to gain biological insight through systems biology approaches.

4.1. Quality check and preprocessing of raw reads

Poor-quality read data can arise from problems in the library preparation or from sequencing itself. Additionally, PCR artifacts, untrimmed adapter sequences, sequence-specific bias, and other possible contaminants can also lead to poor data quality. The presence of poor quality or technical sequences can affect the downstream analysis and data interpretation, and thus give inaccurate results. In order to assess quality of raw sequenced data, several tools such as PRINSEQ [24] and FastQC [25] have been developed. FastQC aims to provide a simple way to do some quality control checks on raw sequence data. It provides a modular set of analyses that can be used to give a quick impression of whether raw sequencing data have any problems that one should be aware of before doing any further analysis. Once the data are checked for quality, they should be processed to remove reads with low-quality bases, adapter sequences, and other contaminating sequences. Tools such as Cutadapt [26] and Trimmomatic [27], which trim adapter or other contaminating sequences based upon user-provided parameters, can be used for performing these operations. After going through the aforementioned steps, the sequencing data are ready for downstream analysis.

4.2. Read mapping

Short sequence reads generated by sequencers must first be mapped or aligned to a reference transcriptome or genome assembly to discover their true locations (origins) with respect to that reference. A large number of read mapping algorithms have been developed in recent years, including TopHat2 [28], STAR [29, 30], GSNAP [31], OSA [32], and MapSplice [33]. To assess the performance of current mapping software, Engström et al. [34] compared 26 mapping protocols based on 11 programs and pipelines and found there were major performance differences among different methods on numerous benchmarks, including basewise accuracy, alignment yield, gap placement, mismatches, and exon junction sites.

Indeed, some features of a reference genome such as repetitive regions, assembly errors, and assembly gaps render this objective impossible for a subset of reads. Furthermore, because RNA-seq libraries are constructed from transcribed RNA, intronic sequences are not present in exon-exon spanning reads. Therefore, when aligning the sequences to a reference genome, reads that span exon-exon junctions have to be split across potentially thousands of bases of intronic sequence. Many RNA-seq alignment tools use reference transcriptomes to inform the alignment of junction reads. The benefits of using a reference transcriptome to map RNA-seq reads have been demonstrated clearly in previous reports [35–37] and our own comprehensive evaluation [38] of RefGene (RefSeq Gene) [39], UCSC Known Genes [40], and Ensembl [41] in mapping of RNA-seq reads and gene quantifications.

The benefits of using a reference transcriptome in mapping of RNA-seq reads are illustrated in **Figure 2**. In **Figure 2A**, 19 junction reads can be uniquely mapped to gene HSP90AB1 when RefGene annotation [36] is provided in the alignment step. However, four reads indicated by the red arrow are mapped to the same gene HSP90AB1 as nonjunction reads with mismatches at one end without the assistance of a reference transcriptome. In **Figure 2B**, those exon-exon spanning reads are mapped to gene TCEA3 with the exact same start and end positions regardless of the use of a transcriptome, but spliced differently. Both mappings are equal in terms of alignment scores and gaps between exons. It is therefore difficult, if not impossible, to tell which alignment is correct without the assistance of a reference transcriptome. Collectively, the two examples in **Figure 2** illustrate the importance of appropriate gene annotations in the correct alignment of junction reads.

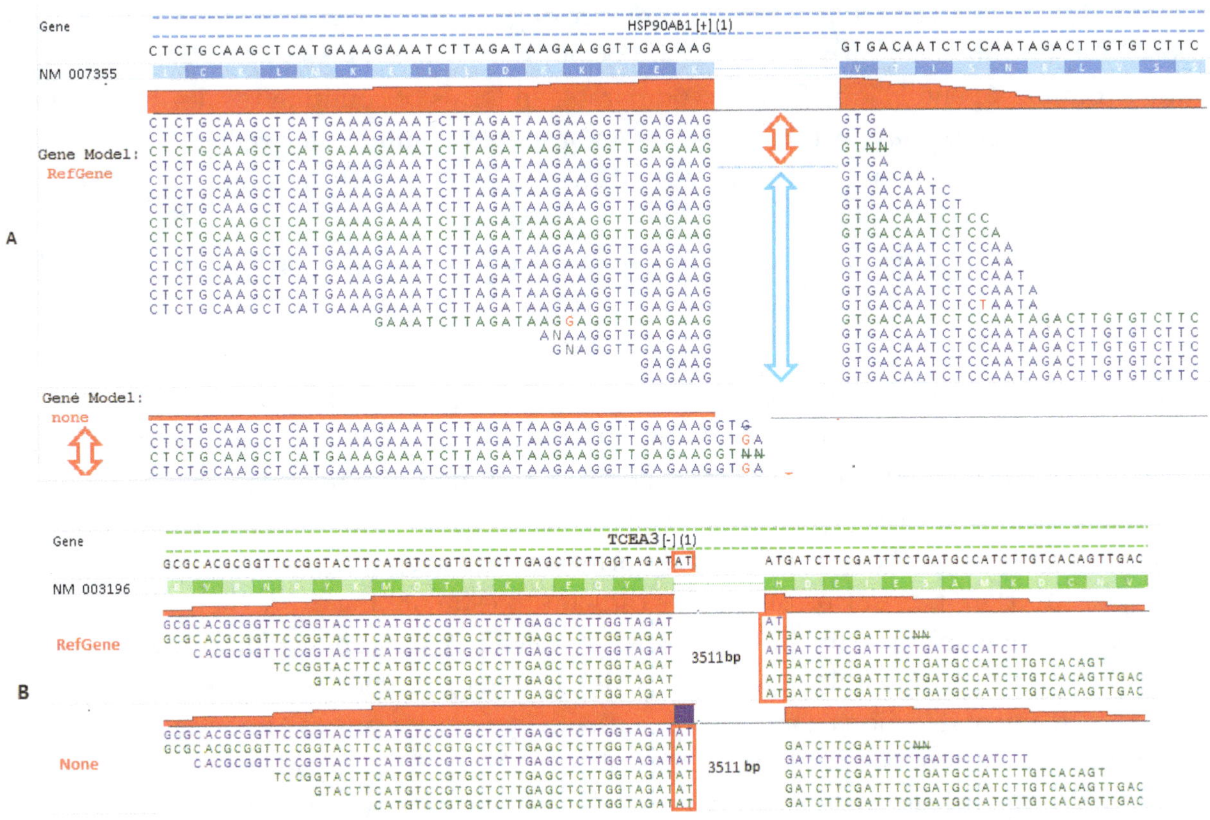

Figure 2. The impact of a reference transcriptome on the mapping of junction reads. Some exon-exon spanning reads are mapped incorrectly without the help of a reference transcriptome. Note the reads colored in blue are mapped to "+" strand, and in green when mapped to "-" strand. The mismatch nucleotide bases are colored in red.

Another problem in read mapping is that of polymorphisms, which occur when sequence reads align to multiple locations of the genome, abbreviated as multireads. Polymorphisms are especially common for large and complex transcriptomes, and the number of multireads for a mammalian genome is estimated to be between 10 and 40% [1, 3]. Generally speaking, there are three common strategies to deal with multireads in practice. The first strategy is to ignore or discard them completely, which we have demonstrated is not ideal for accurate gene quantification [36]. This practice not only discards potentially useful information, but also introduces an underestimation bias in quantifying expression of genes with highly redundant sequences (e.g., young duplicated genes). The second strategy implemented in most mapping software is to randomly assign a position from the possible matches. This practice assumes a multiread can originate from these genomic locations equally, but this assumption is often not valid. The third strategy is to report all mapped locations for a multiread as long as the number of possible matches is below a user-defined cutoff, let's say 10. The problem with this strategy is that the cutoff is somewhat arbitrary. For an accurate detection and quantification of transcripts, it is important to resolve the mapping ambiguity for RNA-seq reads that can be mapped to multiple loci [42].

4.3. Read counting and gene quantification

Since RNA-seq has become a common technology in molecular biology laboratories, a number of methods have been developed for the inference of gene and isoform abundance, including RSEM [43], Cufflinks [44], IsoEM [45], featureCounts [46], and HTSeq [47]. The algorithms featureCounts [46] and HTSeq [47] are comparable in terms of counting results, but featureCounts is considerably faster than HTSeq by an order of magnitude for gene-level summarization and requires far less compute memory. However, neither featureCounts nor HTSeq can count reads at the transcript level due to their implementation. Andelini et al. [48] carried out a simulation study to assess the performance of five widely used counting tools and concluded that performance was heavily dependent upon the true abundance of the isoforms. Lowly expressed isoforms are poorly detected regardless of the methods.

Most recently, Kanitz et al. [49] have evaluated the accuracy of 14 methods for estimating isoform abundance and found that these tools vary widely in memory and runtime requirements. The algorithms for gene quantification can be broadly divided into two categories: transcript-based approaches (such as RSEM [43]) and "union-exon"-based approaches (such as featureCounts [46]). Because different isoforms of a gene typically have a high proportion of genomic overlap, it is intrinsically more difficult to estimate the expression of individual isoforms. Union-exon-based methods are much simpler, in which all overlapping exons of the same gene are merged into union exons. A read is counted to the gene as long as it has sufficient overlap with any of its union exons. Compared with isoforms, reads can be assigned to genes with much higher confidence. Therefore, the union-exon-based counting method is commonly used in RNA-seq, though gene-level counts cannot distinguish Isoforms [50].

A gene can be expressed in one or more transcript isoforms; accordingly, its expression level should be represented as the sum of its isoforms. We carried out a side-by-side comparison between union-exon-based approach and transcript-based method in RNA-seq gene quantification [40], and found that gene expression levels were significantly underestimated when the union-exon-based approach was used. We also discovered that the quantification of gene expression is more accurate if gene expression levels are computed by cumulating expression levels of transcript isoforms than by ignoring the transcript structures.

A gene model that hypothesizes the structure of transcripts produced by a gene also affects the analysis. Among multiple genome annotation databases, RefGene, Ensembl, and the UCSC annotation databases are the most popular. The choice of genome annotation directly affects gene expression estimation. Recently, we systematically characterized the impact of genome annotation on read mapping and transcriptome quantification [38]. Surprisingly, among the 21,958 common genes shared between RefGene and Ensembl annotations, only 16.3% of genes obtained identical quantification results. Approximately 28.1% of genes' expression levels differed by >5% when using different annotation, and of those, the relative expression levels for 9.3% of genes differed by at least 50%. Our study revealed that the difference in gene definition frequently results in inconsistency in gene quantification (**Figure 3**). In Ensembl, the annotation of PIK3CA is much longer than its corresponding definition in RefGene. As a result, much more reads are counted toward PIK3CA if Ensembl annotation is used. According to the mapping profile of RNA-seq reads in **Figure 3**, the PIK3CA gene definition in Ensembl should be more accurate than the one in RefGene.

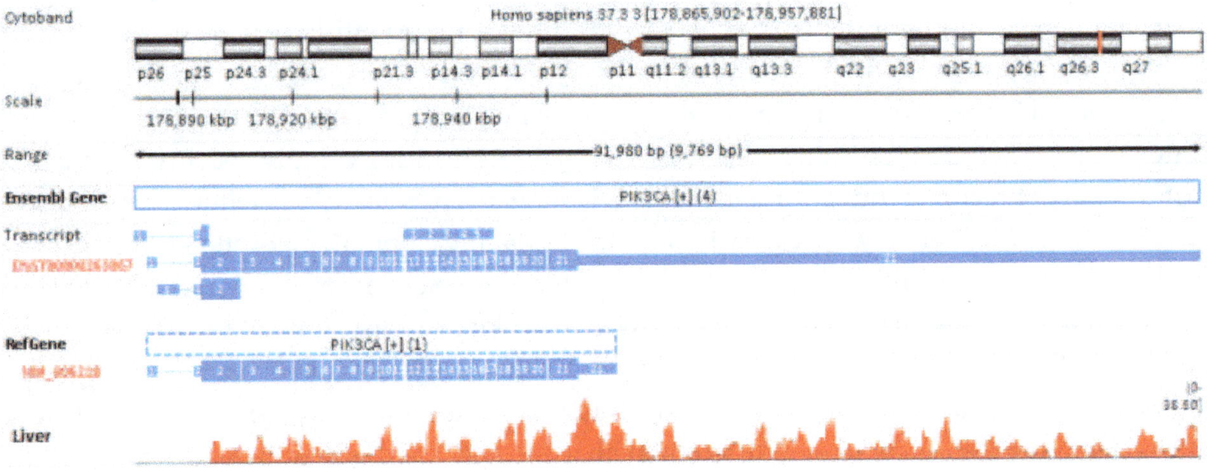

Figure 3. Different gene definitions for PIK3CA give rise to differences in gene quantification. PIK3CA in the Ensembl annotation is much longer than its definition in RefGene, which explains why 1094 reads are mapped to PIK3CA in Ensembl while only 492 reads are mapped using RefGene.

4.4. Data normalization and differential analysis

After calculating the read counts, data normalization is one of the most crucial steps of data processing, and this process must be carefully considered, as it is essential to ensure accurate inference of gene expression and subsequent analyses thereof. First, the *sequencing depths* or *library sizes* (the total number of mapped reads) typically vary for different samples, which means that the observed counts are not directly comparable between samples. The most straightforward way of normalizing the difference in sequencing library sizes is to rescale the total read counts. However, such normalization is quite often not enough because RNA-seq counts inherently represent *relative* abundances of genes in a sample. The number of reads mapped to a gene is not only dependent on the expression level and length of the gene, but also the composition of the RNA population that is being sampled. A few highly expressed genes may consume a very large portion of the total reads in a sample, and accordingly, the counts for all other genes are repressed. As a result, in comparison to a sample where the reads are more evenly distributed, those repressed genes seem to have lower expression which could give rise to a lot of falsely "differentially expressed" genes.

A fundamental research aim in many RNA-seq studies is to identify differentially expressed genes between distinct sample groups. Many algorithms have recently been introduced specifically for the identification of differentially expressed genes (DEGs) from RNA-seq data, including DESeq [51], edgeR [52, 53], GENE-Counter [54], NOISeq [55], NBPSeq [56], and Cuffdiff2 [57]. However, there is a lack of consensus on how to approach an optimal study design and choice of suitable software for the analysis of an RNA-seq data sets. Recently, numerous groups [58–66] have performed a variety of comprehensive comparisons of different statistical methods for differential RNA-seq data analysis. Still, no general consensus has been reached after so many head to head comparisons and evaluations.

Zhang et al. [61] evaluated the performance of three widely used software tools: DESeq, edgeR and Cuffdiff2. They took a number of important metrics into consideration, including sequencing depth and the number of replicates, and the set of identified DEGs was evaluated with ground truths from either quantitative RT-PCR or microarray. They concluded that no single method is always superior in all DE analyses. It was noted that edgeR performs slightly better than DESeq and Cuffdiff2 in terms of the ability to uncover true positives and that Cuffdiff2 is not recommended for gene-level DE analysis, particularly if sequencing depth is low. Seyednasrollah et al. [65] also carried out a systematic comparison of the state-of-the-art methods in RNA-seq differential analysis to guide the selection of a suitable package. In general, there can be large differences between the algorithms. Similar to the evaluation performed by Zhang et al. [63], it was observed that no single method is likely to be optimal under all circumstances. They also demonstrated how the data analysis tool utilized can markedly affect the outcome of a differential analysis and highlighted the importance of the choice of software. Soneson and Delorenzi [61] have conducted extensive comparisons of 11 methods for DE analysis of RNA-seq data and concluded that very small sample size, which is still common in RNA-seq experiments, imposes problems for all evaluated methods and any results obtained under such conditions should be interpreted with caution.

We have applied edgeR to several whole blood RNA-seq data sets (unpublished results), and the calculated normalization factors might range from 0.4 to 1.6. We were puzzled by such an unreasonably low or high scaling factor. The normalization method implemented in edgeR is based on the premise that most genes are not differentially expressed. Instead of normalizing the raw counts directly, edgeR scales the library size by a factor so that the adjusted abundance (i.e., the ratio of read counts divided by the scaled library size) for many genes is not DE. TMM (Trimmed Mean of M-values) is calculated for each sample in a data set with one sample being considered as the reference sample and the others as test samples. For each test sample, TMM is computed between this test and the reference after exclusion of the most highly expressed genes (5% by default) and genes with the largest fold-changes (30% on each side of up and down by default). Ideally, TMM should be close to 1, but in cases where it is not, its value provides an estimate of the correction factor that must be applied to the library sizes (but not the raw counts) for normalization. However, TMM becomes problematic when the library composition between test and reference samples differs significantly. On the other hand, whether the default parameters in edgeR are appropriate for a given RNA-seq data set is difficult to determine. Moreover, different sets of genes are used for calculation of scaling factors across the entire data set since the genes excluded can vary from sample to sample.

4.5. Pathway enrichment analysis

To obtain a list of DE genes is only the starting point of gaining biological insights into experimental systems, developmental stages, or understanding of disease or molecular mechanisms. To understand the biological context of DE genes, pathway enrichment analysis ensues. Functional enrichment analyses rely upon annotation databases such as Gene Ontology (GO) [67], Kyoto Encyclopedia of Genes and Genomes (KEGG) pathways [68], DAVID [69], and other commercial knowledge systems such as Ingenuity Pathway Analysis (IPA). One traditional analysis starts with a gene list of interest, identified from differential RNA-seq or microarray analyses, and applies statistical methods, such as the Fisher's exact test to test for enrichment of each annotated gene set, network, and pathway.

Gene set enrichment (GSE) analysis transforms information from gene expression profiling into a pathway summary. DE genes are quite often involved in the same biological pathways, and GSE results offer greater biological interpretability over individual gene analysis. GSVA [70] extends the current GSE methods to RNA-seq data, and provides increased power to detect subtle pathway activity changes, and constitutes a starting point to build pathway-centric models of system biology. SeqGSEA [71] is a new open-source Bioconductor package for GSVA and can detect more biologically meaningful gene sets without biases toward longer and highly expressed genes.

Previous pathway analysis methods had been developed based on algorithms considering pathways as simple gene lists and ignoring pathway structures. Recently methods have been developed to incorporate various aspects of pathway topology. For example SPIA captures pathway topology through its scoring system, in which the positions and the interactions of the genes in the pathway are considered [72]. Accordingly, interacting DEG pairs are preferentially weighted over two noninteracting genes. Similarly, TAPPA [73] is a scoring method in which higher weights are automatically assigned to hub genes and interacting gene pairs. DE analysis for pathways (DEAP) [74] makes significant improvements over existing approaches by including information about pathway topological structure. It was demonstrated [74] DEAP identified 14 more important chronic obstructive pulmonary disease–related pathways that existing approaches omitted.

4.6. QuickRNASeq—an integrated pipeline for efficient RNA-seq data analysis and interactive visualization

Although the time and cost for generating RNA-seq data are decreasing, the analysis of massive amounts of RNA-seq data still remains challenging. Numerous software packages and algorithms have been developed, which has led to the need to apply these tools efficiently to obtain results within a reasonable timeframe, especially for large data sets. Based on our own experience with analyses of multiple in-house RNA-seq data sets of varying size using open source tools, the main challenges, gaps, and bottlenecks for large-scale RNA-seq data analyses can be summarized as follows:

1. It is not trivial to select appropriate software packages and set software-specific parameters since it requires an in-depth understanding of the algorithms.

2. Additional bridging scripts are often necessary to make different components work seamlessly in a pipeline.

3. In general, most algorithms are implemented to process an individual sample, and thus it is necessary to integrate and summarize results from individual samples.

4. Due to sample quality and the complicated multistep processes in RNA-seq, it is required to establish stringent RNA-seq data quality metrics to identify outliers that should be excluded from further downstream data analysis.

5. Nearly all RNA-seq data analyses are performed using Linux workstations; however, analysis results in Linux are often inaccessible to most experimental scientists. Thus, sharing results with scientists is a practical challenge.

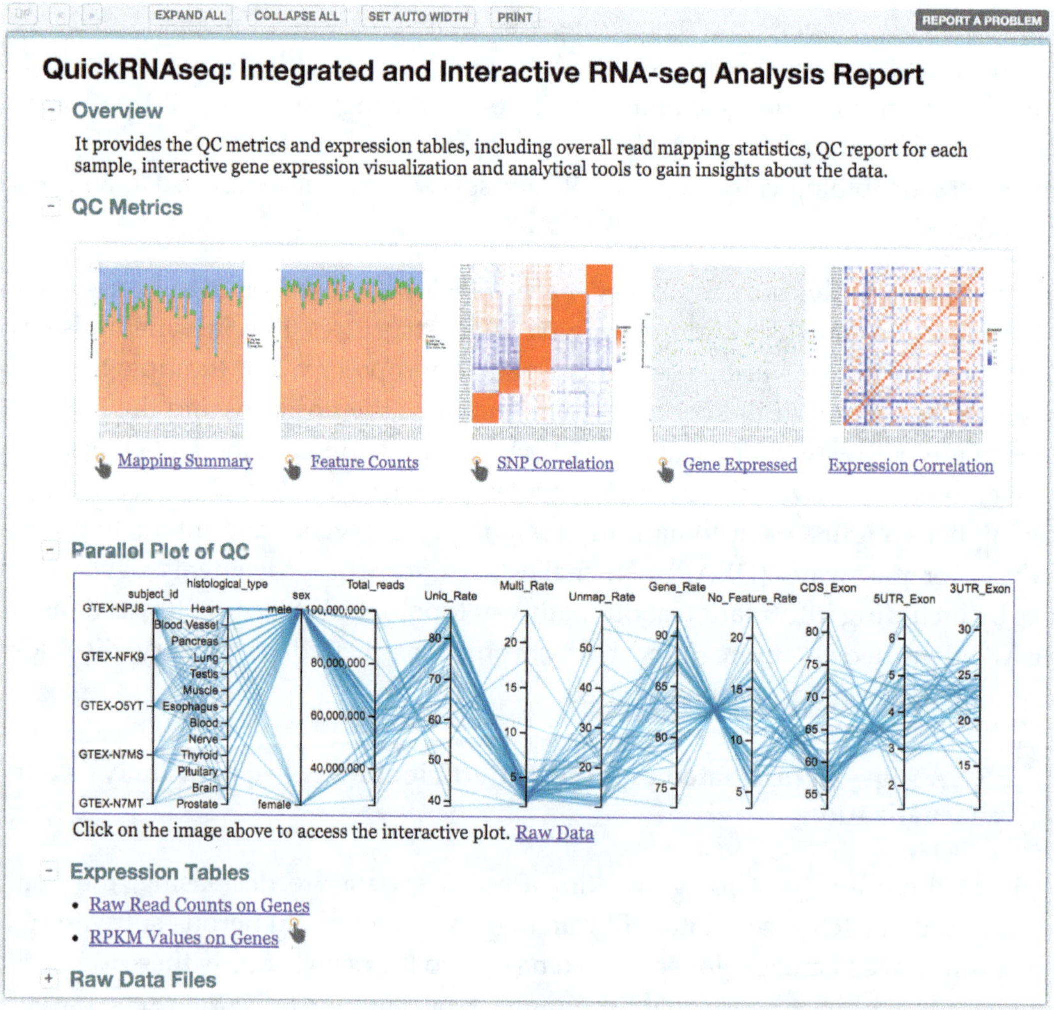

Figure 4. Representative entry webpage for a QuickRNAseq project report. The page layout and printable version of the page can be controlled by the top icons. The QC Metrics section provides QC results in plain text, static plot, and interactive plot formats accessible by clicking on the corresponding hyperlinked texts, the iconized figures, and pointing hand, respectively. The parallel plot of QC offers an integrated view of linked QC measures for a single sample or a group of samples. The expression tables section provides links to raw read counts, a normalized RPKM table, and interactive displays of gene expression levels.

To address these challenges, we have implemented a pipeline named QuickRNASeq to advance the automation and interactive visualizations of RNA-seq data analysis results [75]. QuickRNASeq significantly reduces data analysts' hands-on time, which results in a substantial decrease in the time and effort needed for the primary analyses of RNA-seq data before proceeding to further downstream analysis and interpretation. Additionally, QuickRNASeq provides a dynamic data sharing and interactive visualization environment for end users. All the results are accessible from a web browser without the need to set up a web server and database. The rich visualization features implemented in QuickRNASeq enable nonexpert end users to interact easily with the RNA-seq data analyses results and to drill down into specific aspects to gain insights into often complex data sets simply through a point-and-click approach. A representative entry webpage for a QuickRNASeq project report is shown in **Figure 4**. All the result files and figures are directly accessible by "point and click" from the entry webpage, which makes data navigation and visualization more convenient and intuitive, especially for experimental scientists.

5. New RNA sequencing technologies

5.1. Stranded RNA-seq

One significant shortcoming of the first-generation RNA-seq protocol was that it did not retain the strand information for each transcript. Recently, strand-specific or stranded RNA-seq protocols have been developed [76]. Previous reports [8] demonstrated that data from stranded libraries are more reliable than data from nonstranded libraries and can correctly evaluate the expression of both antisense RNA and overlapping genes. The ability to capture the relative abundance of both sense and antisense expression provides insight into regulatory interactions that might otherwise be missed [9]. With the ability to unlock new information on global gene expression, stranded RNA-seq holds the key to a deeper understanding of the transcriptome. We performed a side-by-side comparison of stranded and nonstranded RNA-seq in our whole blood RNA-seq data set, and demonstrated that stranded RNA-seq provides a more accurate estimate of gene expression compared with nonstranded RNA-seq and is therefore the recommended RNA-seq approach for all future mRNA-seq studies [77].

The advantages of stranded RNA-seq are illustrated in **Figure 5**. ICAM4 (intercellular adhesion molecule 4) shows moderate expression in whole blood. However, nonstranded RNA-seq reports no expression for this gene. As observed in **Figure 5**, ICAM4 is 100% contained within CTD-2369P2.8. In nonstranded RNA-seq, a read mapped to ICAM4 is simultaneously aligned to CTD-2369P2.8 as well. The ambiguous reads in overlapping regions are thus excluded from counting, which explains the lack of expression for ICAM4 with nonstranded RNA-seq. The ambiguous reads in overlapping genes in **Figure 5** can be perfectly resolved using stranded RNA-seq. By considering the read direction, all reads are assigned to ICAM4 (but not CTD-2369P2.8), because they are all reverse complementary to ICAM4. According to a stranded sequencing protocol, it is impossible for such reads to originate from CTD-2369P2.8.

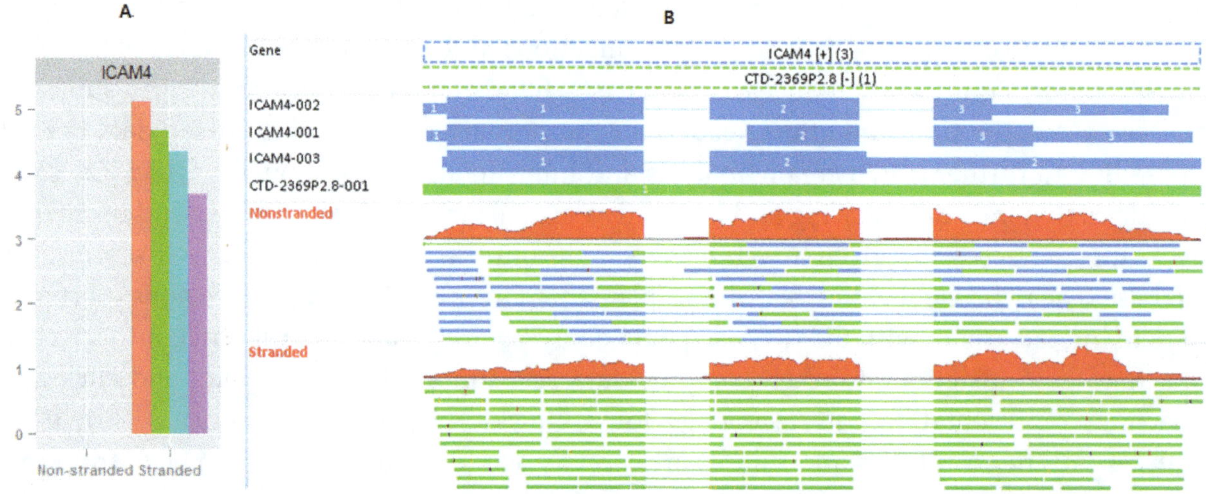

Figure 5. (A) Gene expression of ICAM4 in stranded and nonstranded RNA-seq. (B) Mapping profiles for ICAM4 (intercellular adhesion molecule 4). ICAM4 is on the "+" strand, and 100% contained within CTD-2369P2.8 in the "-" strand. In nonstranded RNA-seq, the ambiguous reads in overlapping regions are excluded from counting, which explains why there is no expression for ICAM4. However, the ambiguous reads can be perfectly resolved in stranded RNA-seq. By considering the read direction, all reads can be counted to ICAM4 but not CTD-2369P2.8. *Note:* (1) RNAs were extracted from pooled whole blood samples, and four replicates were pair-end sequenced using both stranded and nonstranded protocols. The unit of y-axis is RPKM in the plot to the left. (2) All genes, transcripts, and sequence reads are colored in blue if they are in the "+" strand and colored in green if in the "-" strand. According to our stranded sequencing protocol, a sequence read should be reversely complementary to its transcript origin.

5.2. Targeted RNA-seq

RNA-seq can be a powerful tool to measure gene expression, detect novel transcripts, characterize transcript isoforms, and identify sequence polymorphisms. However, this unbiased RNA-seq method can be costly and yields complex data sets that are time consuming to analyze. Often one is interested in only a small subset of genes or the goal is to study only one component of the transcriptome, such as long noncoding RNAs (lncRNAs), which constitute only a small fraction of transcripts in a total RNA sample [78–80]. A targeted quantitative RNA-seq method that is reproducible and reduces the number of sequencing reads required to measure key transcripts would be better suited to these purposes. Most recently, Tan et al. [80] describes a targeted enrichment method for the analysis of lncRNAs. Targeted RNA-seq can measure dozens to hundreds of targets simultaneously. Targeted RNA-seq gives an economical way to focus on genes of interest, and provides enhanced coverage for sensitive gene discovery, robust transcript assembly and accurate gene quantification. Common uses of this method include:

- Profiling expression of select target genes, to assess disease-associated variants and epigenetic alterations.
- Analyzing gene fusions and gene expression alterations to provide a focused view of functionally relevant changes occurring in cancer.
- Studying genes associated with a variety of key signal transduction pathways, such as NF-kB, P450, IFN response, apoptosis pathways, and many more.

There are several approaches for target RNA enrichment, either by hybridization only (Agilent SureSelect and Roche SeqCap system), hybridization followed by extension (Illumina Targeted RNA-seq; **Figure 6**) or PCR amplification (Thermo Fisher Ampliseq and BD Cellular Research Precise Assays). Genes of interest are enriched by custom-designed probes (**Figure 6**). However, these targeted sequencing procedures can potentially introduce biases caused by nonuniform hybridization as well as variation in amplification efficiencies across different genes and transcripts. According to our in-house pilot study (unpublished data), gene expression results are very sensitive to the choice of probes and targeted regions. Cellular Research Precise Assays [81] allow researchers to introduce so-called unique-sequence barcodes into the samples to overcome possible amplification bias and thus produces more accurate estimates of transcript abundance. By counting unique barcodes instead of reads, researchers can determine how many copies of each transcript are expressed with very high accuracy. In the meantime, targeted RNA-seq poses new challenges for data analysis. For example, most differential analysis packages assume the majority of genes are not differentially expressed, but this assumption is most likely not valid in targeted RNA-seq data sets.

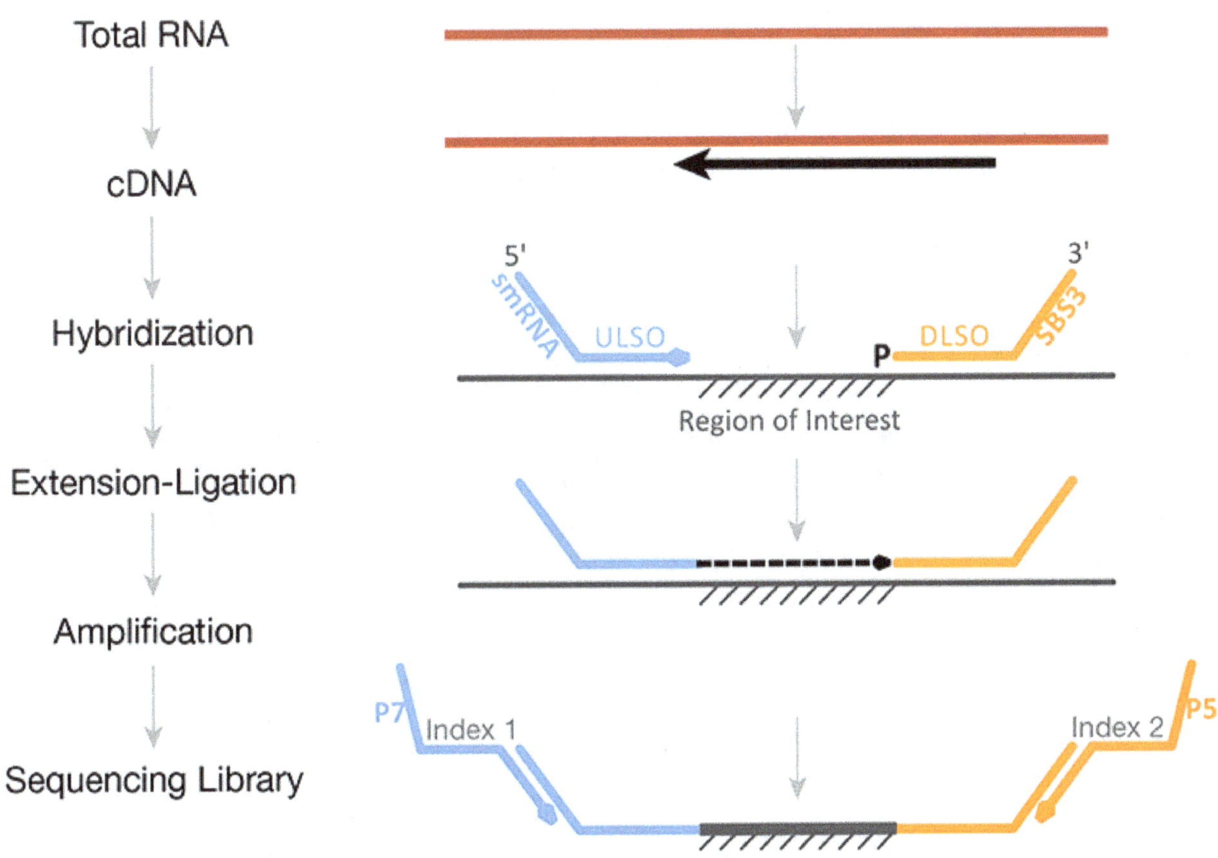

Figure 6. TruSeq targeted RNA expression workflow. Two custom-designed oligonucleotide probes with adapter sequences hybridize up and downstream of the region of interest. ULSO stands for upstream locus-specific oligo and DLSO stands for downstream locus-specific oligo.

5.3. Single-cell RNA-seq

Cell identity and function can be characterized at the molecular level by unique transcriptomic signatures. At the organismal level, different tissues have distinct gene expression profiles, and even cells in consecutive stages of embryonic development have highly divergent transcriptomic landscapes. Until recently, molecular 'fingerprints' were generated using profiling of gene expression levels from bulk populations of millions of input cells. These Ensemble-based approaches meant that the resulting expression value for each gene was an average of its expression levels across a large population of input cells. In many contexts, such bulk expression profiles are sufficient. However, there are also important questions for which bulk measures of gene expression are insufficient, for instance heterogeneity in immune cells. Besides, Ensemble measures do not provide insights into the stochastic nature of gene expression.

Recently, RNA-seq has achieved single-cell resolution, and scRNA-seq enables unbiased, high-throughput, and high-resolution transcriptomic analysis of individual cells [82, 83]. This provides an additional dimension to transcriptomic information relative to traditional methods that profile bulk populations of cells. What is currently still missing is an effective way to routinely isolate and process large numbers of individual cells for quantitative in-depth RNA-seq. Klein et al. [84] and Macosko et al. [85] have independently developed a high-throughput droplet-microfluidic approach for barcoding of RNA from thousands of individual cells for subsequent analysis by next-generation sequencing. Droplet-based scRNA-seq will be an attractive method for many laboratories because of its seemingly unlimited scalability and relatively low cost. By combining sophisticated RNA-seq technology with a new device that isolates single cells and their progeny, MIT researchers can now trace detailed family histories for several generations of cells descended from one "ancestor" [86].

Alongside the large-scale generation of single cell transcriptomic data, it is important to consider the specific computational and analytical challenges that still have to be overcome [87]. Although some tools for bulk RNA-seq analysis can be readily applied to single-cell data, many new computational strategies are required to fully exploit this new data type. For instance, many genes at single cells are expressed in a stochastically-bursting fashion and their abundance exhibits a bimodal distribution in cell populations. Another main problem we face is that each cell can be in a different cell cycle phase, and they might vary in size and RNA content. The traditional RNA-seq data analysis does not take transcriptional bimodality into consideration, and implicitly assume the total RNA amount is the same or at least comparable. Additionally, compared with bulk RNA-seq, scRNA-seq data have much larger technical variations mainly because starting amount of RNA in a single cell is much lower and requires many more cycles of amplification. Recently, Korthauer et al. [88] have developed scDD, a differential analysis method particularly suited for scRNA-seq, to characterize differences in expression in the presence of distinct expression states within and among biological conditions. Using simulated and case study data, they demonstrate that the modeling framework is able to detect DE patterns of interest under a wide range of settings. Compared with existing approaches, scDD has higher power to detect subtle differences in gene expression distributions that are more complex than a mean shift, and is able to characterize those differences.

6. Concluding remarks

In much the same way that the advent of NGS technologies transformed our approaches in DNA-sequencing, RNA-seq approaches have dramatically changed our abilities to analyze the transcriptome of cells and tissues with a new level of detail and sensitivity. But we must also recognize that RNA-seq analysis is vulnerable to the general biases and errors inherent to next-generation sequencing (NGS) technology upon which it is based.

While RNA-seq technology is considered unbiased, it is important to note that the preparation and fragmentation of RNA and the library construction (which includes size selection) can introduce biases. Fragments are not uniformly sampled and sequenced, as there is variability in sequencing depth across the transcriptome due to preferential sites of fragmentation, variable primer, and tag nucleotide composition effects [89, 90]. RNA-seq is a complicated multistep process that involves sample collection and stabilization, RNA extraction, fragmentation, cDNA synthesis, adapter ligation, amplification, purification, and sequencing. Any step in this complex sequence of protocols can result in biased data.

Data normalization is one of the most crucial steps of RNA-seq data processing and has a profound effect on the results of the analysis. In practice, normalization of high-throughput data still remains an important topic and has received a lot of attention in the literature. The increasing number of normalization methods makes it difficult for scientists to decide which method should be used for which particular data set [58–66]. Even worse, different research groups draw contradictory conclusions. For example, Zyprych-Walczak et al. [66] concluded that the use of TMM method in most cases is displayed poorly, whereas the study by Dillies et al. [64] indicated that the use of TMM method led to good performance on simulated RNA-seq data sets. Considering the confusion and the numbers of methods available, we believe RNA-seq data normalization is a fundamental question that is far from being solved, and RNA-seq community will have to develop more sophisticated and robust algorithms to tackle this problem.

Additionally, for RNA-seq data collected with multiple sequencing platforms or at multiple sites, other normalization methods are required to remove site- or technology-specific effects [91]. Several methods for such normalization have been developed, including sva [92], RUV2 [93], and PEER [94], but they vary in their ability to remove these systematic biases [91]. As the cost of NGS continues to decrease, it is likely that additional studies will be conducted to readdress the same biological question. Therefore, there is an increasing need for analyzing data from multiple studies and different labs [95].

In this chapter, we focused on data analysis workflow for mRNA-seq, but have not covered RNA-seq experimental design. However, a crucial prerequisite for any successful RNA-seq study is a good experimental design, making sure the data generated have the potential to answer the biological questions of interest. For other aspects of RNA-seq data analysis, including experimental design, alternative splicing, gene fusion, and eQTL mapping, please refer to the most recent review by Conesa et al. [96]. In practice, RNA-seq is not used as a stand-alone technology platform in research. The integration of RNA-seq data with other types of genome-wide data allows us to connect the regulation of gene expression with specific aspects of molecular physiology and functional genomics. Integrative analyses that incorporate RNA-seq data with other genomic or proteomic experiments are becoming increasingly prevalent. For instance, the combination of RNA and DNA sequencing can be used to explore RNA-editing or expression quantitative trait loci (eQTL) mapping. Pairwise DNA-methylation and RNA-seq integration can reveal the correlation between gene expression and methylation patterns. Additionally, integration of RNA-seq and miRNA-seq data has the potential to unravel the regulatory effects of miRNAs on transcript steady-state levels. All these integrative analyses pose additional challenges that are beyond the scope of this chapter.

Author details

Shanrong Zhao*, Baohong Zhang, Ying Zhang, William Gordon, Sarah Du, Theresa Paradis, Michael Vincent and David von Schack

*Address all correspondence to: shanrong.zhao@pfizer.com

PharmaTherapeutics Clinical R&D, Pfizer Worldwide Research and Development, Cambridge, MA, USA

References

[1] Mortazavi A, Williams BA, McCue K, Schaeffer L, Wold B. Mapping and quantifying mammalian transcriptomes by RNA-seq. Nat Methods. 2008;5:621–628. DOI: 10.1038/nmeth.1226

[2] Wang Z, Gerstein M, Snyder M. RNA-seq: a revolutionary tool for transcriptomics. Nat Rev Genet. 2009;10:57–63. DOI: 10.1038/nrg2484

[3] Costa V, Angelini C, De Feis I, Ciccodicola A. Uncovering the complexity of transcriptomes with RNA-seq. J Biomed Biotechnol. 2010;2010:853916. DOI: 10.1155/2010/853916

[4] Mutz KO, Heilkenbrinker A, Lönne M, Walter JG, Stahl F. Transcriptome analysis using next-generation sequencing. Curr Opin Biotechnol. 2013;24:22–30. DOI: 10.1016/j.copbio.2012.09.004

[5] Garber M, Grabherr MG, Guttman M, Trapnell C. Computational methods for transcriptome annotation and quantification using RNA-seq. Nat Methods. 2011;8:469–477. DOI: 10.1038/nmeth.1613

[6] Capobianco E. RNA-seq data: a complexity journey. Comput Struct Biotechnol J. 2014;11:123–130. DOI: 10.1016/j.csbj.2014.09.004

[7] Han Y, Gao S, Muegge K, Zhang W, Zhou B. Advanced applications of RNA sequencing and challenges. Bioinform Biol Insights. 2015;9(Suppl 1):29–46. DOI: 10.4137/BBI.S28991

[8] Mills JD, Kawahara Y, Janitz M. Stranded RNA-seq provides greater resolution of transcriptome profiling. Curr Genomics. 2013;14:173–181. DOI: 10.2174/1389202911314030003

[9] Sigurgeirsson B, Emanuelsson O, Lundeberg J. Analysis of stranded information using an automated procedure for strand specific RNA sequencing. BMC Genomics. 2014;15:631. DOI: 10.1186/1471-2164-15-631

[10] Junker JP, van Oudenaarden A. Single-cell transcriptomics enters the age of mass production. Mol Cell. 2015;58:563–564. DOI: 10.1016/j.molcel.2015.05.019

[11] Saliba AE, Westermann AJ, Gorski SA, Vogel J. Single-cell RNA-seq: advances and future challenges. Nucleic Acids Res. 2014;42:8845–8860. DOI: 10.1093/nar/gku555

[12] Wilson NK, Kent DG, Buettner F, Shehata M, Macaulay IC, Calero-Nieto FJ et al. Combined single-cell functional and gene expression analysis resolves heterogeneity within stem cell populations. Cell Stem Cell. 2015;16:712–724. DOI: 10.1016/j.stem.2015.04.004

[13] Mantione KJ, Kream RM, Kuzelova H, Ptacek R, Raboch J, Samuel JM, Stefano GB. Comparing bioinformatic gene expression profiling methods: microarray and RNA-seq. Med Sci Monit Basic Res. 2014;20:138–142. DOI: 10.12659/MSMBR.892101

[14] Hurd PJ, Nelson CJ. Advantages of next-generation sequencing versus the microarray in epigenetic research. Brief Funct Genomic Proteomic. 2009;8:174–183. DOI: 10.1093/bfgp/elp013

[15] McIntyre LM, Lopiano KK, Morse AM, Amin V, Oberg AL, Young LJ, Nuzhdin SV. RNA-seq: technical variability and sampling. BMC Genomics. 2011;12:293. DOI: 10.1186/1471-2164-12-293

[16] Malone J, Oliver B. Microarrays, deep sequencing and the true measure of the transcriptome. BMC Biol. 2011;9:34. DOI: 10.1186/1741-7007-9-34

[17] Fu X, Fu N, Guo S, Yan Z, Xu Y, Hu H, Menzel C, et al. Estimating accuracy of RNA-seq and microarrays with proteomics. BMC Genomics 2009;10:161. DOI: 10.1186/1471-2164-10-161

[18] Zhao S, Fung-Leung W-P, Bittner A, Ngo K, Liu X. Comparison of RNA-seq and microarray in transcriptome profiling of activated T cells. PLoS One. 2014;9:e78644. DOI: 10.1371/journal.pone.0078644

[19] Hangauer MJ, Vaughn IW, McManus MT. Pervasive transcription of the human genome produces thousands of previously unidentified long intergenic noncoding RNAs. PLoS Genet. 2013;9:e1003569. DOI: 10.1371/journal.pgen.1003569

[20] Wang ET, Sandberg R, Luo S, et al. Alternative isoform regulation in human tissue transcriptomes. Nature. 2008;456:470–476. DOI: 10.1038/nature07509

[21] Griffith M, Griffith OL, Mwenifumbo J, et al. Alternative expression analysis by RNA sequencing. Nat Methods. 2010;7:843–847. DOI: 10.1038/nmeth.1503

[22] Picardi E, Horner DS, Chiara M, Schiavon R, Valle G, Pesole G. Large-scale detection and analysis of RNA editing in grape mtDNA by RNA deep-sequencing. Nucleic Acids Res. 2010;38:4755–4767. DOI: 10.1093/nar/gkq202

[23] Maher CA, Kumar-Sinha C, Cao X, Kalyana-Sundaram S, Han B, Jing X et al. Transcriptome sequencing to detect gene fusions in cancer. Nature. 2009;458:97–101. DOI: 10.1038/nature07638

[24] Schmieder R, Edwards R. Quality control and preprocessing of metagenomic datasets. Bioinformatics. 2011;27:863–864. DOI: 10.1093/bioinformatics/btr026

[25] FastQC [Internet]. 2016. Available from: http://www.bioinformatics.babraham.ac.uk/projects/fastqc/ [Accessed: 2016-02-23].

[26] Martin M. Cutadapt removes adapter sequences from high-throughput sequencing reads. EMBnet. J. 2011;17:10–12. DOI: 10.14806/ej.17.1.200

[27] Bolger AM, Lohse M, Usadel B. Trimmomatic: a flexible trimmer for illumina sequence data. Bioinformatics. 2014;30:2114–2120. DOI: 10.1093/bioinformatics/btu170

[28] Kim D, Pertea G, Trapnell C, Pimentel H, Kelley R, Salzberg SL. TopHat2: accurate alignment of transcriptomes in the presence of insertions, deletions and gene fusions. Genome Biol. 2013;14:R36. DOI: 10.1186/gb-2013-14-4-r36

[29] Dobin A, Davis CA, Schlesinger F, Drenkow J, Zaleski C, Jha S, et al. STAR: ultrafast universal RNA-seq aligner. Bioinformatics. 2013;29:15–21. DOI: 10.1093/bioinformatics/bts635

[30] Dobin A, Gingeras TR. Mapping RNA-seq Reads with STAR. Curr Protoc Bioinformatics. 2015;51:11.14.1–11.14.19. DOI: 10.1002/0471250953.bi1114s51

[31] Wu TD, Nacu S. Fast and SNP-tolerant detection of complex variants and splicing in short reads. Bioinformatics. 2010;26:873–881. DOI: 10.1093/bioinformatics/btq057

[32] Hu J, Ge H, Newman M, Liu K. OSA: a fast and accurate alignment tool for RNA-seq. Bioinformatics. 2012;28:1933–1934. DOI: 10.1093/bioinformatics/bts294

[33] Wang K, Singh D, Zeng Z, Coleman SJ, Huang Y, Savich GL, et al. MapSplice: accurate mapping of RNA-seq reads for splice junction discovery. Nucl Acids Res. 2010;38:e178. DOI: 10.1093/nar/gkq622

[34] Engström PG, Steijger T, Sipos B, Grant GR, Kahles A, et al. Systematic evaluation of spliced alignment programs for RNA-seq data. Nat Methods. 2013;10:1185–1191. DOI: 10.1038/nmeth.2722

[35] Wu P-Y, Phan JH, Wang MD. Assessing the impact of human genome annotation choice on RNA-seq expression estimates. BMC Bioinformatics. 2013;14:S8. DOI: 10.1186/1471-2105-14-S11-S8

[36] Chen G, Wang C, Shi L, Qu X, Chen J, Yang J, et al. Incorporating the human gene annotations in different databases significantly improved transcriptomic and genetic analyses. RNA. 2013;19:479–489. DOI: 10.1261/rna.037473.112

[37] Zhao S. Assessment of the impact of using a reference transcriptome in mapping short RNA-seq reads. PLoS One. 2014;9:e101374. DOI: 10.1371/journal.pone.0101374

[38] Zhao S, Zhang B. A comprehensive evaluation of Ensembl, RefSeq, and UCSC annotations in the context of RNA-seq read mapping and gene quantification. BMC Genomics. 2015;16:97. DOI: 10.1186/s12864-015-1308-8

[39] Pruitt KD, Tatusova T, Maglott DR. NCBI reference sequences (RefSeq): a curated non-redundant sequence database of genomes, transcripts and proteins. Nucl Acids Res. 2007;35(Database):D61–D65. DOI: 10.1093/nar/gkl842

[40] Hsu F, Kent WJ, Clawson H, Kuhn RM, Diekhans M, Haussler D. The UCSC known genes. Bioinformatics. 2006;22:1036–1046. DOI: 10.1093/bioinformatics/btl048

[41] Flicek P, Amode MR, Barrell D, Beal K, Billis K, Brent S, et al. Ensembl 2014. Nucl Acids Res. 2014;42(Database issue):D749–D55. DOI: 10.1093/nar/gkt1196

[42] Dao P, Numanagić I, Lin YY, Hach F, Karakoc E, Donmez N, et al. ORMAN: optimal resolution of ambiguous RNA-seq multimappings in the presence of novel isoforms. Bioinformatics. 2014;30:644–651. DOI: 10.1093/bioinformatics/btt591

[43] Li B, Dewey CN. RSEM: accurate transcript quantification from RNA-seq data with or without a reference genome. BMC Bioinformatics. 2011;12:323. DOI: 10.1186/1471-2105-12-323

[44] Trapnell C, Williams BA, Pertea G, Mortazavi A, Kwan G, van Baren MJ, et al. Transcript assembly and quantification by RNA-seq reveals unannotated transcripts and isoform switching during cell differentiation. Nat Biotechnol. 2010;28:511–515. DOI: 10.1038/nbt.1621

[45] Nicolae M, Mangul S, Măndoiu II, Zelikovsky A. Estimation of alternative splicing isoform frequencies from RNA-seq data. Algorithms Mol Biol. 2011;6:9. DOI: 10.1186/1748-7188-6-9

[46] Liao Y, Smyth GK, Shi W. featureCounts: an efficient general purpose program for assigning sequence reads to genomic features. Bioinformatics. 2014;30:923–930. DOI: 10.1093/bioinformatics/btt656

[47] Anders S, Theodor P, Huber W. HTSeq—A Python framework to work with high-throughput sequencing data. Bioinformatics. 2014;31:166–169.DOI: 10.1093/bioinformatics/btu638

[48] Angelini C, De Canditiis D, De Feis I. Computational approaches for isoform detection and estimation: good and bad news. BMC Bioinformatics. 2014;15:135. DOI: 10.1186/1471-2105-15-135

[49] Kanitz A, Gypas F, Gruber AJ, Gruber AR, Martin G, Zavolan M. Comparative assessment of methods for the computational inference of transcript isoform abundance from RNA-seq data. Genome Biol. 2015;16:150. DOI: 10.1186/s13059-015-0702-5

[50] Zhao S, Xi L, Zhang B. Union exon based approach for RNA-seq gene quantification: to be or not to be? PLoS One. 2015;10(11):e0141910. DOI: 10.1371/journal.pone.0141910

[51] Anders S, Huber W. Differential expression analysis for sequence count data. Genome Biol. 2010;11:R106. DOI: 10.1186/gb-2010-11-10-r106

[52] Robinson MD, Oshlack A. A scaling normalization method for differential expression analysis of RNA-seq data. Genome Biol. 2010;11:R25. DOI: 10.1186/gb-2010-11-3-r25

[53] Robinson MD, McCarthy DJ, Smyth GK. edgeR: a bioconductor package for differential expression analysis of digital gene expression data. Bioinformatics. 2010;26:139–140. DOI: 10.1093/bioinformatics/btp616

[54] Cumbie JS, Kimbrel JA, Di Y, Schafer DW, Wilhelm LJ, Fox SE, et al. GENE-Counter: a computational pipeline for the analysis of RNA-seq data for gene expression differences. PLoS One. 2011;6:e25279. DOI: 10.1371/journal.pone.0025279

[55] Tarazona S, Garcia-Alcalde F, Dopazo J, Ferrer A, Conesa A. Differential expression in RNA-seq: a matter of depth. Genome Res. 2011; 21:2213–2223. DOI: 10.1101/gr.124321.111

[56] Di Y, Schafer D, Cumbie J, Chang J. The NBP negative binomial model for assessing differential gene expression from RNA-seq. Stat Appl Genet Mol Biol. 2011;10: Article 24. DOI: 10.2202/1544-6115.1637

[57] Trapnell C, Hendrickson DG, Sauvageau M, Goff L, Rinn JL, Pachter L. Differential analysis of gene regulation at transcript resolution with RNA-seq. Nat Biotechnol. 2013;31:46–53. DOI: 10.1038/nbt.2450.

[58] Bullard JH, Purdom E, Hansen KD, Dudoit S. Evaluation of statistical methods for normalization and differential expression in mRNA-seq experiments. BMC Bioinformatics. 2010;11:94. DOI: 10.1186/1471-2105-11-94

[59] Kvam VM, Liu P, Si Y. A comparison of statistical methods for detecting differentially expressed genes from RNA-seq data. Am J Bot 2012;99:248–256. DOI: 10.3732/ajb.1100340

[60] Robles JA, Qureshi SE, Stephen SJ, Wilson SR, Burden CJ, Taylor JM. Efficient experimental design and analysis strategies for the detection of differential expression using RNA-sequencing. BMC Genomics. 2012;13:484. DOI: 10.1186/1471-2164-13-484

[61] Zhang ZH, Jhaveri DJ, Marshall VM, Bauer DC, Edson J, Narayanan RK, Robinson GJ, Lundberg AE, Bartlett PF, Wray NR, Zhao QY. A comparative study of techniques for

differential expression analysis on RNA-seq data. PLoS One. 2014;9:e103207. DOI: 10.1371/journal.pone.0103207

[62] Soneson C, Delorenzi M. A comparison of methods for differential expression analysis of RNA-seq data. BMC Bioinformatics. 2013;14:91. DOI: 10.1186/1471-2105-14-91

[63] Rapaport F, Khanin R, Liang Y, Pirun M, Krek A, Zumbo P, Mason CE, Socci ND, Betel D. Comprehensive evaluation of differential gene expression analysis methods for RNA-seq data. Genome Biol. 2013;14:R95. DOI: 10.1186/gb-2013-14-9-r95

[64] Dillies MA, Rau A, Aubert J, Hennequet-Antier C, et al. A comprehensive evaluation of normalization methods for Illumina high-throughput RNA sequencing data analysis. Brief Bioinform. 2013;14:671–683. DOI: 10.1093/bib/bbs046

[65] Seyednasrollah F, Laiho A, Elo LL. Comparison of software packages for detecting differential expression in RNA-seq studies. Brief Bioinformatics. 2015;16:59–70. DOI: 10.1093/bib/bbt086

[66] Zyprych-Walczak J, Szabelska A, Handschuh L, Górczak K, Klamecka K, Figlerowicz M, Siatkowski I. The Impact of normalization methods on RNA-seq data analysis. Biomed Res Int. 2015;2015:621690. DOI: 10.1155/2015/621690

[67] The Gene Ontology Consortium. Gene Ontology Consortium: going forward. Nucl Acids Res. 2015;43(Database issue):D1049–D1056. DOI: 10.1093/nar/gku1179

[68] Kanehisa M, Goto S, Kawashima S, Okuno Y, Hattori M. The KEGG resource for deciphering the genome. Nucl Acids Res. 2004;32(Database issue):D277–D2780. DOI: 10.1093/nar/gkh063

[69] Huang D, Sherman BT, Lempicki RA. Systematic and integrative analysis of large gene lists using DAVID bioinformatics resources. Nat Protoc. 2009;4:47. DOI: 10.1038/nprot.2008.211

[70] Hänzelmann S, Castelo R, Guinney J. GSVA: gene set variation analysis for microarray and RNA-seq data. BMC Bioinformatics. 2013;14:7. DOI: 10.1186/1471-2105-14-7

[71] Wang X, Cairns MJ. SeqGSEA: a Bioconductor package for gene set enrichment analysis of RNA-seq data integrating differential expression and splicing. Bioinformatics. 2014;30:1777–1779. DOI: 10.1093/bioinformatics/btu090

[72] Tarca AL, Draghici S, Khatri P, Hassan SS, Mittal P, Kim JS et al. A novel signaling pathway impact analysis. Bioinformatics. 2009;25:75–82. DOI: 10.1093/bioinformatics/btn577

[73] Gao S, Wang X. TAPPA: topological analysis of pathway phenotype association. Bioinformatics. 2007;23:3100–3102. DOI: 10.1093/bioinformatics/btm460

[74] Haynes WA, Higdon R, Stanberry L, Collins D, Kolker E. Differential expression analysis for pathways. PLoS Comput Biol. 2013;9:e1002967. DOI: 10.1371/journal.pcbi.1002967

[75] Zhao S, Xi L, Quan J, Xi H, Zhang Y, von Schack D, Vincent M, Zhang B. QuickRNASeq lifts large-scale RNA-seq data analyses to the next level of automation and interactive visualization. BMC Genomics. 2016;17:39. DOI: 10.1186/s12864-015-2356-9

[76] Levin JZ, Yassour M, Adiconis X, Nusbaum C, Thompson DA, Friedman N, Gnirke A, Regev A. Comprehensive comparative analysis of stranded RNA sequencing methods. Nat Methods. 2010;7:709–715. DOI: 10.1038/nmeth.1491

[77] Zhao S, Zhang Y, Gordon W, Quan J, Xi H, Du, S, von Schack D, Zhang B. Comparison of stranded and non-stranded RNA-seq transcriptome profiling and investigation of gene overlap. BMC Genomics. 2015;16:487. DOI: 10.1186/s12864-015-1876-7

[78] Mercer TR, Gerhardt DJ, Dinger ME, Crawford J, Trapnell C, Jeddeloh JA, Mattick JS, Rinn JL. Targeted RNA sequencing reveals the deep complexity of the human transcriptome. Nat Biotechnol. 2011;30:99–104. DOI: 10.1038/nbt.2024

[79] Mercer TR, Clark MB, Crawford J, Brunck ME, Gerhardt DJ, Taft RJ, Nielsen LK, Dinger ME, Mattick JS. Targeted sequencing for gene discovery and quantification using RNA CaptureSeq. Nat Protoc. 2014;9:989–1009. DOI: 10.1038/nprot.2014.058

[80] Tan JC, Bouriakov VD, Feng L, Richmond TA, Burgess D. Targeted lncRNA sequencing with the SeqCap RNA enrichment system. Methods Mol Biol.2016;1402:73–100. DOI: 10.1007/978-1-4939-3378-5_8

[81] Cellular Research Precise Assays [Internet]. 2016. Available from: http://www.cellular-research.com/products/precise-assays.html [Accessed: 2016-02-23].

[82] Chattopadhyay PK, Gierahn TM, Roederer M, Love JC. Single-cell technologies for monitoring immune systems. Nat Immunol. 2014;15:128–135. DOI: 10.1038/ni.2796

[83] Avital G, Hashimshony T, Yanai I. Seeing is believing: new methods for in situ single-cell transcriptomics. Genome Biol. 2014;15:110. DOI: 10.1186/gb4169

[84] Klein AM, Mazutis L, Akartuna I, Tallapragada N, Veres A, Li V, Peshkin L, Weitz DA, Kirschner MW. Droplet barcoding for single-cell transcriptomics applied to embryonic stem cells. Cell. 2015;161:1187–1201. DOI: 10.1016/j.cell.2015.04.044

[85] Macosko EZ, Basu A, Satija R, Nemesh J, Shekhar K, et al. Highly parallel genome-wide expression profiling of individual cells using nanoliter droplets. Cell. 2015;161:1202–1214. DOI: 10.1016/j.cell.2015.05.002

[86] Kimmerling RJ, Lee Szeto G, Li JW, Genshaft AS, Kazer SW, Payer KR, de Riba Borrajo J, Blainey PC, Irvine DJ, Shalek AK, Manalis SR. A microfluidic platform enabling single-cell RNA-seq of multigenerational lineages. Nat Commun. 2016;7:10220. DOI: 10.1038/ncomms10220

[87] Stegle O, Teichmann SA, Marioni JC. Computational and analytical challenges in single-cell transcriptomics. Nat Rev Genet. 2015;16:133–145. DOI: 10.1038/nrg3833

[88] Korthauer KD, Chu LF, Newton MA, Li Y, Thomson J, Stewart R, Kendziorski C. scDD: a statistical approach for identifying differential distributions in single-cell RNA-seq experiments. bioRXiv. 2016 [Epub ahead of print]. DOI: 10.1101/035501

[89] Hansen KD, Brenner SE, Dudoit S. Biases in illumina transcriptome sequencing caused by random hexamer priming. Nucl Acids Res. 2010;38:e131. DOI: 10.1093/nar/gkq224

[90] Risso D, Schwartz K, Sherlock G, Dudoit S. GC-content normalization for RNA-seq data. BMC Bioinformatics. 2011;12:480. DOI: 10.1186/1471-2105-12-480

[91] Li S, łabaj PP, Zumbo P, Sykacek P, Shi W, Shi L, et al. Detecting and correcting systematic variation in large-scale RNA sequencing data. Nat Biotechnol. 2014;32:888–895. DOI: 10.1038/nbt.3000

[92] Leek JT, Johnson WE, Parker HS, Jaffe AE, Storey JD. The sva package for removing batch effects and other unwanted variation in high-throughput experiments. Bioinformatics. 2012;28:882–883. DOI: 10.1093/bioinformatics/bts034

[93] Gagnon-Bartsch JA, Speed TP. Using control genes to correct for unwanted variation in microarray data. Biostatistics. 2013;13:539–552. DOI: 10.1093/biostatistics/kxr034

[94] Stegle O, Parts L, Durbin R, Winn J. A Bayesian framework to account for complex non-genetic factors in gene expression levels greatly increases power in eQTL studies. PLoS Comput Biol. 2010;6:e1000770. DOI: 10.1371/journal.pcbi.1000770

[95] Rau A, Marot G, Jaffrézic F. Differential meta-analysis of RNA-seq data from multiple studies. BMC Bioinformatics. 2014;15:91. DOI: 10.1186/1471-2105-15-91

[96] Conesa A, Madrigal P, Tarazona S, Gomez-Cabrero D, Cervera A, McPherson A, Szcześniak MW, Gaffney DJ, Elo LL, Zhang X, Mortazavi A. A survey of best practices for RNA-seq data analysis. Genome Biol. 2016;17:13. DOI: 10.1186/s13059-016-0881-8

Permissions

All chapters in this book were first published in BUFA, by InTech Open; hereby published with permission under the Creative Commons Attribution License or equivalent. Every chapter published in this book has been scrutinized by our experts. Their significance has been extensively debated. The topics covered herein carry significant findings which will fuel the growth of the discipline. They may even be implemented as practical applications or may be referred to as a beginning point for another development.

The contributors of this book come from diverse backgrounds, making this book a truly international effort. This book will bring forth new frontiers with its revolutionizing research information and detailed analysis of the nascent developments around the world.

We would like to thank all the contributing authors for lending their expertise to make the book truly unique. They have played a crucial role in the development of this book. Without their invaluable contributions this book wouldn't have been possible. They have made vital efforts to compile up to date information on the varied aspects of this subject to make this book a valuable addition to the collection of many professionals and students.

This book was conceptualized with the vision of imparting up-to-date information and advanced data in this field. To ensure the same, a matchless editorial board was set up. Every individual on the board went through rigorous rounds of assessment to prove their worth. After which they invested a large part of their time researching and compiling the most relevant data for our readers.

The editorial board has been involved in producing this book since its inception. They have spent rigorous hours researching and exploring the diverse topics which have resulted in the successful publishing of this book. They have passed on their knowledge of decades through this book. To expedite this challenging task, the publisher supported the team at every step. A small team of assistant editors was also appointed to further simplify the editing procedure and attain best results for the readers.

Apart from the editorial board, the designing team has also invested a significant amount of their time in understanding the subject and creating the most relevant covers. They scrutinized every image to scout for the most suitable representation of the subject and create an appropriate cover for the book.

The publishing team has been an ardent support to the editorial, designing and production team. Their endless efforts to recruit the best for this project, has resulted in the accomplishment of this book. They are a veteran in the field of academics and their pool of knowledge is as vast as their experience in printing. Their expertise and guidance has proved useful at every step. Their uncompromising quality standards have made this book an exceptional effort. Their encouragement from time to time has been an inspiration for everyone.

The publisher and the editorial board hope that this book will prove to be a valuable piece of knowledge for researchers, students, practitioners and scholars across the globe.

List of Contributors

S. W. Leong, T. S. Lim and Y. S. Choong
Institute for Research in Molecular Medicine, Universiti Sains Malaysia, Minden, Malaysia

Sidra Younis
The R.O.Perelman Department of Dermatology and Department of Biochemistry and Molecular Pharmacology, NYU Langone Medical Center, New York, USA
Department of Biochemistry, Quaid-i-Azam University, Islamabad, Pakistan

Valeriia Shnayder and Miroslav Blumenberg
The R.O.Perelman Department of Dermatology and Department of Biochemistry and Molecular Pharmacology, NYU Langone Medical Center, New York, USA

Venkateswara R. Sripathi, Abreeotta J. Williams and Govind C. Sharma
Center for Molecular Biology, Alabama A&M University, Normal, AL, USA

Ramesh Buyyarapu and Siva P. Kumpatla
Dow AgroSciences, Indianapolis, IN, USA

Seloame T. Nyaku
Crop Science Department, College of Basic and Applied Sciences, University of Ghana, Legon-Accra, Ghana

Yonathan Tilahun
Cooperative Extension, School of Agriculture & Applied Sciences, Langston University, Langston, OK, USA

Venu Kalavacharla
Molecular Genetics & EpiGenomics Laboratory, College of Agriculture & Related Sciences, Delaware State University, Dover, DE, USA

Arianna Filntisi
School of Electrical and Computer Engineering, National Technical University of Athens, Athens, Greece
Computational Biology and Medicine Group, Biomedical Research Foundation of the Academy of Athens, Athens, Greece

Dimitrios Vlachakis
Computational Biology and Medicine Group, Biomedical Research Foundation of the Academy of Athens, Athens, Greece

George K. Matsopoulos
School of Electrical and Computer Engineering, National Technical University of Athens, Athens, Greece

Trevor Graham Bell and Anna Kramvis
Hepatitis Virus Diversity Research Unit, Faculty of Health Sciences, University of the Witwatersrand, Johannesburg, South Africa

Kiran Kadam, Sangeeta Sawant and Urmila Kulkarni-Kale
Bioinformatics Centre, Savitribai Phule Pune University, Pune, India

V. K. Jayaraman
Shiv Nadar University, Gautam Buddha Nagar, Uttar Pradesh, India

Daniel A. Bórquez
Biomedical Research Center, School of Medicine, Diego Portales University, Santiago, Chile

Christian González-Billault
Department of Biology, Faculty of Sciences, University of Chile, Santiago, Chile
FONDAP Geroscience Center, Brain Health and Metabolism, Santiago, Chile

Mehboob-ur-Rahman and Muhammad Atif Iqbal
National Institute for Biotechnology and Genetic Engineering, Faisalabad, Pakistan

Tayyaba Shaheen and Mahmood-ur-Rahman
Department of Bioinformatics and Biotechnology, Government College University, Faisalabad, Pakistan

Yusuf Zafar
Department of Technical Co-operation, IAEA, Vienna International Centre, Vienna, Austria

Ibrokhim Y. Abdurakhmonov
Center of Genomics and Bioinformatics, Academy of Science of the Republic of Uzbekistan, Tashkent, Uzbekistan

Shanrong Zhao, Baohong Zhang, Ying Zhang, William Gordon, Sarah Du, Theresa Paradis, Michael Vincent and David von Schack
Pharma Therapeutics Clinical R&D, Pfizer Worldwide Research and Development, Cambridge, MA, USA

Index

www.ingramcontent.com/pod-product-compliance
Lightning Source LLC
LaVergne TN
LVHW061510220926
848485LV00017B/706